AF849120

Mountains, Volcanoes, Coasts and Caves

Mountains, Volcanoes, Coasts and Caves

Origins of Aotearoa New Zealand's Natural Wonders

Bruce W. Hayward

With aerial photography by
Alastair Jamieson and Lloyd Homer

AUCKLAND
UNIVERSITY
PRESS

First published 2022
Reprinted 2023
Auckland University Press
University of Auckland
Private Bag 92019
Auckland 1142
New Zealand
www.aucklanduniversitypress.co.nz

Text © Bruce W. Hayward, 2022
Images © Alastair Jamieson and Lloyd Homer, 2022
All images not credited are by Bruce W. Hayward

ISBN 978 1 86940 967 8

This book is made possible by the generous support of Michael Horton, CNZM, the Royal Society of New Zealand Charles Fleming Fund and GNS Science.

A catalogue record for this book is available from the National Library of New Zealand

This book is copyright. Apart from fair dealing for the purpose of private study, research, criticism or review, as permitted under the Copyright Act, no part may be reproduced by any process without prior permission of the publisher. The moral rights of the author have been asserted.

Every effort has been made to trace copyright holders and obtain their permission. The publisher apologises in advance for any errors or omissions and would be grateful to be notified of any corrections that should be incorporated in future reprints or editions.

Cover image: The Upper Taieri Scroll Plain is one of the world's best examples of this evocative landform type, characterised by meandering river scrolls that develop on the floor of a valley with an extremely low gradient. PHOTO: ALASTAIR JAMIESON.

Image on pp. 6–7: Archway Islands, Wharariki Beach, northwest Nelson. PHOTO: ALASTAIR JAMIESON.

Book design by Carolyn Lewis

Site maps and illustration input by Tim Nolan / Blackant Mapping Solutions

Maps contain some data sourced from LINZ, licensed for reuse under CC-BY-4.0

This product is made of FSC®-certified and other controlled material.

FSC
www.fsc.org
MIX
Paper | Supporting responsible forestry
FSC® C016074

Printed in China through Colorcraft Limited, Hong Kong

Bruce W. Hayward is a retired geologist and marine ecologist based in Auckland. He is a former member of the New Zealand Conservation Authority, former president of the Geological Society of NZ and Honorary Life Member of the NZ Marine Sciences Society and the Geoscience Society of NZ. He is the founder and convenor of the NZ Geopreservation Inventory which documents and seeks greater protection for NZ's geoheritage features like those highlighted in this book. His wide interests in natural and human history have resulted in twenty-two previous books on topics as diverse as archaeology, the kauri timber and gum industries, New Zealand fossils, volcanoes, building stones and geoconservation, and the popular *Out of the Ocean, into the Fire* on the geology of northern New Zealand. He is author of AUP titles *Volcanoes of Auckland: A Field Guide* (2019) and *Volcanoes of Auckland: The Essential Guide* (2011).

Alastair Jamieson is an Auckland-based ecologist and photographer who has documented New Zealand's changing natural environment with aerial photography for three decades. He provided the aerial photography for AUP's two Auckland volcano guides. You can see more of his images at www.wildearthmedia.com

Lloyd Homer is a renowned New Zealand landscape photographer who specialised in aerial photography before the emergence of digital cameras. He worked as photographer for the NZ Geological Survey and its successor GNS for 40 years. He has been the photographer co-author of eight publications and is also the subject of the book *Flying High: The Photography of Lloyd Homer* by Simon Nathan (2019). His vast library of images are accessible at www.gns.cri.nz/Home/Products/Visual-Media-Library

Contents

Introduction

Aotearoa New Zealand is blessed with some of the most spectacular and diverse scenery in the world. This is not a random quirk of nature but a result of the unique and complex geological history of this part of planet Earth. New Zealand owes its very existence to its location over the last 20 million years astride the collisional boundary between two of the world's major tectonic plates. The collisional forces generate earthquakes and have pushed up the land and the axial mountains of both main islands. As the edge of the Pacific Plate has plunged down beneath the edge of the Australian Plate it has generated the magma that has erupted in many places and in many styles in the northern North Island. The land formed by these processes has been further modified by various types of erosion during the Ice Age climate and sea-level cycles of the last 2.5 million years. In the south, ice has carved out fiords and elongate U-shaped valleys and built extensive outwash plains. In the north, the coastline is characterised by eroded valley networks that have been drowned by rising sea level to form intricate harbours and bays. Huge coastal sand-dune barriers and white sand beaches have formed out of volcanic ash thrown out by giant eruptions in the centre of the North Island.

This book highlights 100 of New Zealand's most outstanding natural landforms and geological sites and explains, in relatively simple terms, how each was formed and may have been moulded into the spectacle we see today. When taken together, the accounts of all these sites tell the story of the 500-million-year history of New Zealand and our small, mostly submerged, continent of Zealandia. I hope these stories and images will inspire readers to value more highly New Zealand's geoheritage, both the outstanding and the more common. These features are the nation's irreplaceable archive or heritage library that documents the rich history of Aotearoa – how it was formed and how it has changed over the millennia.

How the sites were selected

I have made a personal selection of 100 of our most outstanding natural features to celebrate here. In making this selection I have been influenced by a number of factors that include equitable geographical coverage of geoheritage features from around the country, the best examples of the widest diversity of landforms, and large geological features that together characterise New Zealand. From the beginning I decided not to include smaller geological features or examples of our mineral deposits and fossils. Where two or more sites have similar features I have chosen the one that is more accessible to the public, but I have not excluded sites on private land or in the wilderness if they are in a class above all the rest.

I have been assembling the list of potential sites over several decades, based mostly on entries in the New Zealand Geopreservation Inventory of outstanding New Zealand

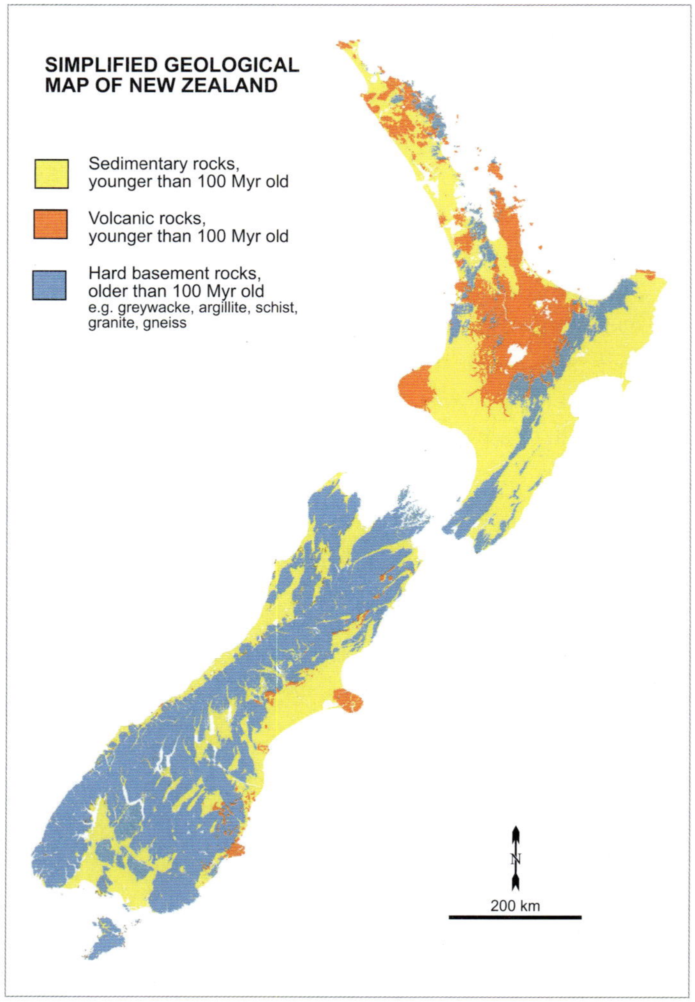

I.1 Simplified geological map of New Zealand

The different colours show where rocks of different kinds occur at the surface. The map shows that volcanic rocks are predominantly found in the northern half of the North Island and that much of the South Island is made of basement rocks (e.g. greywacke, schist and granite).

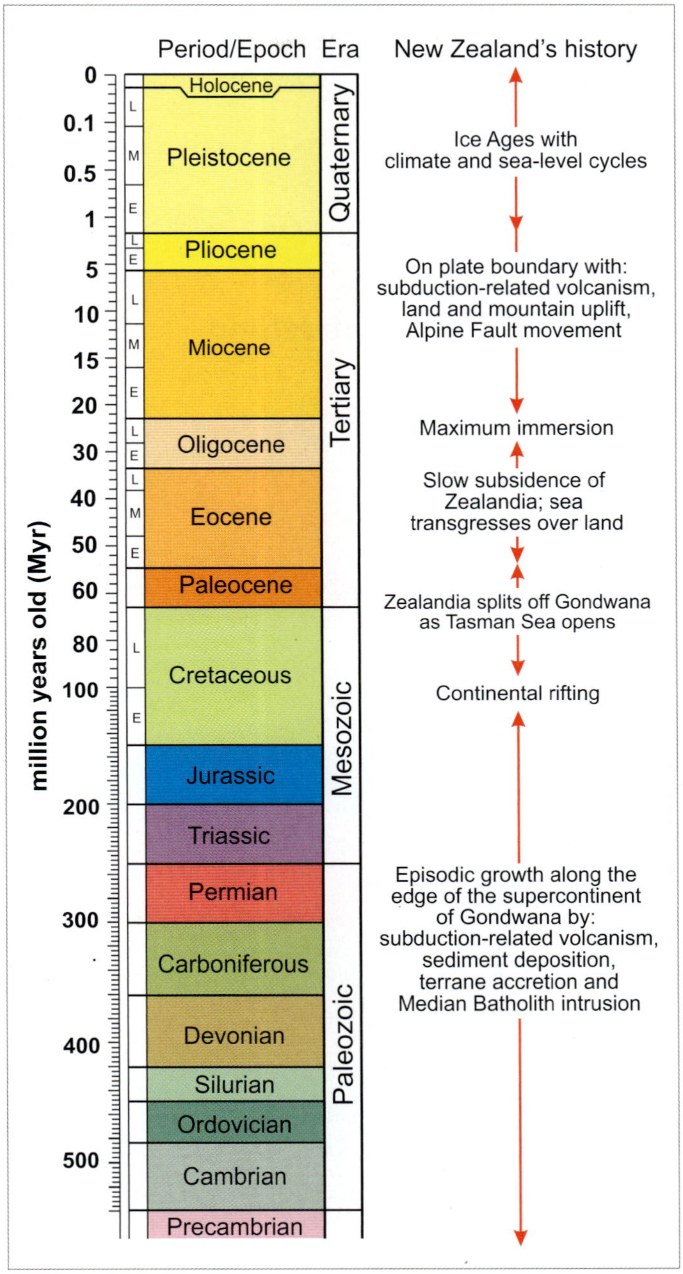

I.2 Geological time scale and a skeleton history of New Zealand

Geological time is a key part of explaining the origins and histories of the sites in this book. The geological eras, epochs and periods are used throughout, and their time ranges are given here (E, M and L denote early, middle and late, respectively). Absolute geological time is usually given in millions of years, which is abbreviated as Myr. The age of younger rocks and events is mostly given in thousands of years, sometimes abbreviated as kyr. How the rocks and events have been dated is summarised in a short section at the end of the book.

geoheritage features (https://services.main.net.nz/geopreservation/) that I have convened since the 1980s. To ensure that I had not missed any obvious front-runners, I also undertook extensive web-based searches for what the public and tourists consider to be the 'must-see' natural wonders, rock formations, waterfalls and scenic sites around the country. I would not expect that anyone will agree entirely with my selection or with the extra 'other contenders' sites in the back of the book, but hopefully all will find new places among my chosen sites and new or additional information about some of those they already know quite well.

How the book is organised

A few of the 100 featured places are not easily accessible to the public. Where they are publicly accessible, simplified instructions on the best places to go to view them or to have a closer encounter are given at the end of each site entry. The book is not written as a field guide, but undoubtedly it may double as such. For more detailed information use the web to find out where to go, how to get there, where to park, and whether or not a visit will be free. Many coastal sites may be best or most safely visited around low tide, so please check the time of this prior to your intented visit.

To explain the origins and history of each site inevitably requires the use of some technical terms, which are usually explained in the text where they first occur but are also defined in the glossary at the end of the book. The book assumes that readers have an understanding of the most basic rock types in New Zealand, but, if not, there is a small section at the back of the book that describes and illustrates them. Similarly a simple explanation of the main plate-tectonic processes and settings behind the creation of many of New Zealand's natural wonders is given at the end of the book to assist the less confident readers. A simplified version of the history of Zealandia and New Zealand is brought together and presented at the end of the book.

Further references for the general reader are listed at the end of the book, but there are also many unlisted web-based resources that can be accessed. The site-specific references listed are a combination of those that are easily accessible and readable (marked with an asterisk *) and those that are more technically advanced but were used as a source for some of the information presented.

The book was written and compiled by Bruce Hayward with many magnificent aerial photos taken by Alastair Jamieson and Lloyd Homer. Figures without a listed source were photographed or drafted by the author, Bruce Hayward.

100 natural wonders of Aotearoa New Zealand

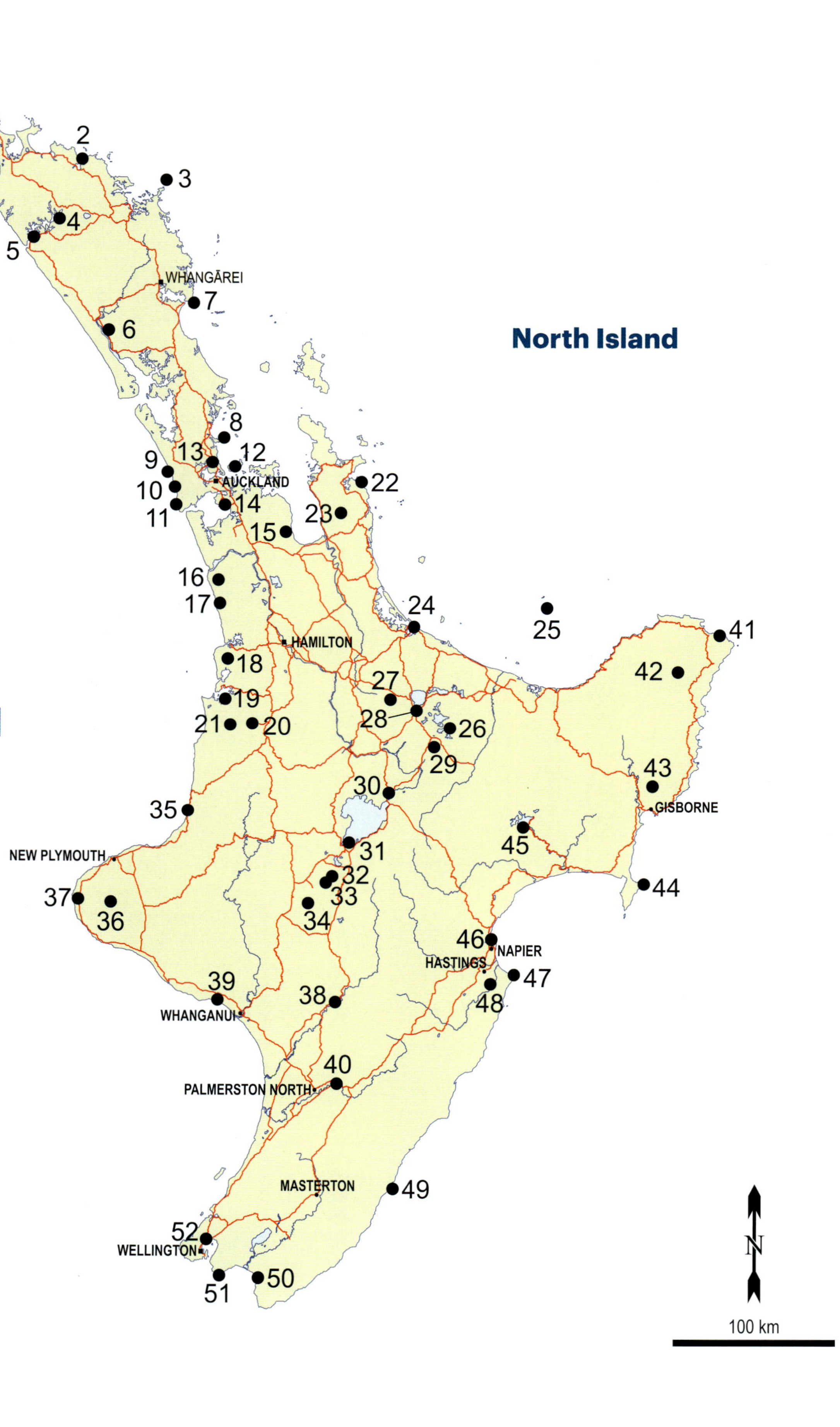

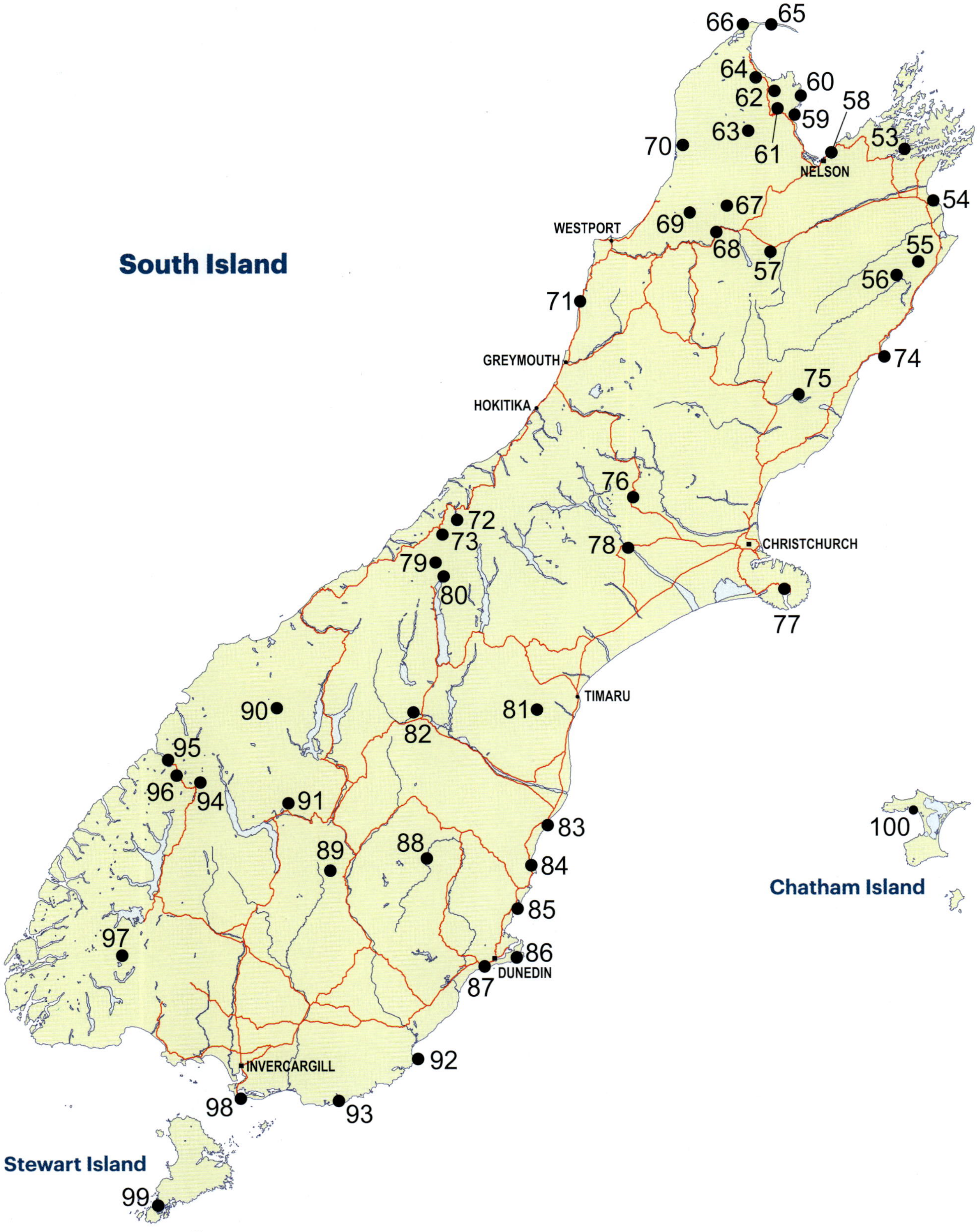

NB: Chatham Island is approx. 800 km east of Christchurch

Ngāuruhoe (foreground) and Ruapehu (distance) volcanoes, central North Island.
PHOTO: ALASTAIR JAMIESON.

North Island

1.

Far North

Giant sand-dune tombolos

The northern tip of New Zealand is the impressive Aupōuri Peninsula, a narrow strip of sand 8–10 km wide extending for 80 km north from Kaitāia. The peninsula consists of three main 'islands' of hard rock joined to the main part of Northland by the growth of sand-dune belts over the last 2–3 Myr to form a giant tombolo. A tombolo is a gravel, sand or shell spit or bar that joins an island to the mainland. The three 'islands' are centred on North Cape, Mt Camel and Cape Reinga-Pārengarenga (1.3). All are mostly composed of the eroded remnants of ancient volcanic rocks that erupted on the deep floor of the Pacific Ocean about 80–70 Myr ago (Cretaceous).

Aupōuri's oldest dune belt is composed of weathered and cemented sand and linked the Cape Reinga-Pārengarenga and Mt Camel islands to the mainland by 3–2 Myr ago. Since then we have been in the midst of the Ice Age cycles, and as sea level went up and down so the tombolo would have cycled in width from a few kilometres during warm periods to 20 km or more during colder glacial periods. Each sand-dune belt was largely deposited as sea level was falling after a warm period, and sea erosion recaptured some of the dunes each time sea level rose again; but over time a sequence of sand belts has been added, mostly along the west side of the tombolo but also some on the east side.

The sand that built the dunes was originally sourced from large eruptions in the centre of the North Island, from the Kaimai Ranges and later from the Rotorua–Taupō area (The Big Picture 6, page 97). Vast quantities of volcanic ash from huge eruptions were carried down rivers into the coastal sediment systems on both coasts. Most of the sand that formed the Aupōuri tombolo was probably transported by northwards longshore drift over tens of thousands of years. As it was moving north, volcanic glass, rock fragments and feldspars were mostly lost through weathering and abrasion, and the heavy black sand grains were left behind because of their density. The majority of the sand that reached Aupōuri was inert, abrasion-resistant quartz grains that are responsible for the brilliant white colour of some of the beaches and dunes (1.5). In other places the quartz is mixed with fragmented shell and lesser amounts of locally derived sediment.

Two smaller hilly and rocky blocks (North Cape and Cape Maria van Diemen) off the east and west sides of the tip of Aupōuri Peninsula were still islands during warm periods when

1.1 Cape Maria van Diemen was three or four small rocky islets on many occasions in the last million or so years during warm interglacial periods with high sea level. Several were probably separate islands during the present (Holocene) interval of high sea level up until a few thousand years ago, when they were linked to Aupōuri Peninsula by small sand tombolos. Motuopao Island (furthest away) is still separate. PHOTO: LLOYD HOMER, GNS SCIENCE, 7795.

1.2 View west to the former islets of Cape Maria van Diemen from Cape Reinga carpark at the end of SH 1. The yellow sand-dune tombolos can be seen linking the rocky islets to the mainland.

sea level was high as recently as the Last Interglacial period (130,000–120,000 years ago) and maybe even as recently as 8000–4000 years ago (Holocene). Both would have been joined by dry land to the peninsula during the lower sea levels of the Last Ice Age (100,000–10,000 years ago). Young sand dunes and trapped wetlands have built new tombolos in the last 4000 years, once again linking these two 'islands' to the peninsula.

Best seen from: Excellent views can be had from Cape Reinga, over the northwest tip of Aupōuri Peninsula with its own small versions of young sand tombolos linking Cape Maria van Diemen's rocky islets to the peninsula. Take an all-day tourist coach trip along Ninety Mile Beach (the west side of the tombolo) to Cape Reinga.

Take a closer look: From Cape Reinga walk down the track to Te Werahi Beach and out to Cape Maria van Diemen over the dune track. Visit the giant Te Paki sand dunes down Te Paki Stream Rd from SH 1.

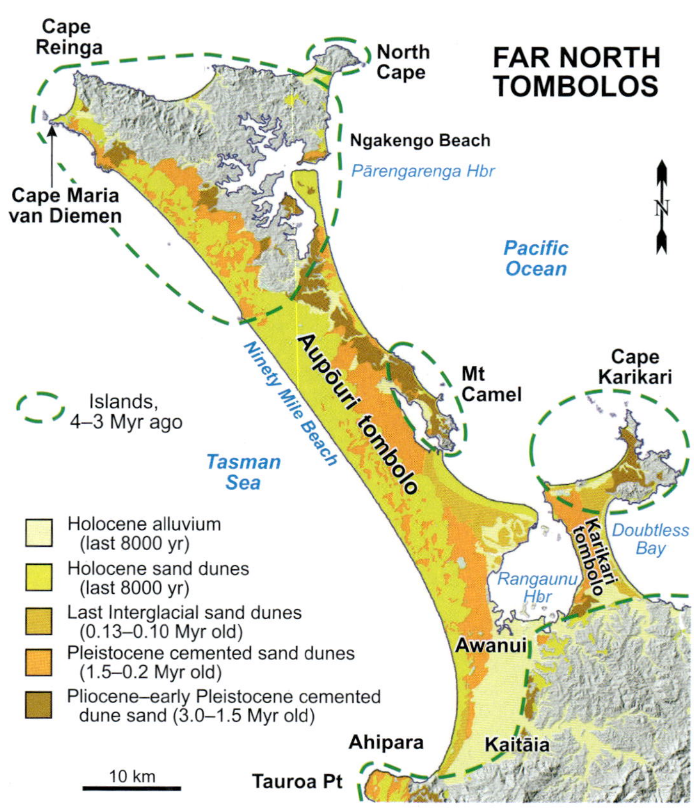

1.3 Map of Aupōuri Peninsula showing the sand-dune belts (orange and yellow) that have built the giant tombolo linking a number of islands to mainland Northland in the last 3–2 Myr. BASE MAP FROM GNS SCIENCE'S QMAP.

1.4 Aupōuri Peninsula is a tombolo composed of numerous sand-dune belts thrown up by the sea on many occasions over the last 2 Myr. Older sand belts are weathered, cemented and partly eroded, whereas the dunes thrown up in the last few thousand years are still fresh, loose sand. Here behind Twilight Beach is an eroded pedestal of slightly older sand dune standing up above the modern wind-blown dune surface.

1.5 The white beach and dune sand of Ngakengo Beach north of Pārengarenga Harbour is almost pure quartz derived from large eruptions in the centre of the North Island. The majority of the sand in the Aupōuri tombolo was derived from these eruptions and has been transported north by longshore drift within the last 2 Myr.

1.6 The steep lee face of Te Paki sand dunes slopes down to Te Paki Stream, which carries much of the advancing sand back to the beach to start its journey inland all over again. This is one of the most active and youngest sand-dune areas that has blown up on to the Aupōuri tombolo within the last few thousand years.

2.

Whangaroa Harbour

Dramatic bluffs made of volcanic ring plain remnants

The picturesque bluffs, rock pinnacles and rounded domes that surround the outer parts of Whangaroa Harbour are the eroded remnants of the Whangaroa Stratovolcano that erupted here 21–18 Myr ago (early Miocene). On land a stratovolcano typically consists of two parts – a central composite cone made of lava flows and volcanic rubble, often completely surrounded by a gently sloping ring plain (TBP 1.3 in The Big Picture 1, page 23). Most of the bush-fringed bluffs around the harbour are made of volcanic conglomerate with indistinct horizontal layering that was deposited by lahars on the ring plain of the volcano. Lahars are dense mud flows of volcanic sediment, including boulders, that periodically sweep down the steep slopes of a volcano's central cone and drop their load on the surrounding plain.

The central cone or cones of Whangaroa Volcano have now eroded away, with their location marked by several areas of exhumed plumbing, boulder-filled volcanic throats and small magma chambers. As the layers of volcanic conglomerate (lahar deposits) accumulated, they were buried and hardened. Stresses within these rocks formed wide-spaced, often vertical, cracks (joint planes) through the conglomerate. Over millions of years most of the volcano and ring plain have been removed by erosion.

Best seen from: Sail into the heart of this extraordinary coastal scenery on a water taxi or charter boat to Pekapeka Arm of Whangaroa Harbour.

Take a closer look: Climb the track from Old Hospital Rd to the top of St Paul's Rock above Whangaroa for panoramic views. The lahar deposits can be seen close-up at the west end of Tauranga Bay and south end of Taupō Bay.

2.1 The outer part of Whangaroa Harbour is surrounded by steep bluffs, cliffs and domes of andesite conglomerate and breccia that was deposited by passing lahars and are remnants of the highly eroded Whangaroa Stratovolcano. PHOTO: LLOYD HOMER, GNS SCIENCE, 2424.

⬆ 2.2 St Paul's Rock above the seaside town of Whangaroa is another eroded remnant of the Whangaroa Volcano's ring plain.

⬈ 2.3 The 300-m-high Taratara, inland from Whangaroa, is one of the best examples in New Zealand of a butte. It is an eroded remnant of the volcanic conglomerate of the Whangaroa Volcano laharic ring plain sitting on top of older rocks. Its flat top is probably the original surface of the ring plain when the volcano stopped erupting 18 Myr ago.

➡ 2.4 Close-up of laharic breccia, Taupō Bay.

⬆ 2.5 The vertical bluffs and overhanging 'Duke's nose' (centre), on the west side of Whangaroa Harbour's Pekapeka Arm, are the result of blocks breaking along joint planes and rolling off downhill.

THE BIG PICTURE 1

Northland Volcanic Arc

IN NORTHERN NEW ZEALAND there are eroded remnants of two northwest–southeast-oriented belts of volcanoes that erupted 22–15 Myr ago (early Miocene) along either side of the Northland Peninsula. Together they form the long-extinct Northland Volcanic Arc.

The western belt consisted of the largest stratovolcano in New Zealand's history – the Waitākere Volcano (The Big Picture 2, page 44), the slightly smaller Kaipara Volcano, and New Zealand's largest basalt shield volcano – the Waipoua Volcano. The western belt also included another four large and at least 35 small volcanoes, most of which are now buried beneath the seafloor off the west coast and have been discovered by remote sensing methods. Remains of three smallish volcanoes (including Tokatoka Volcano, Site 6) occur today along the eastern shores of the Kaipara Harbour. The eastern belt comprises the eroded remains of at least seven andesite stratovolcanoes – including those at Cape Karikari, Whangaroa (Site 2), Whangārei Heads (Site 7), Hen Island, North Great Barrier Island and North Coromandel. The Northland Volcanic Arc became inactive about 15 Myr ago and volcanic activity migrated southeastwards towards its present location as a belt between Mt Ruapehu, through Rotorua and the Bay of Plenty on up to the Kermadec and Tonga islands (The Big Picture 5, page 90).

The onset of a young phase of plate-tectonic collision through New Zealand began about 45 Myr ago. The eruption of the Northland Volcanic Arc was one of the earliest expressions of this young collision phase. The leading edge of the dense oceanic crust and lithosphere of the Pacific Plate pushed under the eastern margin (ancestral Northland and Auckland) of the Australian Plate, which here comprised lower-density, more buoyant, continental crust and lithosphere (TBP 1.2). The volcanic arc erupted above this west-dipping subduction zone. As the edge of the Pacific Plate slowly descended into the mantle (asthenosphere) it carried down with it an upper layer of water-saturated seafloor sediment. With increased temperature and pressure at about 100 km depth, the water was driven upwards out of the subducted rocks into the overlying mantle. The addition of water lowers the melting point of the mantle above the subducting plate, causing partial melting. This magma rises and collects in large chambers in the overlying continental crust. Magma periodically erupts from these as andesite, dacite or rhyolite forming a line of volcanoes above the subducted plate and parallel to the plate boundary. These lines of volcanoes are called volcanic arcs and together they form the well-known Pacific Ring of Fire. The onset of eruptions in the Northland Volcanic Arc was the start of fiery activity along this southwestern end of the Ring of Fire.

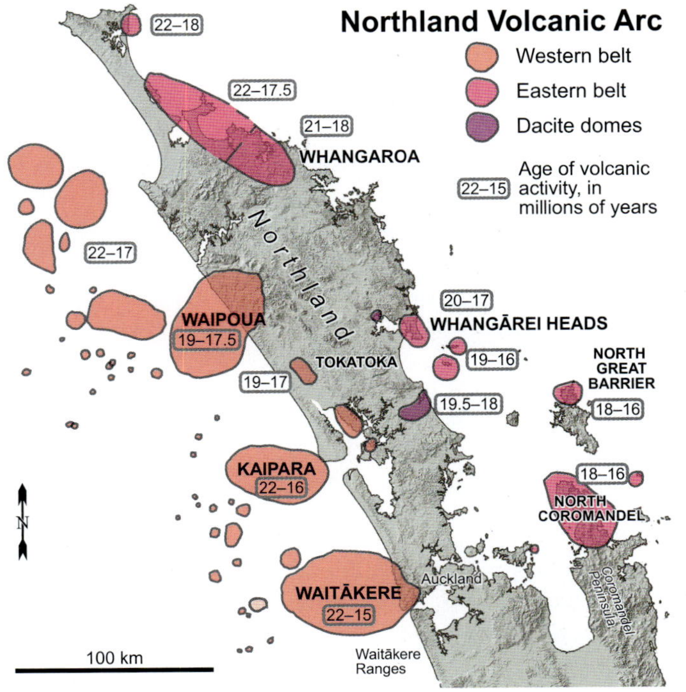

TBP 1.1 Map showing the volcanoes of the early Miocene Northland Volcanic Arc and the time of their eruption in millions of years ago.

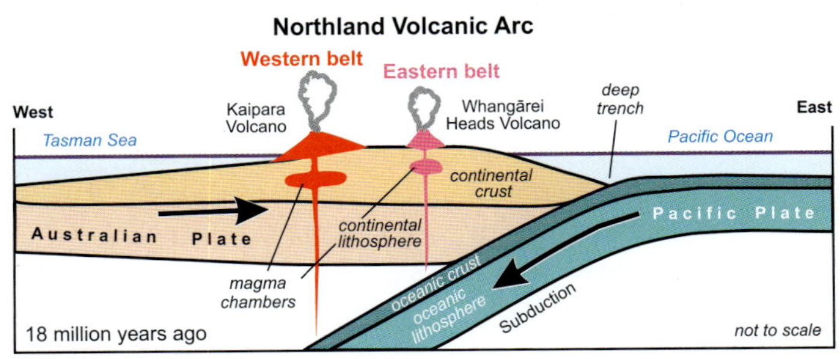

TBP 1.2 East to west cross-section through northern New Zealand about 18 Myr ago, showing the leading edge of the oceanic Pacific Plate subducting beneath the continental Australian Plate, resulting in the generation of magma bodies and the eruption of two linear belts of volcanoes within the Northland Volcanic Arc.

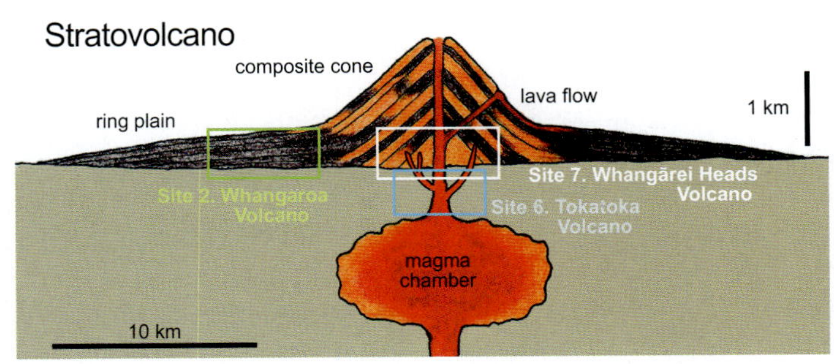

TBP 1.3 Schematic cross-section through a terrestrial stratovolcano, showing the central steep-sided cone surrounded by a ring plain composed of breccia deposited by lahars. The rectangles show the parts of the stratovolcano seen in the eroded remains of three Northland Arc examples featured in this book.

3.

Motu Kōkako / Hole in the Rock, Bay of Islands

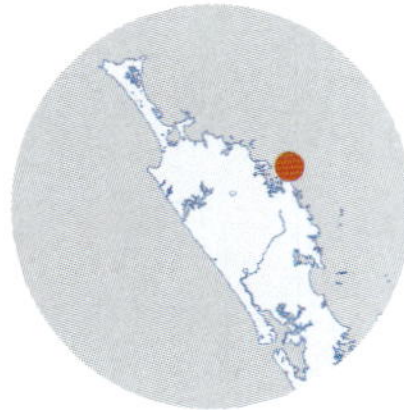

Large arch through greywacke and limestone sea stack

One of the major tourist attractions of the Bay of Islands is to cruise out to view, or in calm weather sail through, the Hole in the Rock of Motu Kōkako / Piercy Island. Motu Kōkako is a 152-m-high, barely vegetated, monolithic rock stack (3.5) off the end of Rakaumangamanga / Cape Brett, forming the southern entrance to the Bay of Islands.

The islands and coast of the Bay of Islands (3.2), like most of eastern Northland, are eroded out of hard greywacke rock (Waipapa Terrane). These rocks have had an eventful life, having been deposited as mud and sand in deep water off the coast of Gondwana 250–150 Myr ago (Triassic–Jurassic). They were then plastered (accreted) onto the east side of Gondwana and deeply buried, hardened and weakly metamorphosed at burial depths of 6–12 km. Subsequently they have been pushed up and the overlying rocks removed by erosion. The eroded surface of these old rocks has been pushed up higher in the east and tilted down to the west, forming the 'basement rocks' beneath the younger rocks of central and western parts of the Northland Peninsula.

During the long periods of lowered sea levels in the colder glacial periods of the Ice Ages (X.12, page 364), rivers and their tributary streams eroded their valleys well below modern sea level along the coast. Rising sea level (after the end of the Last Ice Age) slowly drowned an intricate valley system (3.3) between 18,000 and 7500 years ago, creating the present-day Bay of Islands. Ridges between valleys became peninsulas, and hill tops became islands and rock stacks. Motu Kōkako was a low peak on the northeast end of a narrow, meandering, steeply sloping ridge which when drowned became Cape Brett Peninsula.

Piercy Island differs from all other rock stacks and islands on the Northland coast in that it is mostly composed of hardened and altered limestone sitting on top of the basement greywacke. The northern two-thirds and top of the island consists of dark limestone that accumulated as a sandy shell bank on the floor of the shallow sea about 30 Myr ago (Oligocene). The contact with the underlying greywacke is tilted down to the north, and the different properties of the two rock types can be perceived in the island's shape (3.5). The limestone was buried by further strata and then cooked and altered by extremely hot fluids (250–500°C) passing up along fractures through it and the underlying greywacke. These hydrothermal fluids were heated by a shallow magma chamber about 20–17 Myr ago – part of the Northland Volcanic Arc (The Big Picture 1, page 23). The rising fluids acquired dissolved chemicals from the magma chamber, and as the fluids cooled while flowing up the subvertical fractures they deposited quartz veins containing low concentrations of zinc, lead and copper minerals. Close-up you may be able to see some of these lighter-coloured, and at times irregular, networks of quartz veins (3.4).

The Hole in the Rock is a sea arch that has been eroded along fracture planes of broken-up rock through the greywacke beneath the tilted contact with the overlying limestone. Erosion of the arch could only have been by the action of the sea during rough weather. Thus its erosion only happened when sea level was relatively high (within 10 m or so of the present level) during warm interglacial periods of the Ice Age cycles. It is possible that all, or most, of the erosion that produced this arch occurred during the present period of high sea level (last 8000 years) which followed 100,000 years of lower sea level when the island would have been fully vegetated and surrounded by high scree slopes of weathered rock.

Best seen from: Cruise out to the Hole in the Rock off Cape Brett on one of several possible cruises from Paihia or Russell.

3.1 The Hole in the Rock in Motu Kōkako / Piercy Island is sufficiently large for sizeable catamaran tourist boats to pass through in calm weather. The northern (left) two-thirds and top of this monolithic rock is composed of 30-Myr-old limestone that overlies hard greywacke rock along a north-tilted contact (down-tilted to the left) which passes above the sea arch. The hole has been eroded by the forces of the sea along several near-vertical fracture planes of broken-up rock.

⊙ 3.2 View southeast over the outer southern islands and peninsulas of the Bay of Islands that the boat trip passes en route to the Hole in the Rock. All are composed of hard greywacke rock while the landforms were created by rising sea level drowning a dissected valley and ridge topography (after the end of the Last Ice Age) between 10,000 and 7,500 years ago. PHOTO: ALASTAIR JAMIESON.

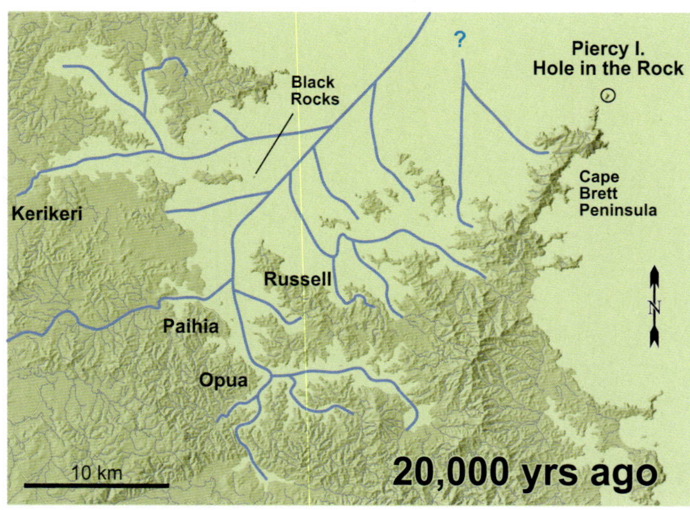

⊙ 3.3 When the sea was 130 m lower than present during the peak of the Last Ice Age, 20,000 years ago, the Bay of Islands was completely land, drained by a network of streams and rivers.

3.4 Close-up view of greywacke rock on the southwest side of Piercy Island, near the Hole in the Rock, showing the complex network of near-vertical, light-coloured quartz veins with numerous narrow, irregular, cross-cutting veinlets. These zinc–lead–copper mineralised veins formed by geothermal fluids penetrating up fractures above a shallow magma heat source about 20–17 Myr ago. Photo 50 cm across.

3.5 Profile, looking northwest, of Piercy Island showing the tilted limestone overlying the older greywacke and some of the vertical fracture planes that have been exploited by sea erosion to carve out the Hole in the Rock through the greywacke.

3.6 Piercy Island sits off the end of Cape Brett on the south side of the wide entrance to the Bay of Islands. PHOTO: LLOYD HOMER, GNS SCIENCE, 2016.

4.

Wairere Boulders, Hokianga

Internationally rare basalt karst

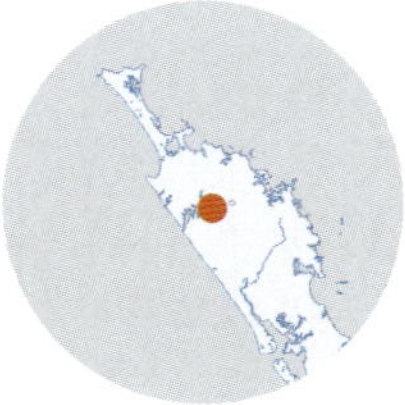

Northern New Zealand has the best-developed basalt karst in the world, and Wairere Boulders is one of the best places to see examples of these small landforms formed by solution of volcanic basalt rock. Karst landforms are usually developed in soluble carbonate rocks such as limestone and marble (The Big Picture 4, page 83), but in a few places in the world much slower solution of rocks such as basalt has formed similar features. Basalt boulders and the edges of basalt flows that are exposed above the ground beneath forest slowly dissolve under the attack of humic acids over thousands of years. Under the forest, the exposed tops of the boulders are often covered in small trees and astelias. These plants shed detritus which accumulates as humus around their roots and breaks down in water into humic acids.

4.1 View down part of the stunning basalt-boulder-filled Wairere Stream. Some of the 5-m boulders retain their vertical fluting, whereas others have rolled over. The stream flows beneath the boulders and only rarely overtops them when in flood.

4.2 A giant boulder in Wairere Valley that has tipped over from the orientation it had for thousands of years when the karst was forming on its surface. The former top with deep dissolved basins and armchair basins is on the left, with the fluting that formed vertically now visible on the right. PHOTO: EGON EBERLE.

4.3 Grassed eastern slope of Wairere Valley with a large block of basalt at the top, possibly still in place where it cooled and solidified as a lava flow 2.7 Myr ago. All the boulders that fill the valley bottom have slid or rolled down from the top of the slopes when they were still forest-covered. Numerous fluted boulders sit on the lower slopes in the middle of the photo.

These weak acids slowly etch out small basins on the tops of the boulders, and where they seep down the sides the boulders gradually develop vertical fluting that can grow to more than 50 cm deep and wide (4.4). The ultimate development is the creation of deep 'armchair basins' on the upper edge of the basalt surfaces where solution has dissolved out deep basins and their overflow seepage has created a deep flute down one side.

At Wairere Boulders, the small valley is chock-full of giant basalt boulders (4.1) with many having karst solution features. The boulders have slid down from the eroding bluffs of a basalt lava flow that forms a high ridge at the top of sloping valley sides, 150 m above the stream bed (4.3). The slopes beneath the flow are made of clay-rich mudstone that becomes very slippery when wet and aids the slide or glide of boulders down the slope through the forest. Some boulders may have come down in a single landslide event, but others possibly took thousands of years to reach the valley floor. There are still some boulders on the deforested slopes today, which are moving a few centimetres downslope each winter. These slowly gliding boulders often retained the same orientation for thousands of years, providing sufficient time for basins and flutes to develop on their surface. In other instances some boulders may have come to a stop at the bottom of the slope or in the stream bed itself, providing the long time required for the karst to form. Some boulders in the valley bottom retain their orientation with vertical fluting but others have been undermined by stream erosion and have toppled over with their fluting now angled (4.2).

Take a closer look: Take the walks among the boulders and up to the lookout at the Wairere Boulders tourist attraction at the end of McDonnell Rd, Horeke.

4.4 The beautifully fluted side of a basalt boulder in Wairere Valley.

4.5 A large basalt boulder with dissolved, elongate, shallow basins on the top at Wairere Boulders.

5.

Koutu Boulders, Hokianga

New Zealand's largest spherical concretions

5.1 This 3.5-m-diameter spherical concretion on the foreshore at Koutu, Hokianga Harbour, is among the largest concretions in New Zealand.

The best-known spherical concretions in New Zealand are the Moeraki Boulders in Otago (Site 84), but those who have visited both there and Northland's Koutu Boulders say they are 'blown away' by the giant size and number of these northern examples. Koutu's spherical concretions occur along a 1-km stretch of the Hokianga Harbour's shoreline.

Concretions are hard, cemented, spherical or ovoid bodies of rock that range in size from a pea to 2 m or more across. The cementing mineral is often calcite (calcium carbonate), but sometimes may be quartz (silicon oxide), siderite (iron carbonate), barite (barium sulfate) or pyrite (iron sulfide). These concretions grow within layers of sedimentary rock (usually sandstone) after they have been buried by additional sediment. In favourable conditions, dissolved minerals in percolating groundwater start to crystallise, often around a small pebble, plant fragment, shell or rip-up clast of mudstone (5.3). The crystals grow in the small spaces between the sandstone grains, cementing them together. They slowly grow outwardly over thousands or hundreds of thousands of years to form hardened concretions. If the minerals crystallise at the same rate in all directions, then a spherical concretion forms. Other times they develop flatter or more elongate shapes. Sometimes the minerals grow around elongate or branching burrows and thus form cylindrical concretions.

The Koutu concretions are mostly cemented by calcite and grew within thick sandstone beds that were deposited as sand on the floor of the deep sea off the coast of Zealandia as it was splitting away from Australia about 70 Myr ago (late Cretaceous). About 25 Myr ago, well after the concretions had grown within the buried rock, the whole region to the northeast of Northland was pushed up out of the sea and giant slabs of rock, including the sandstones at Koutu, slid southwestwards onto a subsiding Northland. Since then, Northland has bobbed back up again to become eroding land and the Koutu rocks have been progressively unearthed by erosion that has removed maybe as much as 1000 m of cover rock in the last 15 Myr or so. The loose concretions that now sit on the Koutu foreshore have all been eroded out of the softer sandstone host rock in the last 7500 years.

Take a closer look: Walk among these concretions on the Hokianga Harbour foreshore (within 2 hours either side of low tide). Access from the carpark on Waione Rd or the end of Cabbage Tree Bay Rd, off Koutu Loop Rd, 4 km east of Opononi.

5.2 At low tide visitors can walk among dozens of large and small concretions over a 1-km stretch of coast at Koutu, Hokianga. The concretions are eroding out of the 70-Myr-old sandstone rocks that form this section of coast. These concretions are 1–2.5 m in diameter.

5.3 The Koutu concretions have grown around pieces of mudstone (rip-up clasts) suspended within the sandstone rocks. The small, rounded holes in the centre of the two halves of this broken, 1-m-diameter concretion are this partly eroded-out chunk of soft mudstone.

5.4 A 0.5-m-diameter concretion on the foreshore at Koutu. Conditions within the host sandstone must have changed over time and the material inside this concretion began to dehydrate and contract. As a result shrinkage cracks formed, which later were filled by further cementing crystals of a different composition. Thus this concretion developed a regular network of erosion-resistant veins that give it the name 'turtle back concretion', because of its resemblance to a turtle shell.

5.5 The most abundant and largest spherical concretions in New Zealand occur along the southern shore of Hokianga Harbour at Koutu. PHOTO: ALASTAIR JAMIESON.

6.

Tokatoka and Maungaraho, North Kaipara

Exhumed plumbing of a vanished volcano

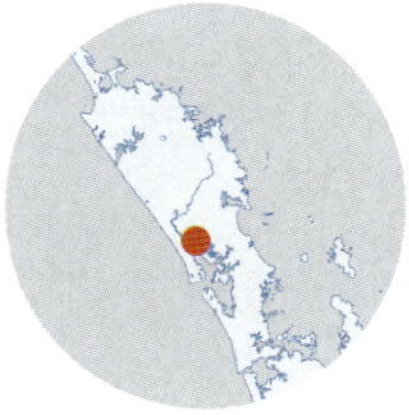

⊙ 6.1 The 180-m-high conical peak of Tokatoka is the deeply eroded plug or small magma chamber that formed in the throat of one of the vents of Tokatoka Volcano about 18 Myr ago. PHOTO: ALASTAIR JAMIESON.

⊙ 6.2 Maungaraho is a 221-m-high blade of andesite rock that protrudes up out of the ground, 4 km north of Tokatoka. Like Tokatoka, it is a deeply eroded remnant of some of the plumbing that fed eruptions of Tokatoka Stratovolcano below, about 18 Myr ago. PHOTO: ALASTAIR JAMIESON.

⊙ 6.3 Schematic cross-section through the Tokatoka area showing the inferred size of Tokatoka Stratovolcano that has been completely eroded away.

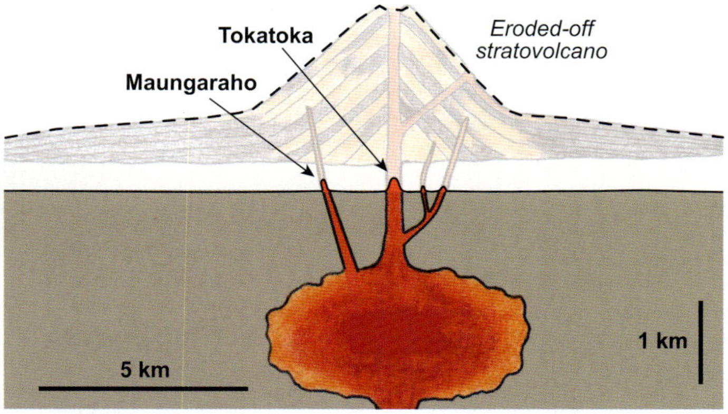

Tokatoka is a conical peak that stands like a sentinel above the kūmara-growing alluvial plain of the Wairoa River, between Dargaville and Ruawai (6.1). It is some of the unearthed plumbing that fed the Tokatoka Stratovolcano that erupted here 19–17 Myr ago. Tokatoka stands high because it is made of hard andesite rock (solidified molten magma), which is more resistant to erosion than the softer, muddy limestone it intruded through beneath the volcano (6.3). There are over 80 eroded remnant plugs and dikes (plumbing) forming small hillocks in the rolling paddocks to the north of Tokatoka. Their distribution indicates that the stratovolcano's cone was at least 8–10 km in diameter and this was probably surrounded by a wide ring plain of lahar breccias. It was slightly smaller than modern Mt Tongariro. Tokatoka Stratovolcano was part of the Northland Volcanic Arc (The Big Picture 1, page 23) that erupted along the length of Northland and Auckland between 22 and 15 Myr ago.

All of Tokatoka Volcano has now entirely eroded away, except for the volcano's plumbing remnants consisting of conduits and fissures now filled with cooled magma, mostly of andesite composition but also some basalt and dacite. Tokatoka itself is eroded from the largest body of andesite in the area and was probably once a small, shallow magma chamber within the throat of one of the volcano's main vents. The highest peak in the district is Maungaraho – a rocky blade of andesite that stands high above the surrounding land (6.2). It is the cooled magma that once filled this vertical fissure and possibly fed lava-fountaining and flows at the surface hundreds of metres above.

Best seen from: The distinctive conical peak of Tokatoka and the steep-sided blade of Maungaraho are most easily seen from SH 12, between Ruawai and Dargaville.

Take a closer look: Both peaks are in signposted reserves on Tokatoka Rd and Maungaraho Rock Rd, respectively. Climb the steep track to the summit of Tokatoka for breathtaking panoramic views (1-hr return).

⊙ 6.4 Tokatoka peak viewed from a harbour cruise on the Wairoa River. PHOTO: EGON EBERLE.

⊙ 6.5 The soft limestone that Maungaraho dike intruded through has been removed by erosion, exposing vertical walls of andesite rock. PHOTO: ALASTAIR JAMIESON.

7.

Whangārei Heads

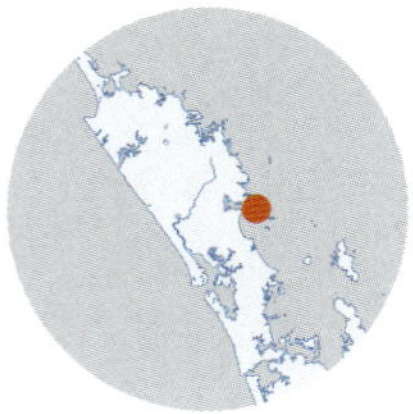

Pinnacled remnants of volcanic cones

Whangārei Heads, at the northern entrance to Whangārei Harbour, owes its existence to the eroded remnants of several small overlapping stratovolcanoes (7.4) that erupted here as part of the Northland Volcanic Arc (The Big Picture 1, page 23) between 20 and 17 Myr ago (early Miocene). The three prominent, rock-capped highland areas at Whangārei Heads – Mt Manaia (420 m), Mt Aubrey (216 m) and Bream Head (476 m) – are all composed of massive to weakly bedded rubbly breccia with angular cobbles and pebbles of andesite interspersed with occasional lava flows. These are the eroded lower remnants of small, steep-sided composite cones. The flat-lying land surface (of uplifted basement greywacke and Northland Allochthon rocks) that they erupted through and over is at about 100 m above sea level today.

In the same area, but usually at a lower elevation, erosion has exposed at least five large bodies of diorite, andesite and dacite that had intruded up into the allochthon and greywacke. These are shallow-level magma chambers, part of the plumbing that lay beneath the stratovolcanoes. Other evidence of the subvolcanic plumbing is numerous andesite dikes, best seen along the coast (7.6). These intrusive sheets of solidified magma have sometimes baked the limestone they pushed their way through to a hard black skarn rock.

Offshore to the southeast, Taranga / Hen Island and Tuturu / Sail Rock are the eroded remnants of a similar 5–8-km-diameter stratovolcano. The rocks of the nearby Marotere / Chicken Islands have been uplifted several hundred metres higher than Hen Island and are mostly composed of basement greywacke intruded by andesite dikes. The two easternmost Chickens are made of a mass of coarsely crystalline diorite and granodiorite that is inferred to have been a small magma chamber at shallow depths beneath yet another stratovolcano.

Best seen from: There are striking views of the Whangārei Heads skyline from the Brynderwyn Hills on SH 1.

Take a closer look: Steep forested tracks lead to the top of Mt Manaia (2 hrs return) or along Bream Head ridge (Te Whara Track, half- and full-day options), or walk along the natural jetty, signposted at Taurikura Bay (5 mins).

7.1 Rugged Bream Head and Mt Manaia (beyond) at Whangārei Heads have been carved by erosion and the weathering out of two 20–17-Myr-old volcanic cones. PHOTO: ALASTAIR JAMIESON.

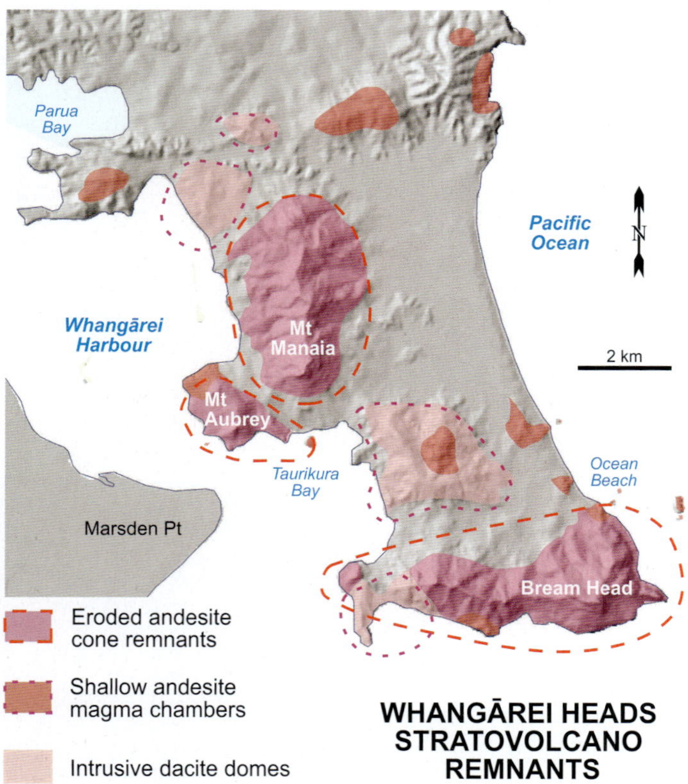

7.2 The huge rock ribs separated by deep guts on the peak of Bream Head are a result of rock faces spalling off along weak fracture plains that cut through the volcanic breccia. PHOTO: ALASTAIR JAMIESON.

7.3 Map showing location of eroded remnants of the 20–17-Myr-old Whangārei Heads Volcano.

➲ 7.4 Profile of Whangārei Heads from the south showing how much of the original volcanic cones has been eroded away. The former land surface that the stratovolcano erupted over is shown by the dashed white line.

➲ 7.5 The top of Mt Manaia is capped by a cluster of tall free-standing pinnacles made of volcanic breccia from the eroded remnant of a volcanic cone. The numerous vertical fracture planes that cut through the rock have been instrumental in producing these rocky fingers.

➲ 7.6 Taurikura Bay has the best example of a natural jetty in New Zealand. It is a dike of andesite – part of the plumbing that fed magma to the erupting volcano above. This dike is a vertical sheet of andesite rock that has horizontal columnar jointing formed by shrinkage as the molten magma cooled and solidified. The softer, muddy limestone that the dike intruded into is eroding faster than the jointed dike, thereby creating the jetty.

8.

Whangaparāoa Head

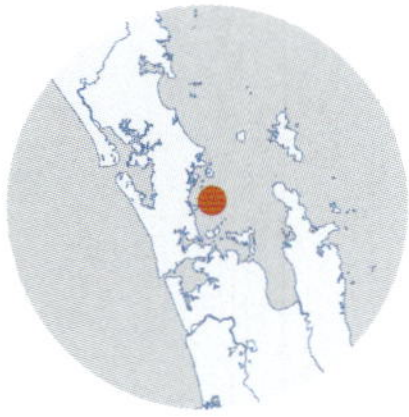

New Zealand's most folded, faulted and tilted strata

8.1 Whangaparāoa Head at the easternmost tip of Whangaparāoa Peninsula is composed of layers of Waitematā Sandstone that have been pushed up to near-vertical from their original horizontal orientation when they accumulated on the deep seafloor about 20 Myr ago.

Some of the most impressively folded and faulted strata to be seen anywhere in New Zealand occur in the Waitematā Sandstones that underlie most of the Auckland region. The best place to see a wide range of mind-boggling deformation is in the cliffs around the eastern end of Whangaparāoa Peninsula. Waitematā Sandstones consist of alternating layers of thick sandstone and thinner mudstone that were deposited on the deep floor (~2000 m deep) of a marine depression, known as the Waitematā Basin, about 20 Myr ago (8.2). At that time most of Northland was eroding land. Rivers carried gravel, sand and mud down to its south coast where it built out a shallow shelf of unstable sediment on the northern edge of the Waitematā Basin. Periodically a large earthquake shook the area and some of this unstable sand and mud flowed off down the submarine slopes as a dense slurry of sediment, known as a turbidity current. The currents probably flowed down submarine canyons and out onto the floor of the Waitematā Basin where they slowed down and deposited their load as a bed of sand.

Sand layers were deposited by turbidity currents once every few hundred years; in the quiet time in between, mud dropped out of suspension from the seawater and slowly built up a thin bed. As the sand and mud layers were buried on the seafloor by the accumulation of more sediment they were progressively compacted and hardened into sandstone and mudstone rocks. Early in this process the beds were cohesive but not solid, and could be deformed plastically into quite tight folds without breaking apart (8.5). With deeper burial the strata became harder and more solid, especially the sandstone beds, and when subjected to bending forces the layers responded in a brittle fashion and broke apart rather than folding.

Walking around the base of the cliffs at the end of Whangaparāoa Peninsula (at low tide) you can see a wide variety of folded, faulted and chaotically deformed strata (see photos) which record landslides and slumps on the seafloor. In these seafloor failures the more plastic sediment layers nearer the surface were folded as they slid downslope, whereas other, deeper, layers broke into giant blocks as they moved. Such were the forces within some of these seafloor slides that originally horizontal strata were at times pushed right up to vertical, as seen in Whangaparāoa Head itself (8.1).

Best seen from: The best examples of deformed strata occur in the cliffs around the end of Whangaparāoa Peninsula and can be seen by walking along the foreshore at low tide between Army Bay and Pink Beach (Shakespear Regional Park). A complete circuit returning overland through Shakespear Regional Park will take 3–4 hours. Similar, slightly less spectacular, deformed strata can be seen in the sea cliffs in a number of other places around Auckland, such as Waiwera, Stanmore Bay, Little Manly, Campbells to Castor Bays, Takapuna Head and Musick Point.

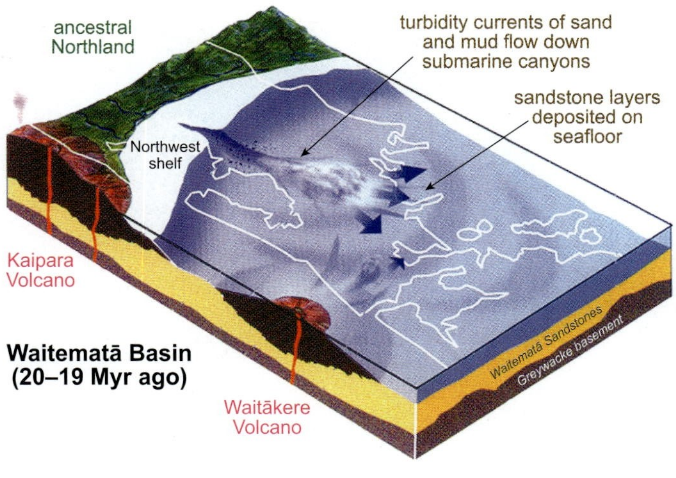

8.2 Three-dimensional diagram showing the geography of Auckland 20 Myr ago when the deep marine Waitematā Basin covered most of the region. Sediment eroded off the ancient Northland land area flowed down the submarine slopes in turbidity currents and was deposited on the basin floor as alternating layers of sand and mud. These were compacted into the Waitematā Sandstones that now underlie the Auckland region.

⬆ 8.3 A large recumbent fold (on its side) in 30-m-high cliffs near Whangaparāoa Head.

⬇ 8.4 A reclined double fold of thick-bedded sandstones and interbedded thin mudstones near Whangaparāoa Head.

➲ 8.5 A tight anticline (upfold) of plastically deformed sandstone and mudstone in an 8-m-high cliff near Whangaparāoa Head.

➲ 8.6 This bed of sandstone has rolled up on itself within the upper layers of the seafloor sediment while it was sliding down the submarine slopes of the Waitematā Basin.

➊ 8.7 This tight syncline (downfold) on the foreshore at Whangaparāoa Head was formed during seafloor landslides about 20 Myr ago. It was subsequently buried by another 500–1000 m of sediment, then uplifted by deep tectonic forces and the overlying layers progressively removed by erosion. This three-dimensional remnant of the syncline has been eroded out by selective marine erosion within the last few hundred years.

➋ 8.8 Gently folded sandstone and mudstone beds in the shore platform at Whangaparāoa Head.

THE BIG PICTURE 2

Waitākere Volcano

THE WAITĀKERE STRATOVOLCANO was the largest cone volcano in New Zealand's history. It erupted off and on for up to 7 Myr between 22 and 15 Myr ago, building up a huge volcano complex with a footprint stretching 50 km east to west and 40 km north to south. It erupted on the floor of the deep sea, perhaps about 2000 m deep, and built up a cone that pierced the waves and at its highest may have been 3000–4000 m high above the seafloor (TBP 2.2). Its volume was at least five times that of modern Ruapehu and Tongariro cone volcanoes combined. The eroded stump of most of the Waitākere Volcano lies offshore beneath the Tasman Sea west of Auckland.

Auckland's western backdrop, the Waitākere Ranges are the uplifted eroded remnants of the eastern submarine slopes of the Waitākere Volcano. There would have been many different volcanic vents, and there is evidence that at times parts of the upper cones collapsed in dramatic slides, sending debris cascading down the submarine slopes. The presence of rare pieces of fossil wood, shallow marine fossils, red-oxidised lumps of volcanic rock (that were erupted hot into the air) and sometimes well-rounded cobbles (rounded in streams or by breaking waves) provide evidence that the top of the volcano was one or more islands for much of its life.

Around 17 Myr ago, tectonic forces related to the collisional plate boundary pushed the Waitākere Volcano and much of the Auckland region up out of the sea. For the last 1–2 Myr of its history, Waitākere Volcano erupted entirely on land with the eroded remains of two lines of young vents recognisable in the Waitākere Ranges (TBP 2.1). One line coincides with the present Tasman Sea coastline and the other is further inland in the vicinity of Scenic Drive. The volcano became extinct 15 Myr ago and since then most of it has been eroded away by the Tasman Sea. Indeed, there have been two phases of erosion. The first phase flattened off the higher portions of the volcano down to a coastal plain by about 5 Myr ago. This was followed by a second period of up to 300 m of uplift and the start of the latest round of erosion. The present flat ridge crests of the Waitākere Ranges are the uplifted and northwest-tilted remains of the 5-Myr-old coastal plain.

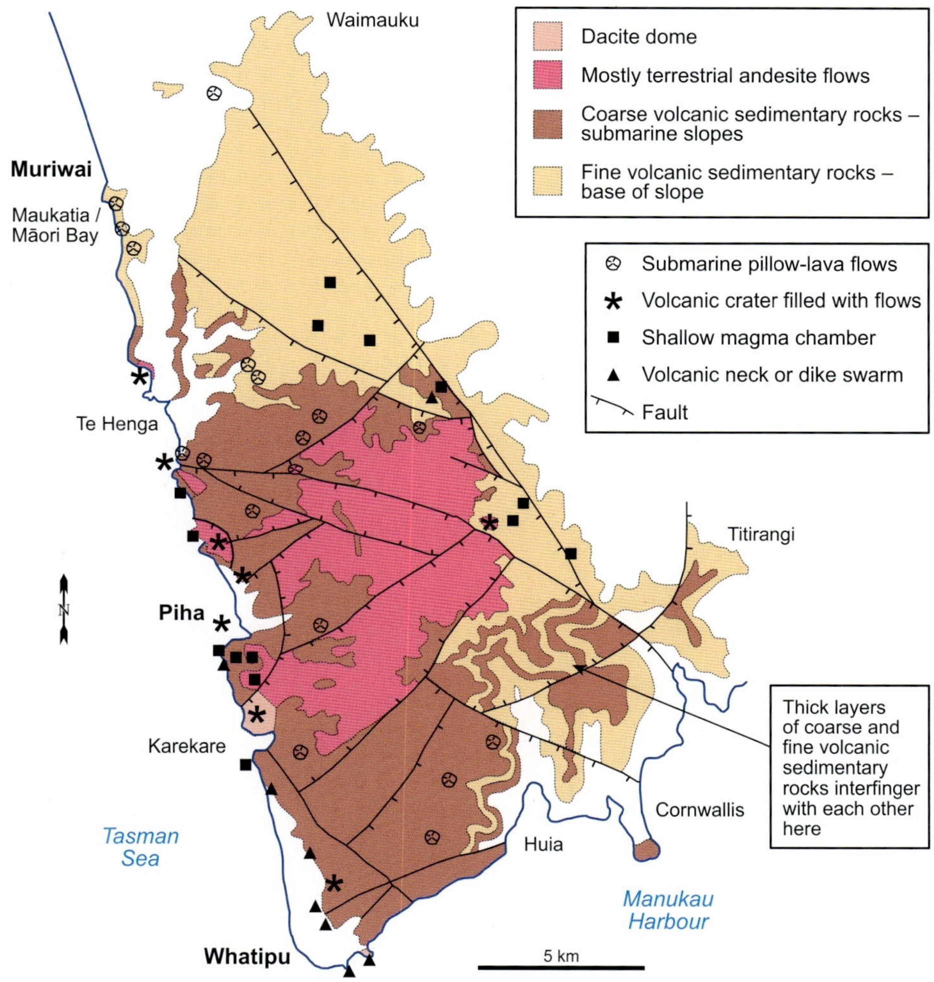

➲ TBP 2.1 Simplified geological map of the Waitākere Ranges, showing the distribution at the surface today of the main rock types that accumulated on the eastern slopes of the Waitākere Volcano, 22–15 Myr ago.

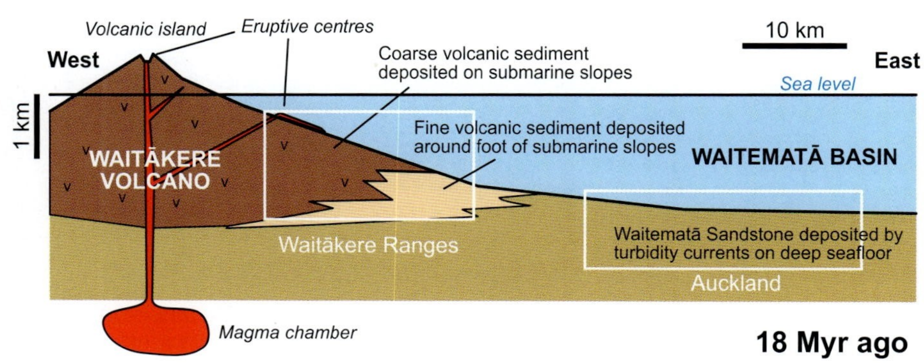

➲ TBP 2.2 Schematic cross-section through the Waitākere Stratovolcano after it had built up on the floor of the ocean and was capped by a volcanic island. It was subsequently pushed up out of the ocean in two phases and all but the eastern slopes (now the Waitākere Ranges) completely eroded away.

9.

Muriwai pillow lava

Undersea lava flows with globally rare fans of huge columnar joints

There are many examples of pillow lava in New Zealand and around the world, but the pillow lavas exposed in the sea cliffs south of Muriwai have an international reputation as the best example anywhere of a large and complex pillow lava flow. Pillow lava forms when molten lava is extruded into water, particularly seawater. Small flows consist of numerous overlapping, often branching lobes of lava (0.2–2 m in diameter; 9.6) that were squeezed out into water on the seafloor. The lobes have distinctive pillow-shaped cross-sections with thin glassy selvedges surrounding solid grey volcanic rock with radiating cooling joints (9.4). An example of one of these smaller flows occurs 1 km south of Maukatia / Māori Bay.

In the cliffs above Māori Bay there is a 20-m-thick pillow lava flow within a sequence of layered volcanic sandstone and conglomerate. The flow is seen in cross-section and has a complex structure. The sides of the flow are composed of pillow lava lobes, whereas the centre of the flow has several 10-m-plus-diameter 'pillows' of solid andesite with fans of huge radiating cooling joints (9.1). This is one of the few places in the world where such detailed internal structure of a thick pillow lava flow can be seen. When this flow was moving down the submarine slopes, the large andesite 'pillows' would have been internal tubes of molten lava that were feeding the front of the flow (9.2). These internal feeders were insulated from the cooling effects of the seawater by the surrounding carapace

9.1 The huge fans of columnar joints within the large Muriwai pillow lava flow are of international significance. The cliff exposes a cross-section cut through a giant internal lava tube with radiating cooling joints (giant 'pillow', left), which fed lava to two 2-m-thick horizontal sheet flows. These sheet flows, at the bottom and top of the flow, have vertical columnar joints. Total thickness of the lava flow is 20 m.

of smaller pillow lobes, and thus the feeder tubes remained molten for some time. When they eventually cooled from the outside inwards, the radiating cooling joints were formed.

The Muriwai pillow lava flow was erupted on the seafloor on the deep northeastern slopes of the giant Waitākere Volcano about 17 Myr ago. The layers of sandstone and mudstone beneath the flow contain rare fossils that indicate that they accumulated on the seafloor at lower bathyal depths of 1000–2000 m.

Best seen from: The track to the gannet colony lookout from the carpark in the old quarry provides a wide view back to the pillow lava flow high in the cliffs above Māori Bay.

Take a closer look: At mid-tide you will see large blocks of columnar andesite on Māori Bay beach. Look up to see the details of the pillow lava flow above. On a spring low tide, walk 1 km south down the coast to see more pillow lava and a dike feeding a pillow roll (9.4, 9.5).

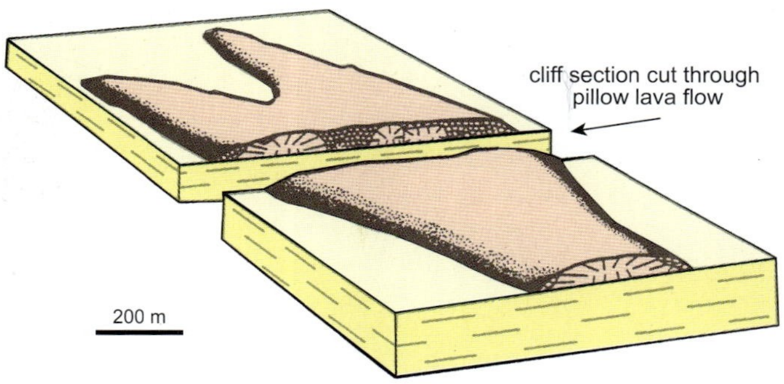

9.2 View looking east of cut-away block diagrams illustrating the inferred shape of the Maukatia / Māori Bay pillow lava flow seen in cross-section in the cliffs (middle of diagram). The giant internal fans of cooling joints are solidified andesite lava that once flowed along as large feeder tubes inside the moving flow on the seafloor.

9.3 View south across Māori Bay showing the large pillow lava flow (dark grey) overlying layered volcanic sandstones and conglomerate forming the cliffs.

9.4 Note the pillow-shape of the lava lobes in the eroded cross-section of this small pillow lava flow in the cliffs, 1 km south of Māori Bay. Each pillow is surrounded by a glassy selvedge 1–2 cm thick with crudely radiating cooling joints of andesite within. The light-coloured material between the pillows is seafloor sediment. Photo 4 m across.

9.5 A sheet-like dike of solid andesite rock, which in its molten state pushed its way through the side of the volcano and extruded a 2-m-diameter lava pillow roll into the unconsolidated seafloor sediment. Here, 1.5 km south of Muriwai, is one of the few places in the world where a dike can be seen attached to and feeding a large pillow roll along its upper edge.

9.6 Part of a small pillow lava flow, 1 km south of Muriwai, showing its internal structure of overlapping lava lobes. Photo 3 m across.

10.

South Piha and The Gap

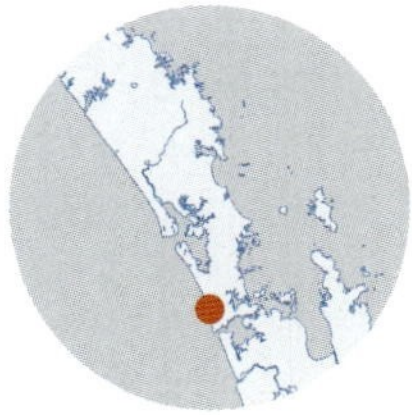

Rocky islets, sea tunnels and caves

The best-known and most easily accessed rocky islets and sea tunnels on the rugged west coast of the Waitākere Ranges are at South Piha. The iconic profile of Whakaari / Lion Rock (10.1) is nationally recognised. Its wave-battered rocky cliffs rise 101 m in a series of steps separated by flax- and pōhutukawa-covered steep slopes. Lion Rock consists of the eroded throat-filling contents of a 16-Myr-old volcanic vent. It is one in a line of vents, running along the present-day west coast, that erupted late in the history of the Waitākere Volcano (The Big Picture 2, page 44). Lion Rock exists because it contains irregular intrusive tongues and dikes of andesite lava that are more resistant to erosion than the surrounding volcanic conglomerate that the volcanic vent erupted through.

The rocky coast to the south of Piha is composed of the greatly eroded remnants of weakly layered volcanic conglomerate that was deposited on the submarine slopes of the growing Waitākere Volcano about 19 Myr ago (The Big Picture 2). Around The Gap (10.3) these conglomerates have been intruded by several steep sheets (dikes) of andesite lava that came up along major joint planes through the rocks about the same time as Lion Rock was erupting.

Today's coastline is the most recent incarnation of a whole series of coastlines that formed here every time sea level was high during the warm periods of the Ice Age cycles. During the long periods when sea level was 20 m or more lower, the coastal islets and embayments of South Piha would have been forested hills, ridges and valleys. The coastal cliffs would have weathered back and fans of vegetated scree built up in front of them. The small Lovett Stream that now flows out through Te Rua o Kaiwhare / The Blowhole would have flowed north through what is now Puaotetai Bay and met Piha Stream off Lion Rock (10.5). As the sea rose towards its present level after the end of the Last Ice Age, the waves rapidly removed the soft soil, weathered rock and scree, exhuming cliffs and islets that last had sea lapping up against them 110,000 years

10.1 Looking north across low-tidal sand to the iconic Whakaari / Lion Rock at South Piha.

10.2 The keyhole-like sea tunnel through Taitomo Island from Puaotetai Bay, South Piha. Breaking waves on the seaward side have blasted their way into the joints of a hard andesite dike, loosening blocks and plucking them out one by one as it eroded its way right through the volcanic conglomerate island. Parts of the uneroded jointed andesite dike can be seen above and on the right side of the tunnel.

earlier. Since then the more exposed cliffs have eroded back about 10 m, creating high-tide platforms, and it is likely that the present sea tunnels, caves and even The Gap itself have all formed in just the last 8000 years of high sea level.

Breaking waves force seawater into small cracks and fractures in the rock and gradually erode them out. The Blowhole is a 50-m-long tunnel that has eroded along a joint plane through a narrow ridge of conglomerate until it skylighted in the small Lovett Stream and has now opened out as a blowhole. Erosion along the horizontal cooling joints and edges of several dikes in Puaotetai Bay has formed the famous sea tunnel that passes right through Taitomo Island (10.2) and also contributed to creation of The Gap and the sea caves that run southwards from it (10.3).

Best seen from: Commanding views of Lion Rock and Piha Beach are obtained from the Piha Rd before descending the hill.

Take a closer look: At low tide, walk south around the rocks from South Piha into Puaotetai Bay. At any tide take the path leading part-way up Lion Rock, or take the Tasman Lookout walk for sweeping views and continue on down into The Gap.

⊙ 10.3 View of the south side of The Gap at Piha, where the tempestuous Tasman Sea has broken a 50-m-wide gap through the conglomerate cliffs to Puaotetai Bay. In the middle of the cliff face are two grey andesite dikes with near-horizontal cooling joints. The thicker dike has intruded through the earlier thinner dike. A sea cave has been eroded along fractures at the right edge of the dikes.

⊙ 10.4 View from Tasman Lookout across the mouth of Puaotetai Bay to the landward-facing cliffs of 63-m-high Taitomo Island. The steep cliff face is a joint plane. Large blocks have been undercut by erosion at high-tide level and then have failed, sliding down the plane to the base of the cliff, where some still sit. The tidal platform on the exposed right side of the island has been cut by the sea at high-tide level over the last 8000 years.

⊙ 10.5 Aerial view looking north up the wave-swept west coast of South Piha. The Gap (middle left) separates Kaiwhare Point (centre) from Taitomo Island (beyond). Lion Rock and South Piha Beach are in the distance. PHOTO: ALASTAIR JAMIESON.

11.

Whatipu, West Auckland

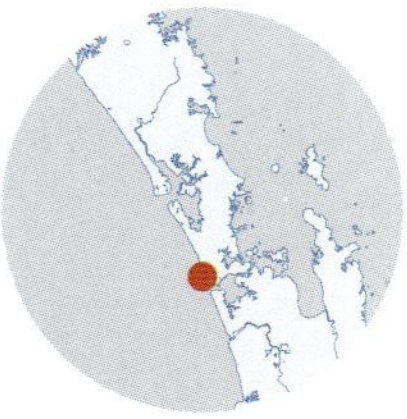

New Zealand's largest and youngest accreting sand flat

The 7000-ha Whatipu sand flat is the largest area of coastal sand build-up (accretion) in New Zealand today. With the slow sea-level rise over the last century or so, most of the country's soft shorelines have been eroding. The few exceptions, where the land is building out into the sea, are where there is an excess of sediment being supplied into the coastal system by rivers. Whatipu is a little different, as it is 40 km from the mouth of the nearest river – the Waikato, which is separated from Whatipu by a stable or eroding coastline of ancient sand-dune cliffs, the Awhitū Peninsula. Longshore drift is from the south, so it is likely that much of Whatipu's excess sand did come down the Waikato River, but some time ago.

Traditional oral histories from pre-European times indicate that during the 18th century and earlier, there was a similar extensive sand flat to the south of Whatipu on the southern side of the Manukau Harbour entrance (11.2). It had dune and wetland vegetation just like the Whatipu sand flat today, and there were shallow freshwater lakes – the source of plentiful food such as eels and waterfowl for early Māori visitors. This land became known as the lost land of Paorae because in the latter half of that century it was eroded away by the sea, although a dry island still existed on the Manukau Bar in the early 19th century. During most of the 19th century much of this excess sand must have been captured in subtidal shoals of an enlarged Manukau Bar, offshore of the harbour entrance.

From the 1930s onwards a new sand flat was thrown up on the north side of the Manukau Harbour mouth. Known as the Whatipu sand flat, it had grown out 1.4 km at its southern end by 1960. Since then, the area of rapid sand accretion has been migrating northwards, building the sand flat out to 1 km wide off Pararaha, 4 km up the beach, by 2000 (11.2). At present the area of greatest sand build-up is 4–8 km north of the Manukau entrance, between Pararaha and Karekare where the high-tidal sand flat has grown to 300–800 m wide. Clearly the pulse of excess sand is still moving northwards, and in the next century or so the Whatipu sand flat may also disappear. By then the excess sand may have built a new sand flat seaward of the cliffs between Karekare and Te Henga.

The Whatipu sand flat has a series of low dunes but is unusual in that large portions are drowned by freshwater

⊙ **11.1 View east over the 1.5-km-wide southern end of the Whatipu sand flat and the Manukau Harbour entrance at low tide. All this flat land up to the abandoned sea cliffs has been thrown up within the last 100 years. PHOTO: ALASTAIR JAMIESON.**

ponds and swamps (11.5). As the sand was accreting it also built up across the mouths of the Whatipu and Pararaha Streams, creating sill-like dams across their courses to the sea. As a result these streams have flooded over the adjacent low-lying flats creating the shallow water bodies.

The sand is dominated by grains of colourless quartz, white plagioclase and creamy-brown pumice. Variable quantities of black sand are also present on the beach and are composed of the heavy and magnetic iron mineral titano-magnetite. Most of the black sand has come down the Waikato River from large eruptions in the centre of the North Island, although some of it may have spilled northwards from erosion of the ring plain around Mt Taranaki, more than 250 km to the south. The light-coloured sand grains are also less dense (lighter weight) than the black ones, and are more readily transported further north by longshore drift than the heavier ones. Thus the highest concentrations of black sand are just north of the Waikato River mouth and they become progressively less abundant northwards beyond Whatipu.

Best seen from: From the Whatipu carpark take the 20-min. walk north to the abandoned sea caves or walk south to see the beach, Manukau Harbour entrance and offshore bar.

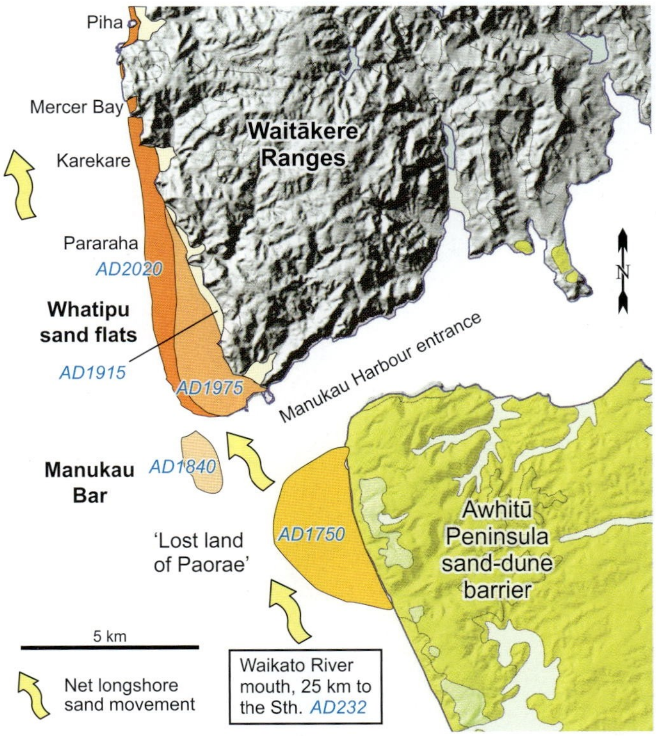

11.2 In the last 250 years, sand that once formed an extensive sand flat attached to Awhitū Peninsula has moved across the Manukau Harbour entrance and built the Whatipu sand flat; it is still moving north. If the rate of northwards movement is backtracked, this pulse of excess sand may have come down the Waikato River soon after the huge AD232 eruption out of Lake Taupō.

11.3 View east from the Whatipu sand flats across the mouth of the Whatipu Stream valley. The peaks on the ridge beyond are rounded because the conglomerate rock is little fractured and weathering is dominated by exfoliation, whereby large, curved sheets of rock detach from the rock face due to slow pressure release after deep burial. The northward-tilted layering of the conglomerate is reflected in the dip slopes (to the left) of two of the peaks. The water in the foreground is one of the freshwater lagoons ponded among the low dunes on the sand flat.

11.4 Just north of the Whatipu Valley mouth there are several old sea caves at the base of the abandoned cliffs, now 1.5 km from the sea. The caves were eroded by pounding waves forced into vertical fracture planes and along the edges of pyroclastic dikes that cut through the weakly layered volcanic conglomerate rock of the cliff. The conglomerate accumulated on the submarine slopes of the Waitākere Volcano (The Big Picture 2, page 44) 20–18 Myr ago. This is the entrance to the largest cave. It was used for dances during the timber-milling days in the 1870s–1880s. Person on right for scale.

11.5 The accumulation of sand that made Whatipu flats partially dammed the outlet of Whatipu Stream and it now spills water over the flat, creating extensive wetland ponds among the low dunes, both vegetated and unvegetated. PHOTO: ALASTAIR JAMIESON.

12.

Rangitoto

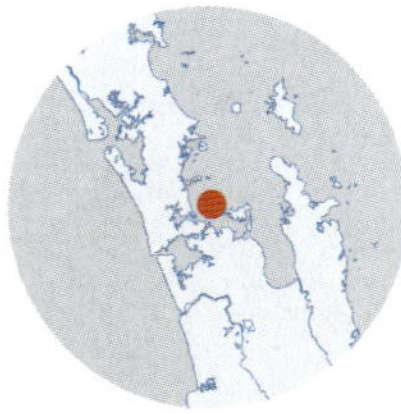

New Zealand's youngest volcano and icon of Auckland

12.1 Oblique aerial view of Auckland's iconic Rangitoto Volcano. PHOTO: ALASTAIR JAMIESON.

12.2 Profile of Rangitoto from North Head showing the gentle lower slopes of the lava flow shield surmounted by the steeper main scoria cones. The two lower, earlier scoria cones (named North and South cones) on either side are clearly visible from this angle.

12.3 The track to the summit from Rangitoto Wharf passes over a long, gentle slope of dark rubbly lava flows that are in the process of being colonised by vegetation. Unlike many lava flows, these ones are not smooth and ropey on the surface (pahoehoe flows). Instead their surface is a clinkery mix of angular blocks and slabs of basalt that had solidified on the lava flow surface upstream. As the flow of lava cooled, it became more viscous and slow-moving and the hard rock surface was broken up and rafted along on top of the molten lava.

Rangitoto Island volcano sits across the entrance to Auckland's Waitematā Harbour and is the icon of New Zealand's largest city, Auckland. It is both the largest and the youngest of the 53 small basalt volcanoes in Auckland's Volcanic Field (The Big Picture 3, page 57). Rangitoto erupted in two phases around 620–600 years ago and is the only eruption so far witnessed by humans in Auckland. A small Māori fishing village on adjacent Motutapu Island was buried by the initial explosive surges of wet volcanic ash that came out of the sea, 4 km away. Human and dog footprints have been found preserved in the wet ash layers. Auckland's second-youngest volcano, Maungarei / Mt Wellington, erupted 10,000 years ago, well before the arrival of humans in Aotearoa.

The first phase of eruption started explosively as rising molten magma (at about 1200°C) encountered cold seawater. Vast quantities of wet volcanic ash were ejected through the water into the air or blasted out sideways as clouds of searing-hot ash and gas. Progressively the volcanic ash built up a ring of land (tuff ring) around the open vent. Once seawater was no longer entering the vent, the eruptions switched from a wet to a dry style and fire-fountaining ensued. Magma is molten rock containing much dissolved gas (mostly carbon dioxide). As the magma surged up inside the throat, the pressure reduced and as a result gas coming out of solution powered a fountaining of frothy lava, in the same way a shaken bottle of Coca-Cola fountains as the pressure is released when the cap is removed. The fountaining froth cooled and solidified as it flew through the air and fell back to the ground around the vent, forming the northern scoria cone.

At about this stage in the eruption, the first batch of magma that had erupted (sourced from about 80–100 km down)

12.4 A 15-minute walk along a sidetrack off the main route to Rangitoto's summit leads to the popular lava caves. There are several caves in the vicinity, but the main one is a 100-m-long, fissure-like cave that runs down inside one of the youngest lava flows on the island. In several places the roof has collapsed allowing access and light into the cave.

⊙ 12.5 As the eruption of Rangitoto Volcano progressed, the lava flows advanced further and further into the sea. When the toe of each flow reached the cold water it was quenched and came to a stop. Upstream of the toe, molten lava continued to move down and on reaching the quenched solid basalt the liquid lava broke out on the side forming a small branch. Here on the southern shore of Rangitoto you see the many branched lobes at the seaward ends of several of the youngest flows that were formed in this way. PHOTO: ALASTAIR JAMIESON.

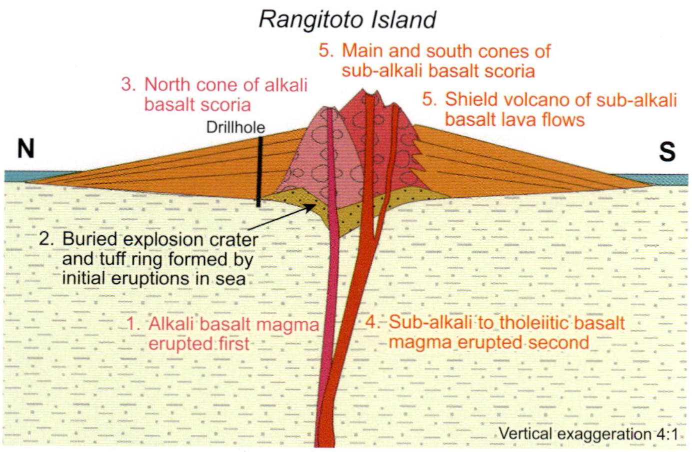

⊙ 12.6 North to south cross-section through Rangitoto Island (vertically exaggerated), showing the eruption sequence from 1 to 5.

ran out. It was replaced just a few years later (12.6) by an even larger batch with a slightly different chemistry (sourced from about 20 km shallower than the first). Dry-fountaining eruptions continued alongside the older, northern cone and built up a succession of several additional scoria cones. The tops of the last two can be seen in Rangitoto's profile as the main cone with crater and the lower south cone (12.6). At the same time as the fountaining was degassing the rising magma, vast quantities of less gaseous lava poured out in all directions from around the base of the scoria cones. Hundreds of small and large flows solidified one on top of the other to create the Rangitoto shield volcano. The almost circular shield, made entirely of basalt lava flows, is 6 km in diameter and has a gentle slope of about 10° away from the centre (12.2).

Best seen from: The profile of Rangitoto Island dominates most seascapes along the east coast of Auckland city, and is clearly seen from Tāmaki Drive, the North Shore beaches or from many high points in the city itself.

Take a closer look: Take a ferry from downtown Auckland or Devonport across to the island. Stroll up the rocky path through the island's lava flows partly clothed in the world's largest pōhutukawa forest. Climb the steeper path to the top of the main scoria cone for panoramic views back to Auckland and over the Hauraki Gulf. Take a torch and join the kids exploring the only publicly accessible lava caves in New Zealand.

⊙ 12.7 The surfaces of some of Rangitoto's larger a'a lava flows are rucked up into arcuate ridges. Pōhutukawa trees are starting to colonise the inhospitable surface of the flows, now 600 years after they were erupted. PHOTO: ALASTAIR JAMIESON.

THE BIG PICTURE 3

The Auckland Volcanic Field

THE AUCKLAND VOLCANIC Field has 53 small basalt volcanoes that have all erupted within the last 200,000 years. With this kind of volcano there is no magma chamber in the crust as there is beneath a stratovolcano. Instead, relatively small quantities of melted mantle periodically rise from depths of 60–100 km and erupt at the surface. Each of these eruptions comes to the surface in a different place and builds a relatively small volcano that never erupts a second time and is therefore called monogenetic. Volcanoes like these erupt for a few days to several years, and are separated from other nearby volcanic eruptions by hundreds to thousands of years of inactivity. They tend to erupt within a discrete area that is called a volcanic field. In these volcanic fields the magma composition is consistently basalt, the composition of the melted mantle. There are over 50 similar monogenetic basalt volcanic fields around the world, including three dormant fields in northern New Zealand, at Kaikohe–Bay of Islands, Whangārei and Auckland. Several extinct monogenetic volcanic fields erupted within the last 10 Myr: south of Auckland, in Northland and around the Mercury Islands off Coromandel Peninsula.

Most of Auckland's volcanoes began life with a wet explosive phase that threw out ash and built a tuff ring around a wide crater. If magma supply was sufficient, the eruption style became dry, and the fountaining of frothy lava built scoria cones often surrounded by lava flow fields. All the highest hills in Auckland are scoria cones and these were used as defended forts (pā) by pre-European Māori. The earthworks remains of these are valued archaeological features. Unfortunately, with the arrival of Europeans many of the city's cones were quarried away or badly damaged to obtain scoria for construction and railway ballast.

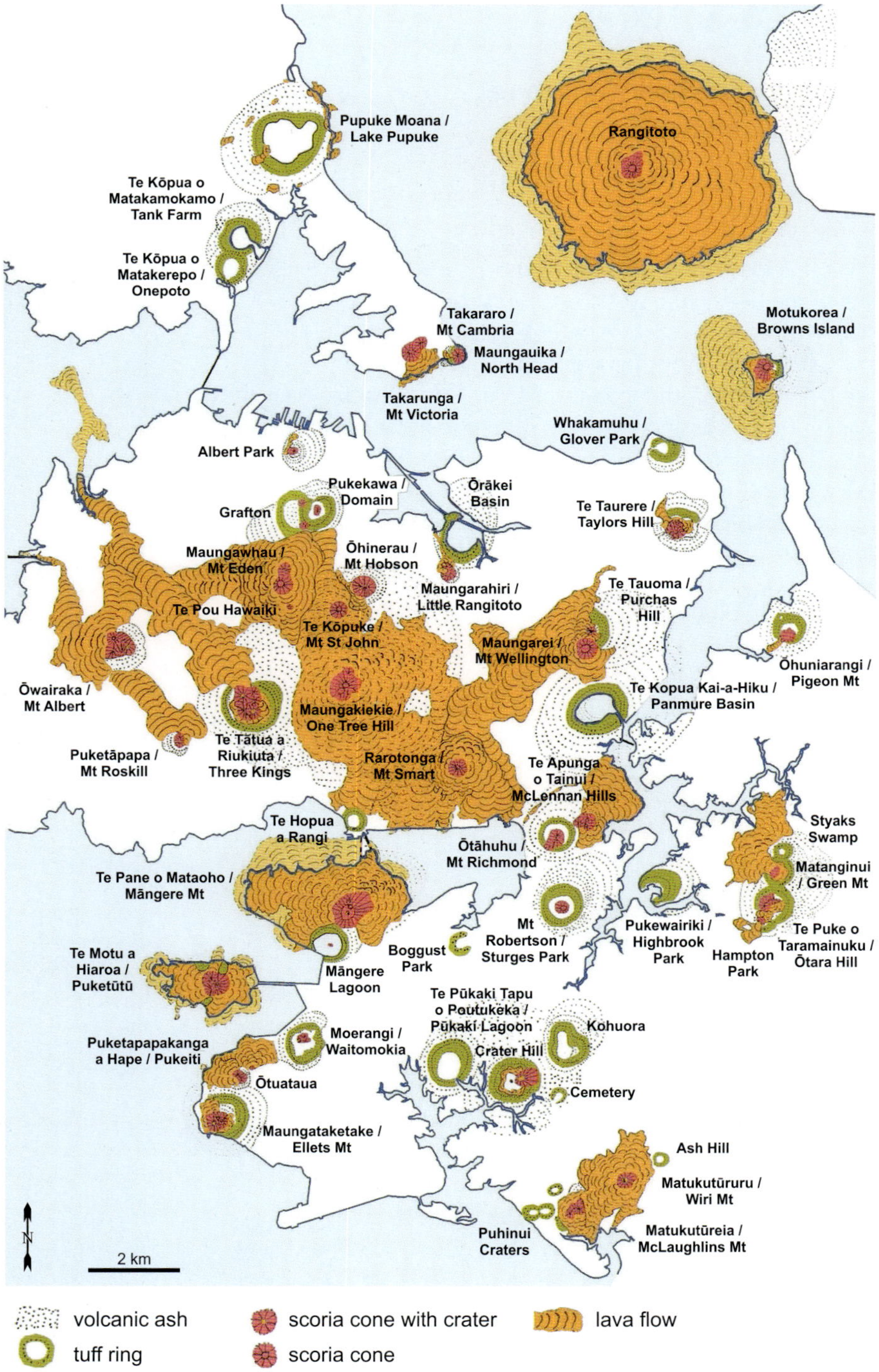

TBP 3.1 Map of the Auckland Volcanic Field.

13.

Takapuna Reef, Auckland's North Shore

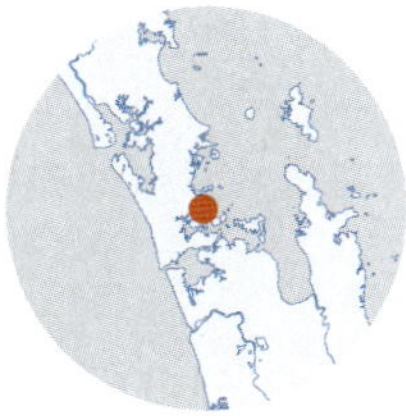

New Zealand's only fossil forest preserved in lava

Takapuna fossil forest is the only example in New Zealand and one of the best in the world of a forest preserved in a lava flow. The fossil forest consists of solidified lava moulds of the stumps and lower portions of the trunks of trees in growth position (13.2) and the moulds of some branches that had fallen into the passing lava flow. Although there is no original wood remaining, the moulds are fossils and this can be called a fossil forest.

The forest, consisting of numerous small trees and a few larger kauri, was growing here around 190,000 years ago when sea level was considerably lower than present. Takapuna Reef is composed of the remains of two thin lava flows from the initial eruptions of Pupuke Volcano that flowed into the forest and set it on fire. The first flowed right through the forest to a depth of 1–2 metres, cooling and congealing around the tree stumps, forming cylindrical moulds composed of basalt 0.3–0.6 m thick (13.1, 13.2). Most of the molten lava between the moulds then drained away. As the moulds formed, the surface of the flow also crusted over in some places, and some of this is preserved as arches between stumps that are close together. Most of the crust that formed on top of this first flow has disappeared, probably broken up by continued lava movement and rafted off by the flow.

Near high-tide level in front of the campground, there are depressions in the dark grey lava flow surface that have been filled with overlying orange-brown tuff, some of which has obviously been washed down into cracks and caverns within the flows themselves. This is the hardened remains of 5–10 metres of volcanic ash (tuff) that erupted later from Pupuke Volcano and buried the lava flows and tree moulds. These remained buried and protected from erosion and weathering until a few thousand years ago, since when they have been uncovered by coastal erosion which has removed the softer tuff and left the lava flows and tree moulds as a hard intertidal reef.

Alongside the coastal path from Takapuna to Milford Beach there is a cartwheel grille over the top of a 4-m-deep, 1.5-m-diameter cylindrical hollow in the basalt rock. This is the mould of the lower trunk of a once-proud kauri tree that was

13.1 Takapuna fossil forest is New Zealand's only example of a forest that has been fossilised in a lava flow. These cylindrical basalt columns look like casts but are in fact moulds that formed around the lower trunks of numerous small trees.

here when Pupuke erupted. The lava flowed around the kauri's trunk and solidified, preserving its shape before the wood had time to burn away. The flow was thicker here than at Takapuna Reef and it solidified in place as one solid flow without some draining away.

Take a closer look: Best seen beside the carpark at the end of The Promenade at the north end of Takapuna Beach. Walk from the carpark on the high-tide track towards Milford to see more branch and tree moulds in the basalt rock as far as sandy Thorne Bay, where water pours out of cracks in the basalt rock – the underground outlet from Lake Pupuke.

⊙ 13.2 One of the tree moulds still in place where the tree was growing when engulfed and set on fire by a passing lava flow from Pupuke Volcano. The lava cooled in layers around the standing tree trunk and the wood slowly incinerated away, leaving the circular hole down the middle.

⊙ 13.3 This circular pool in a basalt lava flow, off the end of Minnehaha Ave, is the mould of the lower part of the standing trunk of a kauri tree that was living here when the lava flow swept around it. The wood burnt away and the hole was filled not long after by brown volcanic ash (tuff). Being softer than the basalt, the ash has eroded down about 30 cm below the lip, creating the tidal pool.

⊙ 13.4 Cross-sections through Takapuna Reef illustrating how the forest trees (1) were engulfed by a lava flow (2) from Pupuke Volcano 190,000 years ago, set on fire (3), and their lower trunks preserved as moulds in the solidified basalt (4). The flow and moulds were buried and preserved by thick volcanic ash (5) until this was eroded away by waves just a few thousand years ago (6). ILLUSTRATIONS: HUGH GRENFELL.

◀ 13.5 This 1.2-m-diameter trough is the mould of a kauri tree trunk that fell into the top of the molten lava flow. Its shape was preserved as the lava rapidly cooled around it before the wood itself was burnt away.

▼ 13.6 Moulds of two branches that fell into the lava as it moved along, seen beside the Takapuna–Milford Beach coastal track. The left-hand mould was still hot and plastic when a later lobe of lava flowed above and squashed it.

▲ 13.7 Aerial view down over Takapuna Reef with the tide out. Each circular protrusion on the reef is the basalt mould of a tree created by a passing lava flow. PHOTO: ALASTAIR JAMIESON.

14.

Wiri Lava Cave, South Auckland

New Zealand's best lava cave

Wiri Lava Cave is unquestionably the best example of a lava cave in New Zealand. All of Wiri Mountain volcano has been quarried away except for the land containing the lava cave, which has been set aside as a Scientific Reserve. Lava caves mostly form inside lava flows. As the lava flows across the ground, it cools around the outside forming a crust – a few centimetres to more than a metre thick – of solidified rock along the sides and over the top (14.2). Just as water flows along inside a hose, so the molten lava flows along inside the crust of the flow. When the supply of lava stops, the molten lava inside may flow out and leave an empty tube or lava cave. In some places the thinner roof of solid basalt may collapse and block the cave or provide a way to get into part of the cave. In New Zealand, accessible lava caves are uncommon and all occur in the basalt lava of the Auckland Volcanic Field. The longest known example in New Zealand is the 290-m-long Wiri Lava Cave. There are much larger lava cave examples elsewhere around the world, like in Queensland, Hawaii and Iceland.

Wiri Lava Cave is special because it is one of only two lava caves in New Zealand that is known to have partly formed inside a scoria cone. The upper end of the cave was formed by lava that rose up inside Wiri Volcano's throat and pushed its way through the northern flanks of the scoria cone to emerge as a lava flow. As the lava passed through the scoria, some of it cooled to form a thin lining of glassy basalt on the inside of the conduit and this is now the 5–20-cm-thick walls of the cave. This upper part of the cave has a ceiling 4–6 m high with a unique 'Gothic Arch' cross-section (14.1). This shape was produced when lava began draining out of the tube and the weight of the surrounding scoria caused the still-plastic lining rock to sag inwards. Part-way along Wiri Lava Cave, the ceiling becomes lower and the walls are much thicker and

⊙ 14.1 Wiri Lava Cave is New Zealand's best example of a cave formed within a lava flow. Note the Gothic Arch cross-section. The ropey floor with basalt kerbs is the remnant of the last lava flow that did not fully drain out of the tube. PHOTO: ALASTAIR JAMIESON.

made of more-solid basalt. This is where the lava emerged from the base of the scoria cone, and from here on the cave was the feeder tube inside a lava flow.

The inside of the cave is not just a featureless tube. In places there are short lava stalactites hanging from the roof (14.5) and lava dribbles down the upper walls. These were formed when extremely hot gas, trapped between the top of the flowing lava and the cave ceiling, partially remelted the basalt lining, giving it a glazed appearance and causing parts to dribble. The cave's floor consists of the dregs of the last surge of lava that flowed down inside the tube (14.1). In places the remaining thin flow inside the cave is rucked up into curved ropes of lava as the centre of the flow moved faster than the edges.

Best seen from: Wiri Lava Cave is underground and accessible by permit only.

14.2 Two cross-sections showing how lava caves, like that at Wiri, are formed inside lava flows. The outside cools and solidifies, forming a crust of basalt rock around an internal tube filled with flowing lava. When the lava supply stops, the molten liquid may drain out leaving behind an elongate lava cave. ILLUSTRATION: GEOFF COX.

14.3 Remnants of the last lava flow lie in a trench that has been eroded into the loose scoria beneath the original floor level by the flowing lava. The sloping bench on either side of the cave marks the original floor level. PHOTO: PETER CROSSLEY.

14.4 This is a small cylindrical blowhole in the roof of Wiri Lava Cave. As lava flowed through the conduit, trapped superheated gas under pressure forced a hole upwards through the plastic lava to the surface. Note the dribbles down the inside walls. PHOTO: ALASTAIR JAMIESON.

14.5 Part of the ceiling of Wiri Lava Cave with solidified dribbles forming lava stalactites.

15.

Pūkorokoro / Miranda Chenier Plain, Firth of Thames

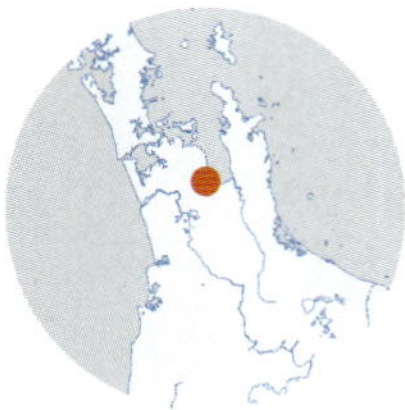

Abandoned shell beaches record changes in sea level

A chenier is a beach ridge made of gravel, shell or sand, resting on top of mud and separated from the shore by a belt of intertidal mudflats. The southwest corner of the Firth of Thames is fringed by a 15-km-long and up to 2-km-wide coastal plain that accumulated in just the last 4000 years (15.3). It is one of the best examples in the world of a combined gravel and shell chenier plain that has built up by the progressive addition of gravel and shell beach ridges on top of tidal mudflats. The gravel beach ridges form the narrower northern part of the plain, with greywacke cobbles and pebbles (15.4, 15.6), eroded from the adjacent Hūnua Ranges, coming down streams to the coast. The gravel is moved progressively southwards down the beach by longshore drift during storms. A succession of storm-beach ridges has built out into the Firth. Moving south, the pebbles became progressively smaller and less numerous, and are replaced by shells, mostly cockle, south of Kaiaua.

The shell ridges originated intertidally during major northeast storms which swept away the mud and concentrated sand and shells into offshore banks. These banks were driven shoreward by later storms and eventually attached to the existing shell beach at their northern ends. From here, they developed into narrow shell spits that grew southwards, parallel to the coast, separated from the previous shell beach by a depression 50–200 m wide (15.1). This elongate depression progressively filled with intertidal mud that was colonised by mangroves and eventually by high-tidal salt marsh, thereby adding new land to the plain. During storms, waves washed shell around the southern tip of the active spit, creating a hook. A succession of these hooks give the inner edges of the shell cheniers a feathery appearance when viewed from the air (15.1) and these can still be seen in many places on the grazed older parts of the chenier plain.

Between 7000 and 4000 years ago, when sea level was 1.5–2 m higher than today, the coastline at Miranda was up to 2 km further west, right up against the foot of the hills. Between 4000 and about 1000 years ago, as sea level dropped to its present level, much of the chenier plain was constructed. Each successive shell or gravel chenier was added seaward of the

preceding one and the plain built out across the intertidal mud. The 4000-year-old chenier up against the hill is 1.5–2 m higher than the 1000-year-old one, and those of today (15.3). Together they document the fall in sea level since 4000 years ago. In the last 1000 years additional shell spits have been added as the chenier plain has grown southwards to Miranda rather than seawards. This is believed to be the result of the stable and more recently rising sea level. What happens to the Miranda Chenier Plain in the coming century or so as sea level rapidly rises again remains to be seen, but active accretion is likely to be replaced by erosion.

Take a closer look: Stop on the foreshore beside East Coast Rd between Miranda and Waharau or take the short walk from the carpark at Robert Findlay Wildlife Reserve. Visit the Pūkorokoro Miranda Shorebird Centre.

15.1 Southern end of the internationally famous Miranda Chenier Plain. These are the two youngest cheniers (shell spits), having been added at the seaward southern end within the last 100 years. The road runs along on top of the third-youngest. The youngest chenier (right) took 40 years to grow the 1.5-km-length shown in the image. Note the feathering (hooks) inside each chenier, where tongues of shell have been swept landwards around the tip of the spit during storms. PHOTO: ALASTAIR JAMIESON.

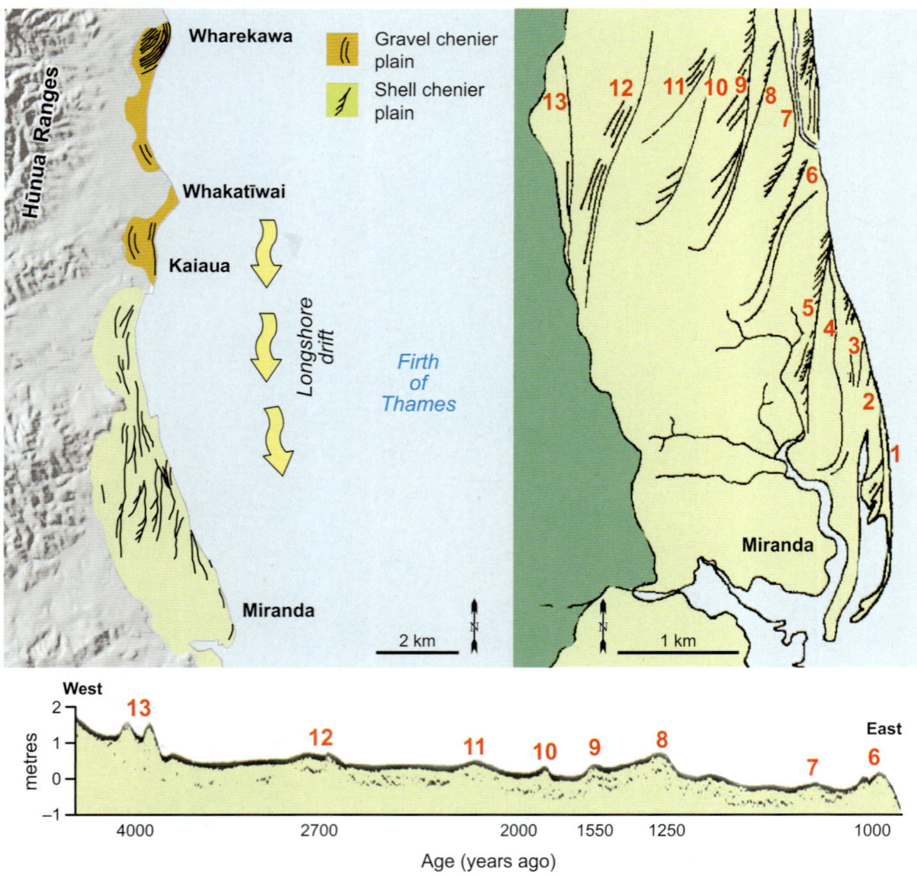

⊙ 15.2 View south over most of the Miranda Chenier Plain. The elongate ridges parallel to the coast and visible in the grassed paddocks are the crests of cheniers (shell spits) that were added on to the edge of the plain between 3000 and 1200 years ago. PHOTO: ALASTAIR JAMIESON.

⊙ 15.3 Left: Map of the Miranda Chenier Plain that has formed along the west coast of the Firth of Thames in the last 4000 years. Right: Map showing numbered shell chenier ridges around Miranda: 13 is the oldest (4000 years old) and 1 the youngest (less than 50 years); FROM SCHOFIELD (1960). Bottom: West to east cross-section through the 4000–1000-year-old portion of the shell chenier plain, showing the drop in sea level (chenier crests) of 1.5–2 m; FROM DOUGHERTY & DICKSON (2012).

⊙ 15.4 The youngest gravel chenier (beach ridge) at the northern end of the Miranda Chenier Plain, north of Kaiaua. The gravel cheniers are made of rounded pebbles of greywacke eroded out of the Hūnua Ranges and transported southwards by longshore drift.

⊙ 15.5 The Miranda cheniers (shell spits deposited over mudflats) are composed of the dead shells of mostly cockles that lived intertidally seaward of the spits. When dead, the shells are driven shoreward by storm waves and currents and rapidly accumulate along the shoreline.

⊙ 15.6 Part of the Whakatīwai Gravel Chenier Plain composed of gravel spits that built out from the coast between 3000 and 1000 years ago. Some crests of the older shore-parallel spits can be seen in the grassed paddock just north of Kaiaua. The excavations left by quarrying of the gravel spits (now stopped) have filled with water, creating the artificial ponds in the distance. PHOTO: ALASTAIR JAMIESON.

16.

Waikawau Valley, Port Waikato

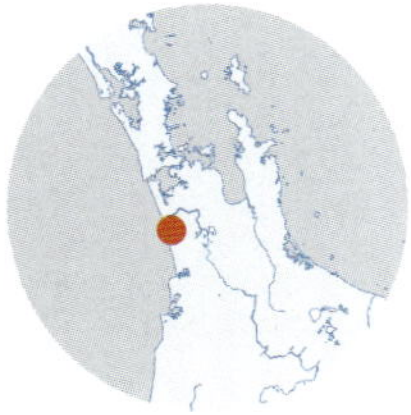

Excellent example of limestone karst (solution landforms)

One of the best examples of limestone karst in New Zealand occurs in the Waikawau Valley and its small tributary valleys a few kilometres south of Port Waikato. The karst landforms (The Big Picture 4, page 83) have formed by solution along cracks in a 10–15-m-thick layer of densely recrystallised, flaggy limestone. The limestone occurs within a sequence of sedimentary rocks that was deposited on the seafloor at relatively shallow depths (shallower than 100 m) during the Oligocene (34–23 Myr ago).

This sequence of strata overlies much older and harder sedimentary rocks (greywacke) that were deposited on the seafloor between 200 and 150 Myr ago, during the Jurassic. Later these greywacke rocks were pushed upwards to form mountains that were slowly eroded down to a flat coastal plain – known around New Zealand as the Waipounamu Erosion Surface (The Big Picture 18, page 309). From about 35 Myr ago, most of this erosion surface slowly subsided and was submerged beneath the sea.

The limestone layer was deposited as a current-swept shell bank sandwiched between layers of shelly sand and mud. This heap was buried by a further 500-m thickness of mud and sand, and under this depth of burial the shell bank was recrystallised into the dense limestone that the karst later developed in. Similar shell banks were deposited in the shallow seas that covered most of New Zealand at this time, and produced the widespread limestones we see today.

Over the most recent 15 Myr, the Port Waikato region has been uplifted progressively and erosion has stripped off most of the overlying younger sedimentary layers, exposing the limestone. In the last million or so years, slightly acidic runoff water from beneath forest has flowed into and along several sets of parallel cracks in the exhumed limestone. Over time these cracks have been dissolved out into deep trenches. At depth, solution along cracks has produced caves, some of which have collapsed to become small stream valleys. The exposed faces of limestone bluffs and walls have developed solution runnels under attack from the slightly acidic seepage that ran down the faces when they were under forest (16.2). In places these runnels are shallow and close together (16.3); elsewhere they have eaten deeper into the rock, creating large fluting.

Best seen from: Waikawau Valley karst is on private farmland. It can be viewed as you drive along Port Waikato-Waikaretu Rd, between 3 and 13 km south of Port Waikato.

⬇ 16.1 An amazing area composed of deep trenches (called grikes) separated by intervening vertical-sided narrow limestone ridges (clints) occurs in the Waikawau Valley, near Port Waikato. The trenches have formed by solution of the rock by water passing along a number of parallel cracks (joints) through the rock. The people provide a scale.

↑ 16.2 Beautifully fluted limestone faces in a tributary of the Waikawau Valley, Port Waikato. The fluting would have developed by solution by slightly acidic seepage running over the rock faces when the area was clothed in native forest.

↓ 16.3 Limestone face with intricate solution runnels, Waikawau.

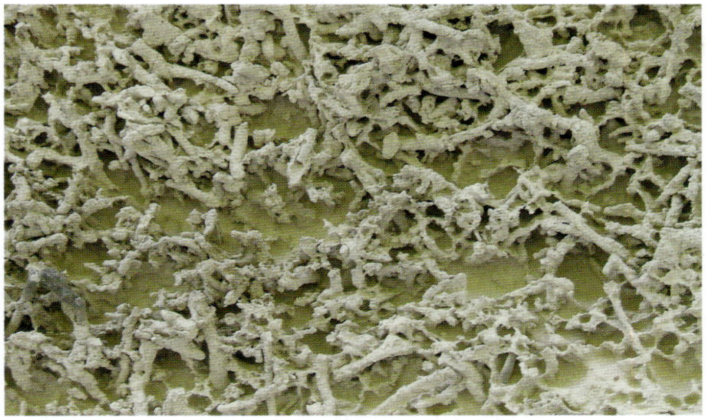

16.4 This intricate network of branching tubes occurs within the Waikawau Valley limestone. The tubes are the sandy limestone-filled burrows of marine organisms that lived in the sediments when they were being deposited on the seafloor about 25 Myr ago. The burrows were excavated into a thin mud layer which is now softer than the sand-filled burrows and has been eroded away from around the burrows.

16.5 Ngapuriri Natural Arch was formed by wetting and drying and wind erosion of soft and broken rock within the harder limestone, Waikawau Valley.

16.6 The Waikawau Valley and many of its tributaries are bordered by gently west-tilted limestone bluffs. The limestone is the most erosion-resistant rock layer within the tilted sequence, with softer calcareous sandstone below and calcareous siltstone above.

17.

Ngatutura Cliffs, Port Waikato coast

Eroded remains of a thick lava flow on a stormy coast

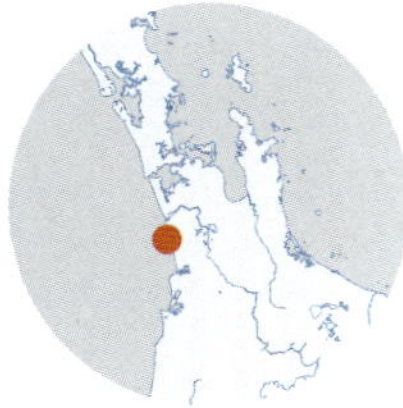

Amazing vertical and conical sea stacks and adjacent castellated cliffs occur at Ngatutura Point (17.1, 17.2) on a remote section of Waikato's west coast about halfway between Port Waikato and Raglan Harbour. These are the eroded remains of a 50-m-thick lava flow from a basalt volcano that erupted near here 3.3 Myr ago (Pliocene). The basalt flow is of limited extent and appears to have ponded in a deep valley. As the lava slowly cooled and solidified, it shrank and regular columnar cooling joints (cracks) were formed through it. These joints are usually perpendicular to the cooling fronts at the top, bottom and sides of the flow. In the centre and top of the flow the columns are largely vertical, having formed during cooling from the top of the flow down and bottom up, At the northern end of the flow, in the cliffs around Shag Rock (17.3), the columns are curved around to be perpendicular to the valley side. Thin basalt flows, breccias and a volcanic neck occur in the cliffs of Ngatutura Bay, just north of the thick lava flow. These are the products of another, younger, basalt volcano that erupted here around 1.7 Myr ago.

17.1 Shag Rock off Ngatutura Point is a superb 50-m-high vertical-sided stack made of columnar-jointed basalt. It is part of the eroded remains of a thick lava flow that ponded in a coastal valley when it was erupted here by a small volcano about 3.3 Myr ago. The gap between the coastal cliffs and stack would have started as a sea arch, eroded through a narrow point by the crashing waves. Slowly, block by block, the jointed basalt above the arch would have been undercut and dropped out, until finally it completely collapsed leaving the separated stack.

17.2 View from the north over Ngatutura Bay and Point. The reef in the left foreground is made of a 20-Myr-old (Miocene) sandy limestone. The three stacks off the point are eroded out of a Pliocene (3.3-Myr-old) basalt lava flow. Pliocene Mt Karioi Stratovolcano at Raglan is on the skyline. PHOTO: ALASTAIR JAMIESON.

17.3 The curved columnar jointing of the basalt flow in Shag Rock is believed to have been produced by shrinkage of the rock associated with slow cooling and solidifying of the lava close to the sloping side of a small valley in which the lava was ponded.

17.4 Two 20-m-high conical basalt stacks are all that now remains of the southwestern edge of the Ngatutura Point lava flow.

Ngatutura Volcano is the oldest of dozens of small intra-plate basalt volcanoes that erupted through the continental crust of the Australian tectonic plate between Kāwhia Harbour and Auckland in the last 3.3 Myr. These basalt volcanoes were clustered together in four volcanic fields that show a northwards migration through time. The youngest and northernmost field is the currently dormant Auckland Volcanic Field (The Big Picture 3, page 57) which has been erupting periodically for the last 200,000 years. The second-youngest is the South Auckland Volcanic Field (centred around Pukekohe and the lower Waikato River), which has over 80 small volcanoes that erupted between 1.6 and 0.5 Myr ago. Further south, located on either side of Raglan Harbour, were two more volcanic fields that had 20–30 volcanoes each and erupted during the time span 3.3–1.6 Myr ago.

All four of these intra-plate basalt fields were at least 100 km west of where the subduction-related volcanic arc of dominantly andesite and rhyolite volcanoes was erupting. The basalt magma that erupted at the surface came directly from the partly molten upper mantle and is not believed to have been influenced by the subducting edge of the Pacific Plate.

Best seen from: There is no easy public access to anywhere near this eroded volcano. The sea stacks and coastal cliffs can be seen in the distance from commercial flights off the west coast south of Auckland.

17.5 View north over Ngatutura Bay from the steep slopes behind. The cliffs contain rocks of a variety of ages that were intruded and overlain by a volcanic neck and small lava flows from a second basalt volcano that erupted here 1.7 Myr ago. The sloping reef at the north end of the bay is 20-Myr-old sandy limestone (Miocene). The light rock at the back of the beach is 30-Myr-old calcareous siltstone (Oligocene). The orange-coloured rocks in the middle of the cliff are 4–3-Myr-old marine sediments (Pliocene). They are covered by white rhyolitic ash erupted from the centre of the North Island and dark orange rocks, which are weathered black dune sands (both younger than 2 Myr old; Pleistocene).

18.

Wairēinga / Bridal Veil Falls, Raglan

Exquisite plunge and punchbowl waterfall

One of New Zealand's most beautiful waterfalls, Wairēinga / Bridal Veil Falls is a combination of plunge and punchbowl waterfall types. A plunge waterfall has fast-moving water that descends vertically and loses complete contact with the bedrock on its way down. In a punchbowl waterfall the water descends in a constricted flow and then spreads out into a wider pool at the bottom (18.2).

The falls have a single drop of 55 m and plunge over the edge of an eroding basalt lava flow. The flow erupted from a small unnamed volcano located 1.5 km away to the northwest, about 2.6 Myr ago (early Pleistocene). This lava appears to have flowed southeast across the middle section of the Pakoka River and initially dammed it to form an elongate lake upstream above Bridal Veil Falls. The river soon overtopped the flow and has progressively eroded the hard basalt rock back upstream by nearly 1 km.

18.1 The delicate Wairēinga / Bridal Veil Falls, inland from Raglan, cascade 55 m into a 5-m-deep plunge pool. The top of the falls are held up by the hard grey rock of a 2.6-Myr-old basalt lava flow. The lower half of the face is composed of older, softer, sandstone, which erodes back more quickly undercutting the fractured basalt. With their support removed the basalt columns progressively become loose, fall out of the face and slowly the falls recede.

Initially the river cascaded over the edge of the basalt flow and eroded down into the underlying, softer, 30-Myr-old (Oligocene) sandstone. Splashback and the swirling waters at the foot of the falls has eroded into the older sandstone, undercutting the more resistant basalt above (18.1). As the support beneath them was removed, columnar blocks of basalt broke away along their weak joint planes and collapsed into the plunge pool. Over time these blocks became smaller by abrasion and the smaller pieces were swept away downstream by flood waters. In 2.6 Myr the waterfall has receded back upstream creating the 1-km-long gorge, while the upper surface of the basalt flow, which forms the lip of the falls, has not eroded down much at all.

Best seen from: Attractive vistas are obtainable from the two viewing platforms near the top of the falls ~10 mins walk each way from the carpark on Kawhia Rd, south of Raglan.

Take a closer look: Steep steps lead from the top to two lower viewing platforms for different perspectives of the falls and plunge pool (a further 20-min. return).

18.2 Bridal Veil Falls plunge over the eroding edge of a basalt flow in a dramatic single column of water.

18.3 The basalt lava flow consists of a dense array of close-spaced columns. As molten lava cools and solidifies into rock, it contracts and cooling cracks are formed within it. They often form columns oriented perpendicular to the cooling surface. In a lava flow the columns are usually oriented vertically, as here, because the cooling surfaces are the horizontal top and bottom of the flow.

18.4 In part of the rock face next to the waterfall the close-spaced basalt columns have been bent around to slope at a low angle. This is believed to have been produced by the hottest lava, deep within the flow, still moving slowly as the columns formed. This rotated the columns from vertical to this low angle, sloping down in the downflow direction.

19.

Rākaunui Peninsula, Kāwhia

Drowned limestone karst — rare in New Zealand

⓿ 19.1 Kāwhia Harbour's Rākaunui Peninsula is composed of an unusual landscape of drowned coastal karst. Dolines (sinkholes), formed by limestone solution when sea level was lower during colder times, have been partly drowned as sea level rose after the end of the Last Ice Age. These two dolines are now mud-filled tidal inlets. PHOTO: PAUL WILLIAMS.

The best New Zealand example of drowned coastal karst (The Big Picture 4, page 83) occurs around the shore of Rākaunui Peninsula, southern Kāwhia Harbour (19.1, 19.4). Limestone pinnacles, caves and solution sinkholes (dolines), formed under forest when sea level was lower during the long cold periods of the Ice Ages, have been partially submerged by the rising sea after the end of the Last Ice Age. In the last 1 Myr, sea level has been 20 m or more lower than present for 90% of the time. Prior to the high sea level of the present warm interglacial period of the last 8000 years, there had been an interval of 110,000 years with lower sea levels when much of the drowned karst would have been formed, although it would have started developing during earlier periods of low sea level.

Like many other places in New Zealand the karst at Rākaunui Peninsula developed in dense limestone that was deposited as shell banks on the current-swept shallow seafloor between 30 and 25 Myr ago (Oligocene). The surface of this limestone was dissolved into a dense network of dolines creating polygonal karst as at Mangapōhue (Site 21). At the same time the Rākaunui River dissolved and eroded itself a narrow, steep-sided gorge through the limestone. Cave streams that drained the polygonal karst flowed out into this stream valley. Around the gorge, solution along several sets of cracks had created limestone pinnacles and bluffs (19.3, 19.5) that were fluted by seepage running down their sides.

When sea level rose, it advanced up the surrounding valleys and branches creating the Kāwhia Harbour. The Rākaunui Gorge was flooded, forming a narrow inlet with the tide flowing around the base of the bluffs and some of the pinnacles on either bank. The water partly flooded the lower-elevation dolines. Dolines nearest the coast filled with mud carried in by each incoming tide and are now tidal lagoons or high-tide salt marshes (19.1, 19.4). Inland, the subterranean drainage system is now flooded as the groundwater level is controlled by sea level. As a result, the bottoms of some of these inland dolines are now shallow ponds and wetlands (19.2).

Best seen from: The drowned karst is mostly in private farmland and is best seen using boat transport from Kāwhia township. Kayak across on a calm day (5 km in direct line at high tide).

⬆ 19.2 The cave drainage system beneath the inland dolines on Rākaunui Peninsula is now submerged because of sea-level rise. As a result many of the dolines are partly filled with freshwater ponds and wetlands. PHOTO: ALASTAIR JAMIESON.

⬅ 19.3 The narrow, winding Rākaunui Inlet on the south side of Kāwhia Harbour is a former river gorge that has been partly submerged by sea-level rise after the end of the Last Ice Age. The gorge had eroded through a band of hard limestone karst; these stunning pinnacles of flaggy limestone are part of the display that has not been drowned.

↑ 19.4 Full tide on the Kāwhia Harbour clearly shows how sea-level rise after the end of the Last Ice Age has drowned the polygonal karst and underlying cave drainage systems that were developed during lower sea levels. PHOTO: ALASTAIR JAMIESON.

← 19.5 A 20-m-high flaggy limestone pinnacle on the southern shore of Kāwhia Harbour.

↓ 19.6 A limestone mushroom rock on the Kāwhia Harbour shoreline of Rākaunui Peninsula. PHOTO: NICOLAS BARTH.

20.

Waitomo Caves

Limestone caves, stalactites and stalagmites

⊙ 20.1 Limestone caves around Waitomo contain some of the best and most easily accessible examples of speleothem decorations in New Zealand. The most attractive are the highly decorated stalactites, stalagmites and columns such as these in Hollow Hill Cave. PHOTO: PETER CROSSLEY.

The most accessible and dazzling displays of stalactites and stalagmites (speleothems) in New Zealand caves occur in the limestone country around Waitomo. Here the karst has developed over the last 3 Myr or so, in 60–100-m-thick limestone, the origins of which can be traced back to shell banks that accumulated on the seafloor 30–25 Myr ago.

As described in The Big Picture 4 (page 83), slightly acidified rainwater infiltrates along cracks in the dense limestone and over time dissolves them out into underground passageways. When many of these small passageways join together they may form an underground stream flowing along within a cave. Over the last 3 Myr or so, as the Waitomo area has risen, the surrounding stream valleys eroded down and the underground streams made new caves at lower levels. This left behind dry abandoned caverns which have become decorated with speleothems on their roofs, walls and floors (20.1).

As carbonate-rich groundwater drips from cracks in the cave roof or trickles down the walls it loses carbon dioxide to the air inside the cave, becoming less acidic and unable to retain in solution all the carbonate it had dissolved while passing through the limestone above. Thus tiny crystals of calcite (calcium carbonate) are precipitated and grow very slowly into a variety of cave decorations that may be tens to hundreds of thousands of years old. The best-known speleothems are stalactites, which hang vertically down from the cave ceiling. The simplest form of stalactite is a thin, hollow straw with water dripping down inside it. When this central cavity becomes blocked, straws develop into conventional solid stalactites with water dribbling down the outside and dripping off the end (20.3). Stalagmites grow upwards from the cave floor where the drips land from the stalactites above. If downwards-growing stalactites join up with the upwards-growing stalagmites beneath them, they form a solid column from ceiling to floor (20.1).

Water reaching the roof of a cave does not always form droplets. Sometimes it may dribble along the gently sloping roof or down the wall. Water constantly trickling along this dribble path may deposit a narrow strip of calcite which eventually grows out at an angle from the wall or ceiling to form a curtain or shawl (20.2). Folds in the shawl may result from the irregular original path of the dribble. A more extensive sheet of carbonate-rich water may flow down the wall and over rock or mud deposits on the cave floor. As it passes over these obstacles it deposits a thin film of calcite which over time grows thicker and thicker into a blanket-like speleothem called flowstone.

Sometimes small knobbly clusters of calcite crystals are formed in the cave and these may be referred to misleadingly as coralloids or cave popcorn (20.5). Some of the caves around Waitomo also contain speleothem decorations made of calcium sulfate (gypsum). In caves the most common of these gypsum speleothems is satin spar – a soft, silky, fibrous ribbon often referred to as selenite.

⊙ 20.2 The sloping growth lines in this shawl in Ruakuri Cave show how it has progressively grown downwards from the roof. The folds in the shawl reproduce the meandering path of the dribble on the gently sloping ceiling.

The other attraction of the Waitomo Caves is the presence of an unusual insect that lives in wet caves. These are glow-worms – the larvae of a fungus gnat (insect) that lives only in New Zealand. To catch food, the carnivorous larvae dangle fine threads or 'fishing lines', containing sticky mucus droplets, from their silky nests on the cave ceiling. Their prey are flying insects that are lured into the lines by the glowworms' bright blue-green light, produced by chemical reactions in their excretory organs.

Take a closer look: For most people, a guided tour of a tourist cave (Waitomo Glowworm, Ruakuri or Aranui) is the best way to see Waitomo's sensational cave speleothems. Glowworm and Ruakuri Caves have stream passages and glowworms. Glowworm Cave tours include a boat ride. Ruakuri is the largest cave with high boardwalks. Aranui has unforgettable pale brown, pink and white-coloured speleothems. Near the entrances to Ruakuri and Aranui Caves take the Ruakuri Natural Bridge forest walk (free) for surface karst.

20.3 A spectacular display of super-slender straws interspersed with a few conventional stalactites hangs from the roof of Hollow Hill Cave, near Waitomo. PHOTO: PETER CROSSLEY.

20.4 This 10-cm-diameter gypsum flower is a speleothem that has grown on the wall of Pukeiti Flower Cave, south of Waitomo. PHOTO: PETER CROSSLEY.

20.5 These 'coralloid' speleothems with 'popcorn' behind probably grew on the floor and side of a shallow pool of water supersaturated with calcium carbonate in Waipuna Cave, near Waitomo. Photo 50 cm across. PHOTO: PETER CROSSLEY.

20.6 The ceiling of Ruakuri Cave is a dense mass of stalactites and straws.

THE BIG PICTURE 4

Limestone and marble karst and caves

KARST IS THE TERM for landforms produced by solution of water-soluble rock, usually limestone or marble, which are composed of calcium carbonate. Karst landforms (TBP 4.1) are characterised by solution caves, subterranean drainage, sinkholes, springs, pinnacles, natural bridges and fluted rock surfaces, sometimes with quite bizarre shapes. Dense, hard, recrystallised limestone and marble dissolve most readily to develop the best karst features. In New Zealand this kind of limestone is usually of Oligocene age (33–24 Myr old). In the North Island it occurs most commonly in pockets in eastern Northland, extensive plateaux of limestone from Port Waikato to the King Country (best known in the Waitomo District), and long ridges of tilted limestone strata from Hawke's Bay to southern Wairarapa (TBP 4.2). In the South Island limestone remnants occur in northwest Nelson, northern Westland, and in pockets in Mid and North Canterbury, North Otago and Southland. Some of the best karst occurs in the Waitomo and Punakaiki areas, where limestone strata reach thicknesses of 100 m and 700 m, respectively. Many limestones elsewhere are only a few tens of metres thick but nevertheless support extensive karst. Karst landforms in marble in northwest Nelson, however, are in rocks approaching 1 km thick.

Muddy and sandy limestones with more than minor amounts of mud or sand do not usually develop good karst because voids produced by solution are soon clogged with insoluble mud and sand residues. To produce good karst the limestone rock also needs to be dense, with networks of joints that water can flow through, not soft and porous like many of the young limestones of Pliocene and Pleistocene age (younger than 5 Myr) that occur along the east side of the North Island.

Rainwater that passes through soil and decaying organic matter is enriched with carbon dioxide, becoming slightly acidic, thereby enhancing rock solution. Joints are slowly widened, forming caves at depth or deep trenches on the surface. Solution along criss-crossing vertical joints will, over time, corrode limestone into impressive pillars and eventually isolated pinnacles, as found in some places.

Over thousands of years the water percolating through networks of cracks in limestone or marble dissolves out larger and larger cavities, eventually forming a labyrinth of caves with flowing water through the lower levels. Over time, those parts of caves that aren't filled with moving water may become decorated with fantastic displays of speleothems, formed by deposition of minerals. People are most familiar with stalactites (which grow downwards from the ceiling) and stalagmites (which grow upwards from the cave floor). The weakly acidic water passing through the cracks dissolves the limestone and becomes saturated in calcium bicarbonate. When this solution reaches a cave, carbon dioxide comes out of solution and this drives precipitation of calcium carbonate and growth of these fabulous cave speleothems.

LIMESTONE AND MARBLE KARST DEVELOPMENT

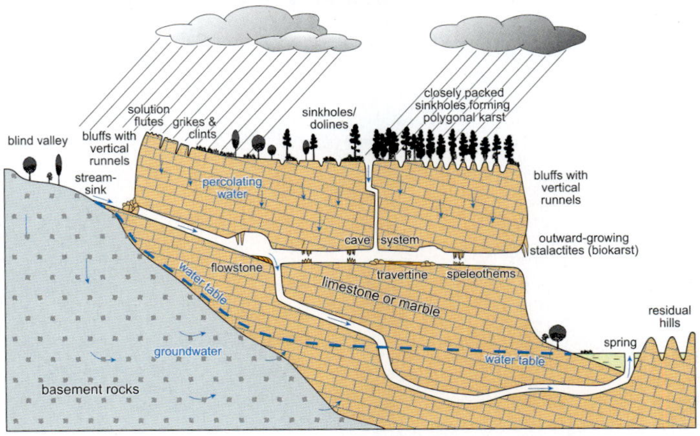

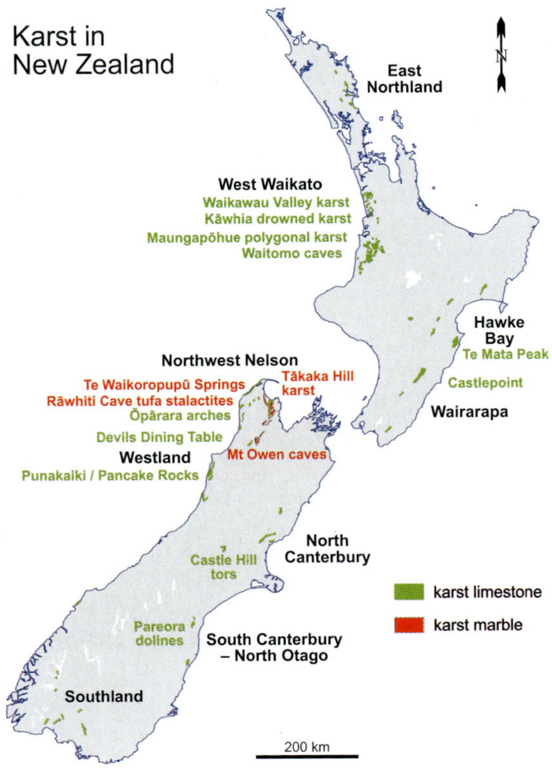

➊ TBP 4.1 Cross-section through an area of limestone sitting on greywacke basement, illustrating how water solution along joints produces the variety of karst landforms.

➋ TBP 4.2 Main locations where karst landforms and caves can be seen in New Zealand.

21.

Mangapōhue, Waitomo

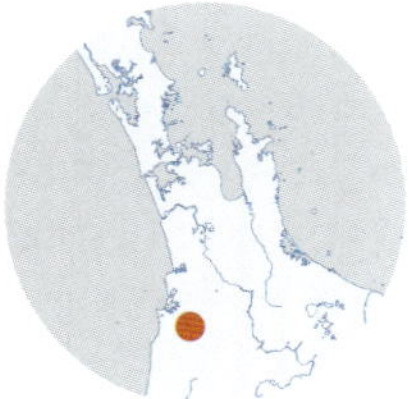

World's best-developed polygonal karst in a temperate climate

In the hills west of Waitomo is some of the best-developed polygonal karst in a humid temperate climate in the world. In this limestone environment, solution dolines (sinkholes) have formed across most of the land surface and the countryside looks like a giant egg tray with intervening sharp ridges and small conical hills between deep saucer-like basins (21.1). Such a landscape is known as polygonal karst.

This karst has developed in 60–100-m-thick limestone that was deposited as shell banks in shallow water around 30–25 Myr ago. The sandy shell layers were subsequently buried by further sediment and recrystallised into a dense limestone. In the last few Myr the region has been rising and the softer layers stripped off by erosion, exposing some of the harder limestone to solution by weakly acidic rainwater flowing into and along fractures in the rock. In this area this solution karst development has been active for about the last 3 Myr.

Branching drainage networks of underground passages have developed through much of the limestone layer. Overhead at the surface, progressive solution of the limestone has dissolved out basins 50–200 m across. Water containing the dissolved limestone seeps down through the rocky soil layers to the deepest part of the funnel-shaped basin and drains

⊙ 21.1 Some of the best polygonal karst developed in a humid temperate climate anywhere in the world occurs in the hills west of Waitomo. Over hundreds of thousands of years, slightly acidic rain water has drained down from the surface through cracks in a thick layer of limestone and dissolved out this dense network of conical solution dolines (sinkholes). All the dolines have tomos (vertical shafts) in the bottom that drain down into underground cave streams. PHOTO: ALASTAIR JAMIESON.

out through widened fractures in the rock (dolines) to the underground stream network beneath. When polygonal karst is well developed, as here, the dolines form a honeycomb-like cellular network with each cell having a pentagonal or hexagonal outline.

The Waitomo polygonal karst is unique because it has been buried on several occasions in the last 1.2 Myr by up to 10-m thickness of ignimbrite and volcanic ash erupted from caldera volcanoes in the centre of the North Island (The Big Picture 7, page 105). After being buried it probably took tens to hundreds of thousands of years for most of the soft pumice ignimbrite and ash to be flushed away through the karst drainage and exhume the dolines again. Today, some parts of the polygonal karst are still blanketed in smooth ash (22.2) and other parts are largely washed clean with the underlying rocky limestone exposed in places at the surface (21.1, 21.3).

Further down the Mangapōhue Valley to the west, the limestone is more eroded. At Mangapōhue Natural Bridge (21.4), a large cave through which the river flows has been skylighted at both ends by roof collapses, creating the bridge. The 17-m-high tunnel has two levels marking the present and a much older and higher riverbed that was abandoned and became a dry cave.

Best seen from: The most easily seen polygonal karst is on private farmland and is best viewed by driving through it on Hauturu Rd 10 km west of Waitomo, or on Waipuna Rd 10–25 km southwest of Waitomo.

Take a closer look: Take the 15–30-minute loop track and walkway from the Mangapōhue Natural Bridge carpark on Te Anga Rd, 25 km west of Waitomo. It will take you right through the bridge and back over farmland past giant fossil oysters in the limestone.

⊙ 21.2 Over the last 300–500 kyr the whole Waitomo district has been blanketed (smoothed) by thick volcanic ash and ignimbrite, erupted from caldera volcanoes in the centre of the North Island. In some places, like here, the originally rocky polygonal karst is still covered by this ash but in other places (21.1, 21.3) it has been eroded off. PHOTO: PAUL WILLIAMS.

⊙ 21.3 In this locality the blanket of volcanic ash that once buried the karst has been flushed away through the cave streams, exhuming this perfect five-sided doline and those around it. As a result there are rocky limestone outcrops exposed at the surface, forming the reticulate ridge crests between the dolines. PHOTO: LLOYD HOMER, GNS SCIENCE, 10464.

⊙ 21.4 A wooden walkway passes through the highest, 17-m-high, part of Mangapōhue Natural Bridge. A lower bridge occurs within the larger, higher one. Both were originally solution caves where the roofs at each end have collapsed and been washed away.

22.

Te Whanganui-a-Hei / Cathedral Cove, Coromandel Peninsula

Iconic Gothic Arch eroded through volcanic ignimbrite cliffs

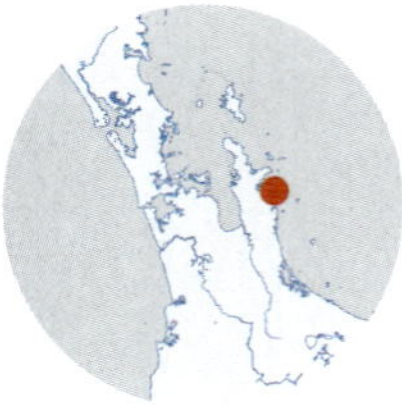

⊙ 22.1 The iconic sea arch at Cathedral Cove has been eroded along a near-vertical fracture plane (visible in the apex of the V-shaped roof) through the massive ignimbrite rock. Its Gothic Arch profile is the basis for the name of this internationally famous tourist attraction.

Cathedral Cove with its sea arch is the centrepiece of Whanganui-a-Hei (Cathedral Cove) Marine Reserve, which stretches along the coast for 4 km north of Hahei. The cliffs of this section of coastline are composed of ignimbrite rock that was erupted out of the Whitianga Caldera about 8 Myr ago. The Whitianga Caldera is a 12-km-diameter giant crater that collapsed into an emptying shallow magma chamber as the hot pyroclastic flows and ash were violently erupted. The pyroclastic flow of extremely hot volcanic glass, pumice, crystals and gas cooled and solidified to form the ignimbrite rock we see at Cathedral Cove. The caldera, which stretches from Whitianga Harbour across to Hahei and the north end of Hot Water Beach, was a smaller version of the large Rotorua and Taupō caldera craters (The Big Picture 7, page 105).

Following the ignimbrite eruptions and caldera formation, degassed viscous rhyolite magma squeezed up along at least eight different subterranean conduits or pipes and was extruded as rhyolite domes into the crater. Each dome is surrounded or separated from the others by the earlier erupted ignimbrite. These eroded domes form the hills between Whitianga Harbour and Hahei (22.3).

⬇ 22.2 Aerial view to the south of Cathedral Cove's sea arch and the larger sea stack that have been eroded by wave action out of the ignimbrite rock over the last 7500 years or so. PHOTO: ALASTAIR JAMIESON.

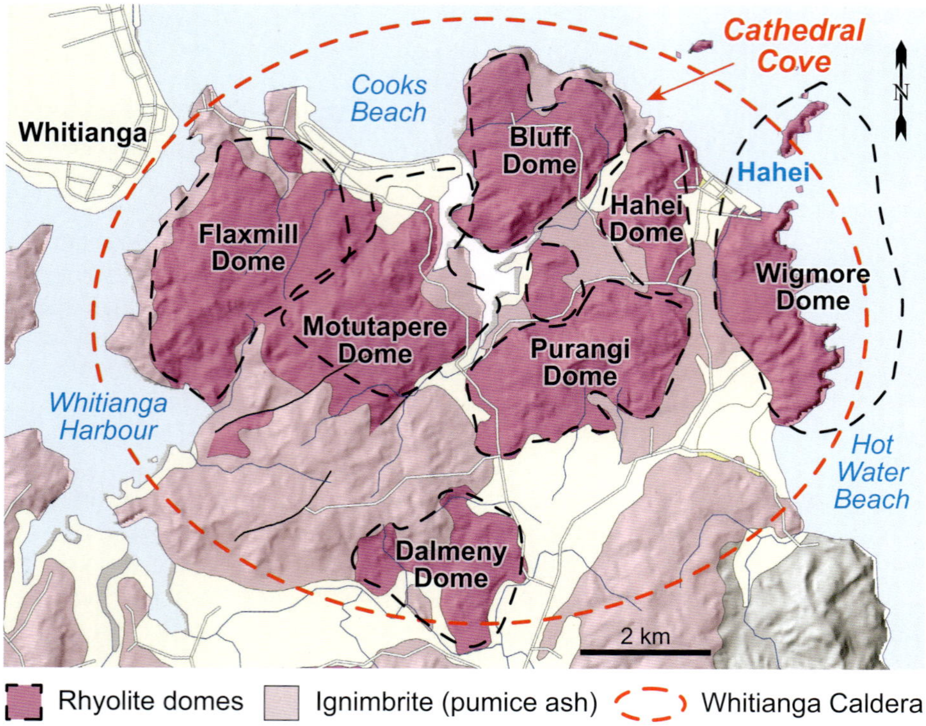

22.3 Map showing the inferred extent of the collapsed giant crater of Whitianga Caldera and the eight partly eroded rhyolite domes that were squeezed out like hot toothpaste into it.

22.4 This yacht-shaped sea stack (Te Hoho), in the low-tide beach on the north side of Cathedral Cove, has been eroded out of the relatively soft ignimbrite rock. Some of the irregular honeycomb surface textures (tafone) higher up the stack are the result of salt-growth erosion as sea spray dries above the reach of the waves.

The ignimbrite that we see along the coast is cut by a few widely spaced near-vertical fracture planes. In the last 7500 years, since sea level reached its present height after the low sea level of the Last Ice Age, storm waves have managed to erode along some of these fractures and then widen them out to sizeable caves. One of these is the 20-m-wide, 15-m-high sea arch that passes right through a short promontory in the middle of Cathedral Cove (22.1). The vertical fracture can still be seen above the 'V' at the apex of the cave's ceiling. Elsewhere along the coast around Hahei there are a number of other sea caves and arches that can be entered in a small boat or kayak when the sea is calm. The waves entering these caves compress the air, which may blast out the fractured rock along the joint plane. By this process the backs of several caves have been eroded upwards to the top of the cliffs creating an arch and 'blowhole' that smaller tourist boats will explore.

Best seen from: There are panoramic views of the coast and islands of Mercury Bay from the lookout at the end of Grange Rd, Hahei. Walk or catch the shuttle bus up the hill from Hahei to the lookout and start of Cathedral Cove Walkway. Alternatively take a tour boat ride or water taxi from Whitianga for views of this embayed coastline or hire a kayak from Hahei and paddle there.

Take a closer look: Take the Cathedral Cove Walkway from the end of Grange Rd along the cliff top and down to the beach (30 mins each way).

22.5 This toadstool rock stack sticks out of the low-tide beach on the south side of Cathedral Cove. The zone of fastest erosion is around high-tide level where the rock alternates between water-saturated and dry on most tidal cycles.

22.6 Spherical accretionary lapilli within ignimbrite at Stingray Bay, immediately south of Cathedral Cove. These 'volcanic hailstones' were formed as ash grains stuck together inside a hot, wet pyroclastic ignimbrite flow erupted from nearby Whitianga Caldera about 8 Myr ago. Photo 20 cm across.

22.7 View to the north over Cathedral Cove. PHOTO: ALASTAIR JAMIESON.

THE BIG PICTURE 5

Coromandel Volcanic Zone

THE PRESENT-DAY COLLISIONAL boundary between the Australian and Pacific tectonic plates has run through New Zealand for at least the last 45 Myr. Evidence for the Pacific Plate subducting down beneath the edge of the Australian Plate under the northern North Island is provided by the eroded remnants of subduction-related volcanoes (andesite and rhyolite) forming a line (volcanic arc) parallel to the boundary above the down-dipping oceanic crust. The outbreak of such volcanism began in the Northland Volcanic Arc 22 Myr ago (The Big Picture 1, page 23). The Coromandel Volcanic Zone was a part of a continuum of southeast-migrating volcanic arc activity from Northland to the central North Island over the last 22 Myr (TBP 5.1).

Volcanoes broke out on what is now the northern tip of Coromandel Peninsula and on Great Barrier Island 18 Myr ago, and volcanic activity extended north to the Poor Knights Islands and south to Thames by 9 Myr ago. Volcanism progressively ceased in the north but migrated southwards, and by 5–2 Myr ago (Pliocene), was confined to the Waihī-Kaimai Ranges. Between 2 and 1.6 Myr ago, volcanism jumped across to the present northeast-trending Taupō Volcanic Zone and the Coromandel Volcanic Zone became extinct.

The early phase of Coromandel Zone volcanism was almost entirely andesitic, producing numerous overlapping stratovolcanoes with surrounding laharic ring plains (TBP 5.2). Rhyolite magma started erupting 12 Myr ago and continued throughout the rest of the Coromandel Zone's history. At most centres, initial rhyolitic eruptions were of the large explosive kind (The Big Picture 7, page 105) producing pyroclastic flows (ignimbrite) and belching voluminous ash high into the atmosphere. Over time, at least six large caldera collapse craters were created, each 5–15 km across and 1.5–3 km deep, and now filled with ignimbrite and rhyolite and partly eroded away. Dozens of lava domes (e.g. Site 24), composed of viscous rhyolite lava, were extruded into and around the rims of these calderas, as well as independent of them. A small intra-plate monogenetic field erupted a cluster of small basalt scoria cones and shield volcanoes in the vicinity of Kūaotunu Peninsula and the Mercury Islands between 10 and 5 Myr ago.

Hydrothermal fluids, heated by shallow magma chambers within the Coromandel Volcanic Zone, dissolved silicon and metals from the rocks they passed up through and deposited quartz and calcite veins and reefs in major rock fractures. Pulses of rising hot fluids released from the magma chambers precipitated sulfide minerals and electrum (silver–gold alloy) in some of the reefs – the target of the Coromandel gold rushes.

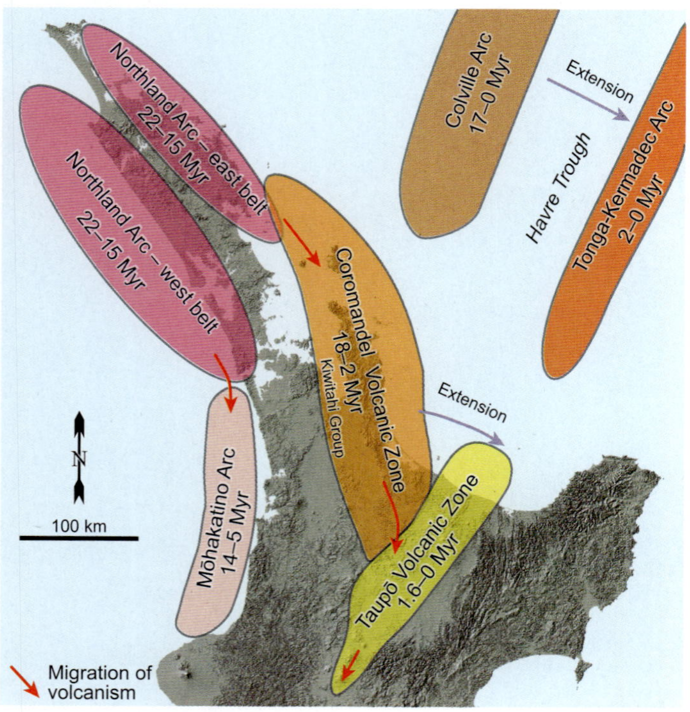

TBP 5.1 Map showing migration of subduction-related arc volcanism from a northwest–southeast-oriented Northland Volcanic Arc 20 Myr ago, via the Coromandel Volcanic Zone (18–2 Myr ago), to its present northeast-oriented location (Taupō Volcanic Zone, last 1.6 Myr) at the southern end of the Tonga-Kermadec Volcanic Arc.

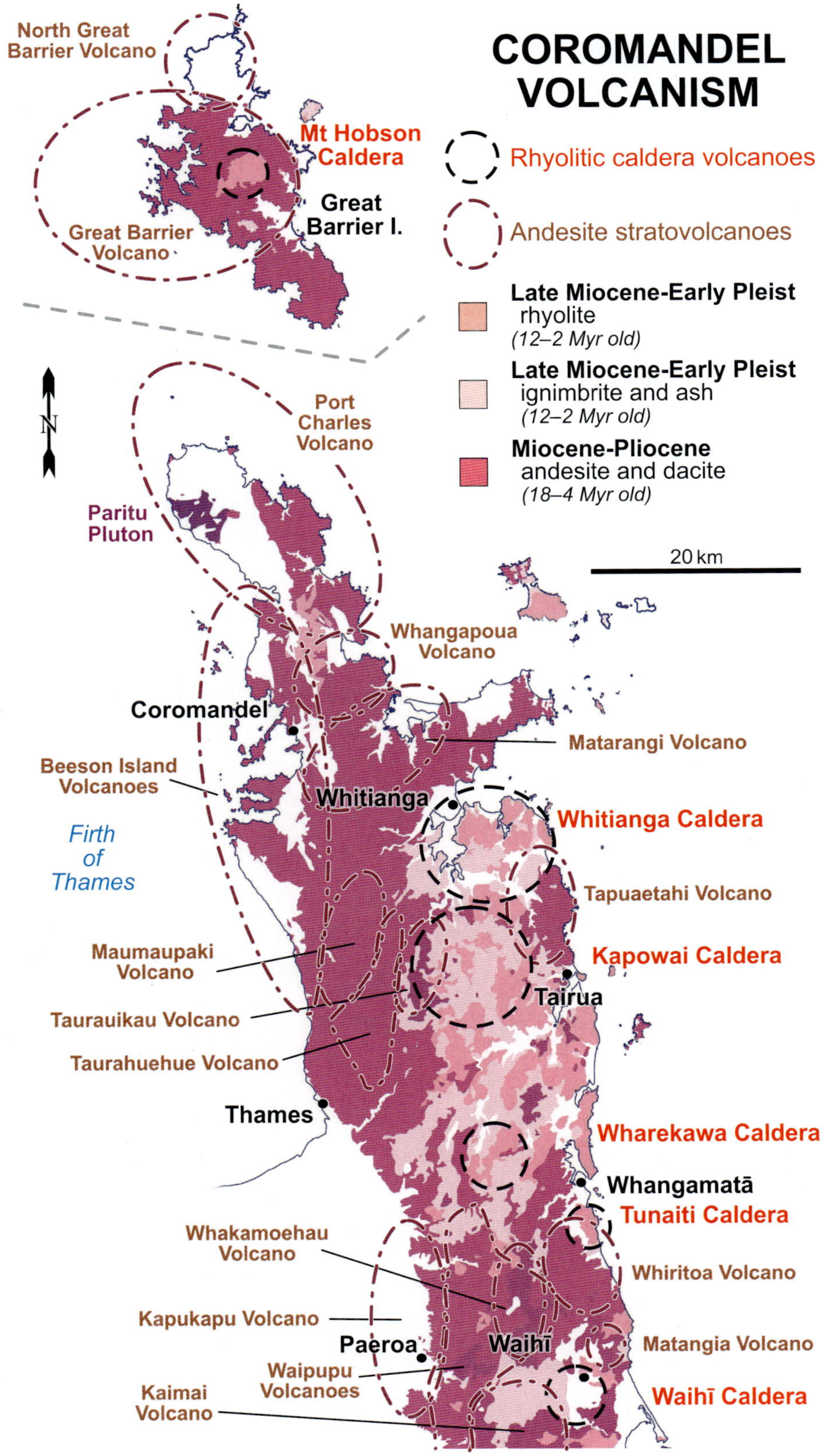

TBP 5.2 Map showing the extent of the Coromandel Volcanic Zone which was intermittently active between 18 and 2 Myr ago. Also shown is the distribution of andesitic and rhyolitic volcanic rocks and the inferred locations of andesite stratovolcanoes and large rhyolitic caldera volcanoes.

23.

Te Kōhatu-Whakairi-a-Ngātoroirangi / Table Mountain, Thames

Surface of an ancient, solidified lava lake

As the name suggests, Table Mountain is a prominent flat-topped plateau on the crest of the Coromandel Range (23.1). Lying inland from Thames, at the head of the Kauaeranga Valley, 800-m-high Te Kōhatu-Whakairi-a-Ngātoroirangi / Table Mountain can easily be recognised by its profile when viewed from the Hauraki Plains (23.2). The plateau is almost completely surrounded by 150-m-high vertical bluffs of columnar-jointed andesite. These joints were produced by contraction of the rock as it cooled from liquid lava to solid rock. The joints are oriented at various angles, apparently the result of a large mass of lava slowly cooling from the sides inwards and from the top down. Thus it is inferred that Table Mountain is a cooled and solidified lava lake that was erupted into and filled a 1-km-diameter elongate crater about 8 Myr ago. This crater is near the western margin of the 15-km-diameter Kapowai Caldera and could have been partly formed by caldera subsidence.

23.1 View east over flat-topped Table Mountain with The Pinnacles eroded rhyolite domes in the left distance. Table Mountain is a former andesite lava lake that solidified about 8 Myr ago inside a volcanic crater. It overlies and intrudes softer rhyolitic sediment and ignimbrite that has been eroded away from around the uplifted plateau – the surface of which was once the surface of the lava lake. PHOTO: ALASTAIR JAMIESON.

23.2 Table Mountain is one of the highest parts of the Coromandel Ranges. Its distinctive flat top is easily recognised on the main divide skyline, up the Kauaeranga Valley inland from Thames.

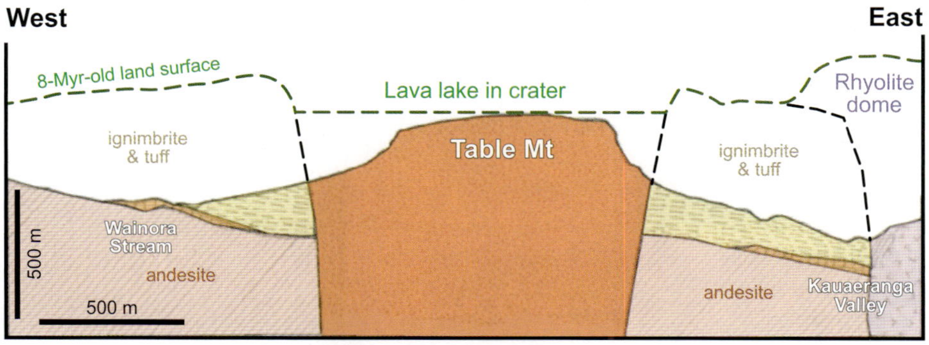

23.3 West to east cross-section through Table Mountain and surrounding areas showing its inferred shape as a lava lake filling a collapsed crater. The dashed line shows the approximate elevation and reconstructed topography that has been eroded away in the 8 Myr since the lake cooled and solidified.

The flat top of Table Mountain is probably close to the original surface of the lava lake, which has not eroded away because it is composed of much harder rock than the surrounding softer ignimbrite and scoriaceous rocks that once formed the crater walls (23.3). Some of the Table Mountain andesite appears to have spilled from the crater lake and flowed over the surrounding land. These flows now cap the high ridge to the northeast. In the 8 Myr since eruption, this part of the Coromandel Ranges has been uplifted and tilted down to the east with erosion removing all the volcanic landforms that once surrounded the lava lake.

The rivers and streams that now drain the slopes of Table Mountain have eroded down well below the level of the original lava lake, into older andesite and rhyolite volcanic rocks, ignimbrites and lake sediments. Thus a wide variety of rock types can be seen if you fossick around in the cobbly bed of the Kauaeranga River.

Best seen from: The flat top of Table Mountain can be seen on the skyline of the Coromandel Range from the Hauraki Plains along SH 25.

Take a closer look: Closer views of its impressive bluffs are obtainable from the end of the Kauaeranga Valley gravel road and from several walking tracks above the valley floor.

23.4 A 6-m-long block of columnar-jointed andesite that broke off the Table Mountain bluffs and now lies in the bed of the Kauaeranga River, where the Billy Goat Track crosses.

24.

Mauao / Mt Maunganui

Volcanic sentinel at entrance to Tauranga Harbour

⊙ 24.1 The 252-m-high Mauao / Mt Maunganui rhyolite dome stands proud at the entrance to Tauranga Harbour. PHOTO: LLOYD HOMER, GNS SCIENCE, 10071.

⊙ 24.2 View east from the top of the Mount over the sand tombolo that has been thrown up by the sea within the last 5000 years and now links the volcanic dome to the rest of the mainland. The north (left) side of the tombolo faces the open Pacific Ocean whereas the south (right) side is inside Tauranga Harbour.

Mauao / Mt Maunganui, commonly known as the Mount, is possibly the most prominent volcanic dome in the country. It was formed, about 2.4 Myr ago (early Pleistocene), by viscous magma rising up a conduit from a shallowly buried rhyolite magma chamber. This magma squeezed out at the surface to form a high dome or mound surrounded by steep-sided lava flows that did not flow very far (24.3). The volcano's original footprint probably extended 1–2 km further out in all directions. This can be envisioned out to the east where Hopukiore / Mt Drury, Moturiki and Motuotau Islands are all less-eroded parts of the distal lava flows from the Mount Volcano. Today Mt Maunganui is a considerably eroded, more-steep-sided version of its original shape.

Since the dome was erupted there have been about 40 Ice Age climate cycles. When sea level was high during warm interglacial times, the Mount was an island with waves eroding its lower slopes. When sea level was lower during colder periods, the Mount was a forest-covered hill and part of the mainland. At the start of the present high-sea-level period, between 8000 and 5000 years ago, the Mount was an island. Since then sand spits and barriers have built up along the coast forming Matakana Island to the west and the tombolo that links the Mount to Pāpāmoa in the east (24.2).

The hot saltwater pool at the foot of Mt Maunganui indicates that there is still some heat left in the old magma chamber beneath the volcano. The magma at depth heats water that penetrates deep into the ground from the sea and then rises up towards the surface. The pools are supplied from a 100-m-deep bore where the water temperature is about 40°C.

Best seen from: The steep-sided profile of Mt Maunganui is Tauranga and Mt Maunganui's most prominent feature.

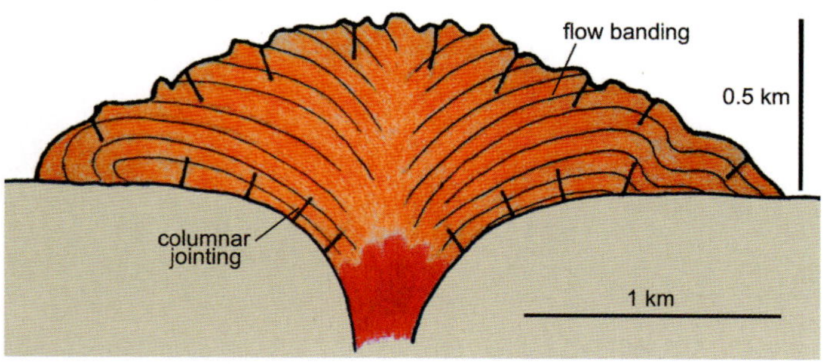

⊙ 24.3 Cross-section through an idealised dacite or rhyolite dome showing the extrusion of the viscous magma through a flaring conduit out onto the surface. The extruded dome with its high, possibly pinnacled, top was surrounded by viscous flows that did not move far from the vent. Lines within the dome represent the general direction of flow-banding, whereas more-solid short lines perpendicular to the outer surface show the orientation of columnar cooling joints that are often formed as magma solidifies. Domes are commonly 1–2 km in diameter. DRAWING: MARGARET MORLEY.

⊙ 24.4 The Mt Maunganui dome consists of flow-banded rhyolite that squeezed out of the ground like molten toothpaste and some of the bands became tightly folded as more lava was pushed into the huge heap. Photo 70 cm across.

Take a closer look: A popular track around its base (45-min. walk) gives views of the coast and flow-banded rhyolite rock. For the more energetic, ascend one of the two tracks to the summit (40-min. climb) for breathtaking panoramic views.

➲ 24.5 White scars in the grass on the lower southern slopes of Mt Maunganui are large heaps of pipi shells. These are the discarded rubbish from numerous pre-European Māori camps. The campsites look over the sheltered waters and tidal flats of the Tauranga Harbour where pipi shellfish were collected.

➲ 24.6 View to the west across the main Maunganui beach to Mt Maunganui volcanic dome.

THE BIG PICTURE 6

Taupō Volcanic Zone

THE TAUPŌ VOLCANIC ZONE is at the southwest end of the Pacific Ring of Fire – a vast near-circular zone of active volcanism and large earthquakes that occurs where dense oceanic lithosphere plunges down into the mantle (subduction) beneath the lithosphere of the surrounding continents (TBP 6.1). The Taupō Volcanic Zone stretches from Ohakune through Lake Taupō to the Bay of Plenty and Whakaari / White Island.

Hydrated basalt magma which forms above the subduction zone rises and collects in magma chambers at the base of or within the crust (TBP 6.2). Here the magma slowly cools for thousands to hundreds of thousands of years. As it cools, high-temperature minerals (olivine, clinopyroxene) crystallise out and, as a result, the composition of the remaining liquid becomes more silica-rich and andesitic. This andesite often rises and erupts as stratovolcanoes, such as Ruapehu, Tongariro, Putauaki / Mt Edgecumbe and White Island (TBP 6.3). The magma has become even more silica-rich in many places by further crystallisation and by crustal extension assisting the addition of melted continental crust (granite). This has produced dacite or rhyolite magma which has erupted explosively to form ignimbrite flows (The Big Picture 7, page 105) or been squeezed out quietly as viscous lava to form volcanic domes. Occasionally small batches of basalt, rising directly from the mantle, have erupted among the andesite and rhyolite volcanoes. Sometimes basalt magma passing through a rhyolite magma chamber has set off a large ignimbrite eruption. Many geothermal systems (The Big Picture 8, page 108), usually heated by rhyolite magma chambers, are present in the Taupō Volcanic Zone.

The Taupō Volcanic Zone has been volcanically active for the last 1.6 Myr. The zone is located within a 50-km-wide northeast–southwest belt of extension that has been pulling apart and widening at a rate of up to 15 mm per year. The belt of extension is bordered by faults which have down-dropped the zone by about 2000–3000 m. On either side are mountain ranges of basement greywacke in places mantled by sheets of ignimbrite that were erupted from the zone. Not only is the belt bounded by young active faults, but many faults also occur within it and are the source of numerous earthquakes.

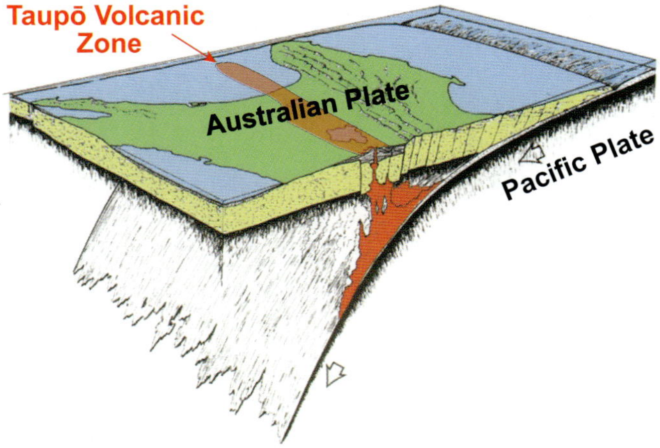

TBP 6.2 Three-dimensional diagram showing the oceanic lithosphere of the Pacific Plate being subducted down beneath less-dense continental lithosphere of the Australian Plate beneath the northern half of New Zealand's North Island. Hydrated basalt magma is generated above the subduction zone and rises in stages to erupt at the surface.

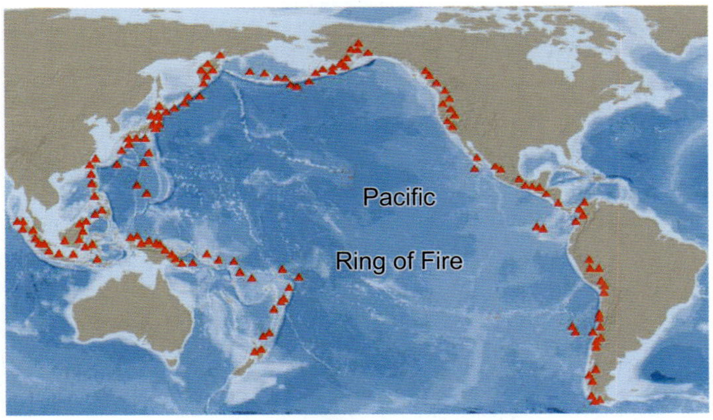

TBP 6.1 The Taupō Volcanic Zone is at the southwest end of the Pacific Ring of Fire.

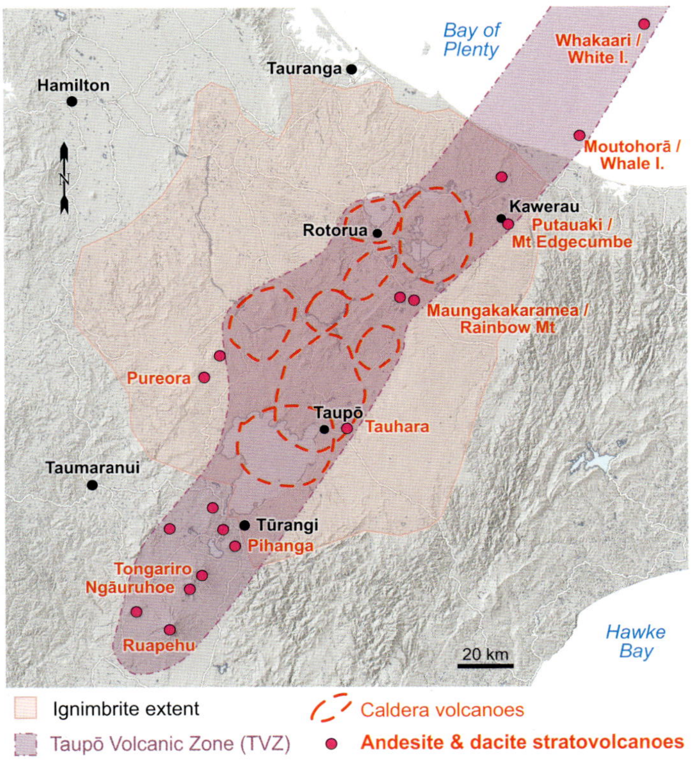

TBP 6.3 Map of the Taupō Volcanic Zone showing named locations of rhyolite centres and the larger andesite stratovolcanoes.

25.

Whakaari / White Island

New Zealand's most frequently active volcano

The periodically active Whakaari / White Island Volcano lies at the northern end of the Taupō Volcanic Zone (The Big Picture 6, page 97) as part of the Pacific Ring of Fire. The island is the upper part of a relatively small andesite/dacite stratovolcano that has been almost continuously active since its inception about 150,000 years ago. It sits 48 km offshore in the middle of the Bay of Plenty just beyond the edge of the continental shelf in ~500 m water depth. Its highest peak, on the crater rim, is 321 m above sea level, which means that its upper third is above water but only about 10% of its volume is.

Above sea level, Whakaari consists of two overlapping composite cones made of lava and breccia. The eroded remnants of the extinct Ngātoro Cone form the high northeastern cliffs, whereas the younger central cone forming the rest of the island is still very active. The elongate amphitheatre-shaped modern crater is inferred to consist of three overlapping sub-craters with only the westernmost having been active historically. Each of the three bays at the southeastern corner of the island (25.1), where the crater opens out to the sea, is believed to have been formed by a phreatic eruption of steam-powered gas and ash.

The mix of seawater, fresh groundwater and hot magmatic fluids produces acidic brines, with a pH as low as 2. These rise to the surface in the western part of the crater forming numerous small hot springs, mud pools, fumaroles, and acid streams and pools. The crater lake of boiling acidic water (25.2) varies in volume due to recent rainfall and fluctuating hydrothermal activity.

Since written records of activity began about 200 years ago, most activity from Whakaari's crater has been phreatic eruptions. This is the technical term for wet eruptions of magma-heated steam and steam-powered ash with no actual magma itself being erupted. The exception was a period between 1976 and 2000, when magma rose inside the crater and there was a series of wet phreatomagmatic and dry strombolian/fountaining eruptions that included the magma itself. Since 2000 there have been only phreatic eruptions with several somewhat larger eruptions of searing gas and ash powered by the sudden release of pressure from an over-pressured underground body of super-hot water within the ash-filled vent.

25.1 View westwards over Whakaari / White Island.
PHOTO: BRAD SCOTT, GNS SCIENCE, 106899.

The most recent of these phreatic eruptions occurred on 9 December 2019 and resulted in the death of 22 tourists and tour guides and horrific burns to 25 more who were also visiting the island at the time. A previous volcano-related catastrophe occurred on Whakaari in September 1914. On that occasion a section of steep crater wall collapsed into the crater lake, blocking the vent for a short while before the built-up pressure threw out the landslide rubble and hot lake water as a small lahar that swept across the crater floor and overwhelmed a sulfur mining camp and the 10 miners who were there at the time. Heaps of debris still present on the eastern part of the crater floor are remnants of that event.

Sulfur was mined on the island during several periods between 1885 and 1914 and again between 1923 and 1933. It was mainly used for agricultural fertiliser as it was not pure enough for more specialised purposes. Because of the frequency of eruptions and the highly corrosive acidic fumes, the island is mostly bare rock, although there are some areas of regenerating pōhutukawa forest on the lower outer slopes well beyond the crater. The outer slopes of the island are also home to several large gannet colonies.

Best seen from: Following the deadly 2019 eruption it is unlikely that tourists will be allowed back on the island, at least for many years. In clear weather the steaming island can be seen offshore from the Bay of Plenty coast.

25.2 View east over the steaming acidic lake at the head of Whakaari's elongate breached crater at a time when it was full of water. The outside of the rim of the crater is in the foreground. PHOTO: KAREN BRITTEN, GNS SCIENCE, 166370.

25.3 View east across Whakaari's breached crater floor covered in mounds of debris left behind by the tragic 1914 lahar that killed 10 sulfur miners.

➔ 25.4 The colourful steaming cliffs on the southern wall of Whakaari's crater. The red-brown colours are relatively fresh volcanic rocks, whereas the white rocks have been chemically altered to varying degrees by highly corrosive and acidic hydrothermal fluids and gases.

➔ 25.5 White and yellow efflorescence of calcium sulfate (gypsum) and some pure sulfur covers parts of Whakaari Volcano's crater floor.

➔ 25.6 Sulfur crystals that have grown around a fumarole on Whakaari.

➔ 25.7 Corroding and rusting remains of the sulfur crushing plant and bagging factory behind Te Awapuia / Crater Bay that was active between 1923 and 1933. The product was used as agricultural fertiliser. PHOTO: BRAD SCOTT, GNS SCIENCE, 156974.

26.

Mt Tarawera

Source of New Zealand's largest historic eruption

Mt Tarawera Volcano fills much of the southeast portion of the large collapsed Okataina Caldera crater. The many lakes in the vicinity are flooded portions of the caldera crater or adjacent smaller collapsed volcanic craters separated by numerous extruded rhyolite domes and flows. Tarawera is the youngest and most recently active volcano in Okataina Caldera. It consists of numerous rhyolite lava domes, extruded during four periods of eruption in the last 25,000 years. The most recent was the Kaharoa Eruption 700 years ago that extruded the three steep-sided domes that now define Tarawera's summit ridge. They are (from northeast to southwest) Wāhanga, Ruawāhia and Tarawera (26.3). Vast quantities of rhyolitic pumice and ash early in this eruption were spewed high into the air and deposited as a thin layer over much of the North Island. This and the earlier eruptions of Tarawera were rhyolitic in composition and sourced from a shallow magma chamber beneath the volcano.

⊙ 26.1 New Zealand's most devastating eruption since the arrival of humans in Aotearoa lasted just 4.5 hours in the early hours of 10 June 1886. View from the south over the northern end of the rift through Mt Tarawera Volcano that opened up during the eruption. Wāhanga rhyolite dome is on the left and Mt Edgecumbe andesite stratovolcano is in the distance. PHOTO: LLOYD HOMER, GNS SCIENCE, 14291.

The historic Tarawera Eruption, however, was quite different and erupted basalt magma sourced directly from the upper mantle, perhaps 60 km or so beneath the surface.

The 1886 Tarawera Eruption has been the largest volcanic eruption in New Zealand since human colonisation, even though it lasted for less than five hours. This eruption, between 1.30 a.m. and 6 a.m. on 10 June 1886, started with violent earthquakes followed by dry scoria fountaining eruptions from craters along a northeast–southwest rift that opened up along the crest of the mountain (26.1, 26.6). Within an hour or two the rift had extended southwest opening up a 17-km-long fissure. Along the southern half of the rift, rising basalt magma encountered the waters of Lake Rotomahana and its geothermal system. This led to explosive phreatomagmatic eruptions ejecting surges of hot mud, blocks, ash and steam (26.4) that resulted in most of the devastation to the surrounding area and the 108 human casualties. Following the eruption, the enlarged crater in the south slowly filled with water to become a much larger Lake Rotomahana and a new geothermal field was formed in Waimangu Valley at the southern end of the fissure (26.2).

Prior to the 1886 eruption numerous tourists had been coming to this rather remote part of New Zealand to see the eighth natural wonder of the world – the Pink and White Terraces on the shores of Lake Rotomahana. Needless to say, these sinter terraces were completely destroyed and their sites buried during the 1886 eruption. If they still existed they would have been featured as 'the jewel in the crown' of this book.

Best seen from: Mt Tarawera can be seen in the distance from a number of places in the Rotorua district. Best views are from the shore of Lake Tarawera or from the Lake Rotomahana boat cruise as part of the Waimangu Volcanic Valley experience.

Take a closer look: The only way to see the crater and top of Mt Tarawera is to pay for a guided walk or helicopter trip. If you can afford it, you will be richly rewarded with the spectacular scenery.

26.2 Aerial view northeastwards along the southern sector of the 17-km-long rift and line of craters that opened up during the 1886 Tarawera Eruption. Waimangu Geothermal Field is in the foreground, with Lake Rotomahana between there and Mt Tarawera. PHOTO: ALASTAIR JAMIESON.

26.3 The three young rhyolite lava domes that form the upper parts of Mt Tarawera. From northeast to southwest (left to right) they are Wāhanga, Ruawāhia and Tarawera. Each was extruded as viscous magma over a matter of weeks near the end of the Kaharoa Eruption event 700 years ago. PHOTO: ALASTAIR JAMIESON.

➲ 26.4 Vast quantities of wet, muddy ash erupted from Lake Rotomahana in 1866 were responsible for the Buried Village. Here to the south of the lake the volcanic mud is slowly moving downslope in wet winters producing the highly uneven topography. Mt Tarawera is in the distance.

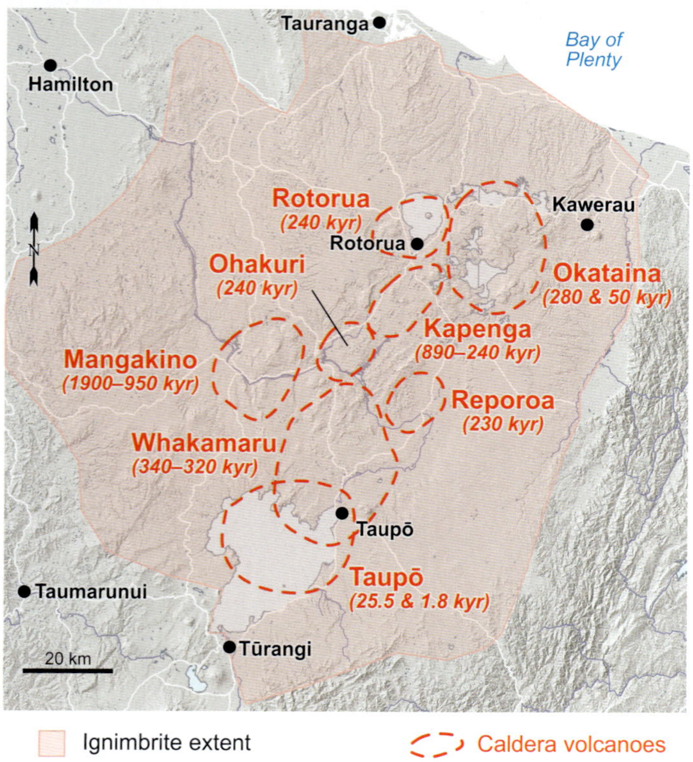

➲ 26.5 In the walls of northern part of the rift that opened up across Mt Tarawera in 1886 we see a 20–60-m-thick layer of red (and black) basalt scoria, from that 1886 eruption, capping pink and white pumice breccia and ash layers erupted during the rhyolitic Kaharoa Eruption 700 years ago. The red scoria was oxidised as the molten froth was fountaining through the air, whereas the unoxidised black scoria was erupted so rapidly it did not have time to be oxidised. A thin, dark, vertical line through the cliffs at the bottom of the 'V' is a basalt dike – one of a number of feeder conduits for the 1886 eruption. PHOTO: ALASTAIR JAMIESON.

➲ 26.6 Aerial view along the crest of Mt Tarawera from the northeast showing the 1886 rift through all the rhyolite domes. Beyond the volcano to the right is Lake Tarawera. PHOTO: ALASTAIR JAMIESON.

➲ TBP 7.1 Location of caldera volcanoes in the Taupō Volcanic Zones (two dormant and likely to erupt again).

THE BIG PICTURE 7

Caldera volcanoes and ignimbrite eruptions

THE RHYOLITE VOLCANOES of the Coromandel and Taupō Volcanic Zones were fed from shallow magma chambers within the continental crust. Rhyolite magma containing little dissolved gas, mostly carbon dioxide, often erupted viscous domes at the surface (see Site 24). As magma with lots of dissolved gas rose to the surface, the confining pressure decreased and the gas came out of solution, often producing an enormous fountaining of hot, frothy lava (TBP 7.2). Volcanic ash sometimes reached 50 km into the air and was blown away as a plume by the wind. The ash eventually mantled a wide area of land and sea in a layer of light-coloured glassy particles, such as fine pumice.

Denser parts of the eruption column of frothy lava, hot gas, crystals and glass collapsed back towards the ground. As it landed, the energy of the collapsing column blasted it out sideways as a searing-hot pyroclastic (or ignimbrite) flow that travelled just above the ground at speeds of up to 500 km/h. It would have incinerated and buried the forests it engulfed.

The largest such eruptions came out from the Taupō Volcanic Zone in the last 2 Myr with at least four super-eruptions, each expelling more than 450 km^3 of magma. If the flow was still extremely hot as it came to a stop, the deposit of pumice, glass, crystals and a few rock fragments welded together into a hard ignimbrite rock as it cooled. If the flow had cooled sufficiently during its journey, then the resulting ignimbrite deposit was an unwelded, relatively soft pumice breccia deposit.

These huge eruptions lasted just a matter of hours to a few days. The land overlying a rapidly emptying shallow magma chamber collapsed into the hole, creating a huge crater known as a caldera. Calderas on the Coromandel Peninsula are 5–15 km in diameter, whereas those in the Taupō–Rotorua area are up to 30 km across. The older calderas are hard to recognise because they are filled, buried or eroded. The three calderas that still retain their recognisable shape are Rotorua (now extinct), Okataina and Taupō (TBP 7.1). The latter two are currently believed to be dormant and could have further eruptions. Viscous rhyolite domes may be extruded from the same centres before or after the caldera-forming eruptions.

TBP 7.2 Cross-section through an idealised caldera volcano eruption showing an enormous cloud of ash and pumice, up to 30–50 km high, with the finer parts blowing away and the denser parts collapsing and blasting out sideways as hot pyroclastic flows that sped across the land and left behind a sheet of ignimbrite rock. MODIFIED FROM DRAWING BY GEOFF COX.

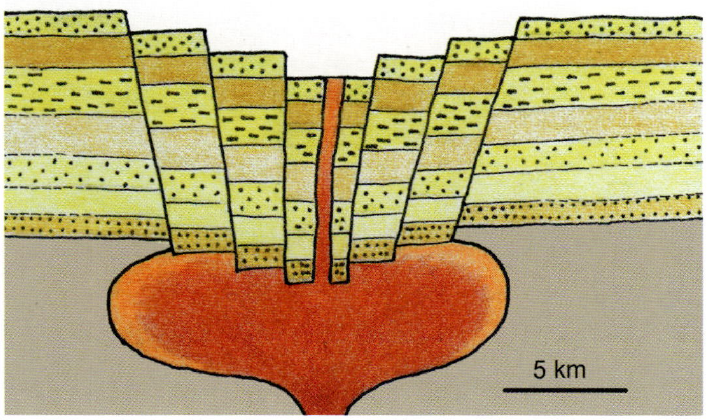

TBP 7.3 Cross-section showing the inwards collapse of a magma chamber's roof creating a large, circular caldera, soon after a huge ignimbrite eruption. Caldera width 5–30 km. DRAWING: MARGARET MORLEY.

27.

Mamaku Plateau, Rotorua

Internationally rare ignimbrite pillars and small conical hills (tors)

Driving into Rotorua over the 600-m-high Mamaku Plateau you pass through a field of conical rocky mounds, some topped with a narrow vertical rock column. This is the only place in New Zealand where globally rare ignimbrite tors can be seen. The near-flat-topped Mamaku Plateau is capped by a sheet of ignimbrite, up to 100 m thick, that was erupted as a huge pyroclastic flow (The Big Picture 7, page 105) from Rotorua Caldera, 240,000 years ago. The centre of the collapsed caldera crater is now occupied by Lake Rotorua. As the searing-hot cloud of volcanic gases and pumice came to a stop they compacted down and cooled into ignimbrite rock. Hot gases escaping from the cooling ignimbrite caused

27.2 A rare single, 10-m-high column of ignimbrite among the Mamaku tors. Person at base of tor for scale.

27.1 This 10-m-high column of rock is one of the two most impressive of the ignimbrite tors that can be seen from SH 5 over Mamaku Plateau. Ignimbrite tors are a rare landform globally.

cementation in gas escape pipes, and the ignimbrite itself shrank and cracked along vertical joints, forming columns.

The top of the highest tors are all about 10 m above the surrounding plateau (27.2) and this is believed to have been approximately the upper surface of the original ignimbrite sheet. There are no large streams or rivers on the top of the plateau where the tors are found. The tors appear to have been formed by wind erosion during the several Ice Age cold (glacial) periods since the eruption. Pollen studies indicate that during these cold periods the top of the Mamaku Plateau was above the bush line and thinly clad in alpine tussock land, but there was no ice sheet nor glaciers.

During these cold times the freeze and thaw of ice would have broken up the softer parts of the ignimbrite sheet. Rainwater infiltrated along the cooling cracks, which expanded on freezing, thereby gradually breaking open the fractures and promoting erosion. Strong winds blew the broken-up small particles away as loess, leaving behind a relatively flat deflation surface between. The cemented pipes and harder parts of the ignimbrite sheet resisted these erosional forces and remained upstanding as pinnacles with flaring conical bases (27.1). Some of the ignimbrite split away from the tors along the vertical cooling cracks, leaving behind the straight-sided columns of rock (27.2).

The floor of the Mamaku Plateau around the tors is fresh ignimbrite rock mantled with layers of volcanic ash (27.4) that have been erupted in the last 15,000 years, indicating that the most recent period of fierce ice and wind erosion of the Last Ice Age came to an end about that time. There is very little alluvium on the top of the plateau, which confirms that fluvial erosion by water was not a major player in carving out the tors.

Best seen from: SH 5 between Tirau and Rotorua passes through the middle of the field of ignimbrite tors on the top of the Mamaku Plateau, about 18 km east of Fitzgerald Glade.

Take a closer look: From SH 5 drive along Maraeroa Rd to Mamaku township through the bizarre landscape.

27.4 This road cutting slices right through one of the Mamaku tors and shows that it has been eroded out of ignimbrite rock (forming the cone). Note the 1-m-wide vertical pipe of cemented ignimbrite in the core of the tor. The fresh rock is draped by wind-blown loess and volcanic ash erupted in the last 15,000 years.

27.5 The Mamaku ignimbrite tors consist of upstanding small cones, sometimes capped with vertical-sided columns of rock. They were eroded by ice and wind on the top of the Mamaku Plateau during the coldest periods of the recent Ice Ages. PHOTO: ALASTAIR JAMIESON.

27.6 The field of ignimbrite tors on the top of the Mamaku Plateau covers nearly 100 km². The ignimbrite was erupted from the Rotorua Caldera Volcano 240,000 years ago. Lake Rotorua (in distance) now partly fills the bottom of the caldera crater that was formed during that eruption. PHOTO: ALASTAIR JAMIESON.

27.3 Three of the larger rocky tors on the Mamaku Plateau. All have vertical sides produced by blocks of ignimbrite rock breaking away along the vertical columnar cooling joints.

THE BIG PICTURE 8

Geothermal systems

HIGH HEAT FLOW from shallow magma chambers, mostly rhyolitic, heats deeply circulating groundwater. Being hotter and less dense than surrounding cold water, it then rises, usually up a zone of fractured rock like a fault, and may discharge at the surface. The deep groundwater at depth is recharged by infiltrating rainwater from above (TBP 8.2).

More than 20 geothermal fields have been recognised in the Taupō Volcanic Zone between Whakaari / White Island and Mt Tongariro (TBP 8.1). Some are sources of superheated water and steam trapped at depth with little discharge at the surface. Others have developed large areas at the surface characterised by geothermal discharge features (TBP 8.6), such as steaming ground and fumaroles (e.g. Craters of the Moon), hot pools, geysers and sinter terraces (e.g. Whakarewarewa, Waiotapu, Ōrākei Kōrako) and boiling mud pools (e.g. Tikitere). Many extinct and eroding geothermal systems are known, especially on the Coromandel Peninsula. Some systems may have been active for tens to hundreds of thousands of years. The youngest geothermal system in the world is in Waimangu Valley, having formed after the Tarawera Eruption in 1886. Several young geothermal systems are associated with active or dormant volcanoes, such as Whakaari, Moutohorā / Whale Island and Ketetahi on Mt Tongariro.

Geysers and hot springs occur where the hot water from depth can rise rapidly to the surface and not lose too much heat on the way up. Water discharged from hot springs is usually weakly alkaline and rich in dissolved chloride – properties acquired at depth by chemical interaction with the rocks it passed through. These chloride pools are usually clear and often boiling. Geysers usually eject hot chloride water.

Where the deep reservoir temperature exceeds 175°C, silica will dissolve in the water. As this silica-rich water cools after discharging at the surface, silica sinter is precipitated out of solution building up sinter rings around some pools or sinter terraces like those around Pōhutu Geyser at Whakarewarewa. Pure silica sinter is white, but the presence of iron impurities may produce grey or pink sinter. Microbial mats growing on the surface of the sinter produce a range of orange, yellow, green and black colours depending on the temperature of the water.

In some places the hot geothermal water may get trapped below the surface by an impermeable cap layer such as mud. This hot water may boil to form steam, which along with the common volcanic gas hydrogen sulfide may escape up narrow cracks to the surface forming fumaroles. Here the rising steam and gas may alter the near-surface rock and clays to form the grey bubbling mud. Around these fumaroles lemon-yellow sulfur crystals often precipitate.

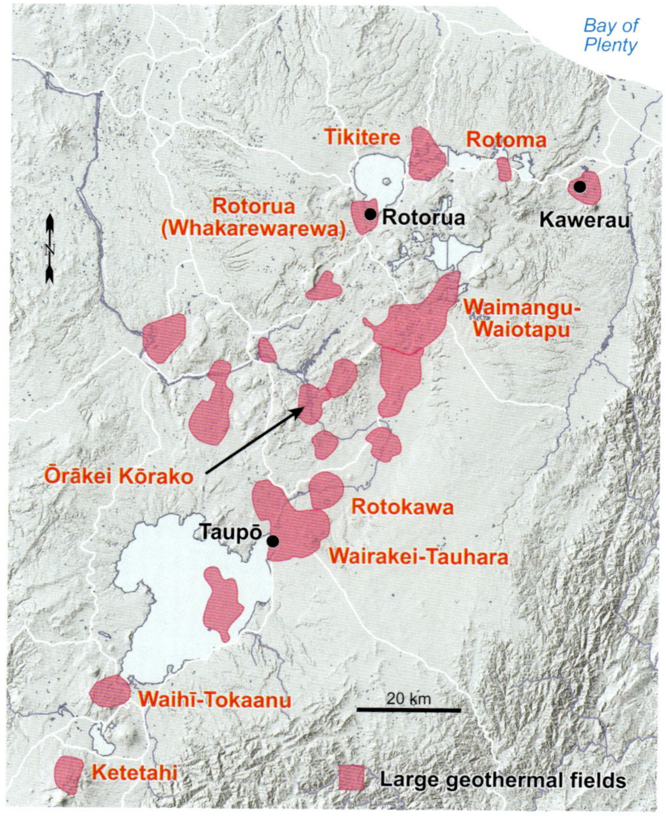

TBP 8.1 Location of geothermal fields in the Taupō Volcanic Zone. Those with significant surface features are named.

TBP 8.2 Cross-section through the ground beneath a surface geothermal field showing how a hot magma chamber heats deep meteoric water which then rises to the surface.

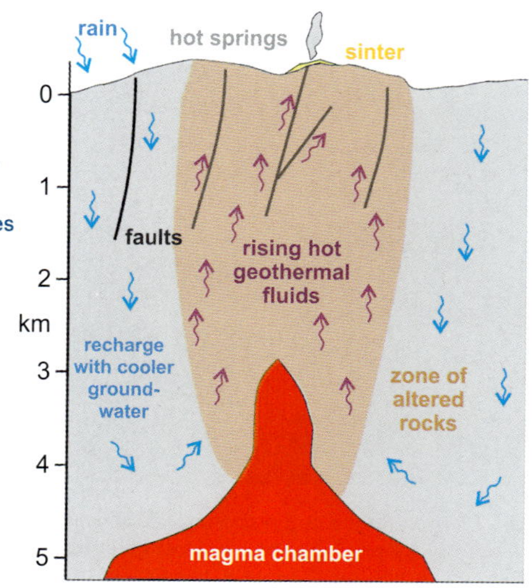

The trapped hot water and volcanic gas may react with the surrounding soil, altering their composition and forming acidic sulfate-rich waters. This acidic steam and gas may escape through waterlogged soil or ponds to form bubbling mud pools, like those at Waiotapu (TBP 8.5). In dry spells, the gas bubbles burst out through dryer and more viscous mud. The blobs of mud they eject may build up into a steep-sided heap called a mud volcano, sometimes with a crater on top (TBP 8.4). When there has been a lot of rain the mud volcanoes wash away and the mud pool becomes more liquid. At these times the escaping gas causes concentric turbulence in the surface of the mud.

TBP 8.4 A large (4 m wide, 1.5 m high) mud volcano erupted from Hauanu boiling mud pool during a dry spell, Whakarewarewa.

TBP 8.3 A boiling alkaline chloride pool at Whakarewarewa, surrounded by textured geyserite sinter tinged yellow and black with sulfur in places. PHOTO: PAUL VAN DER WERFF, GNS SCIENCE, 107416.

TBP 8.5 Gas bubbles rise and burst through a mud pool creating concentric mud ripples, beside Loop Rd, Waiotapu.

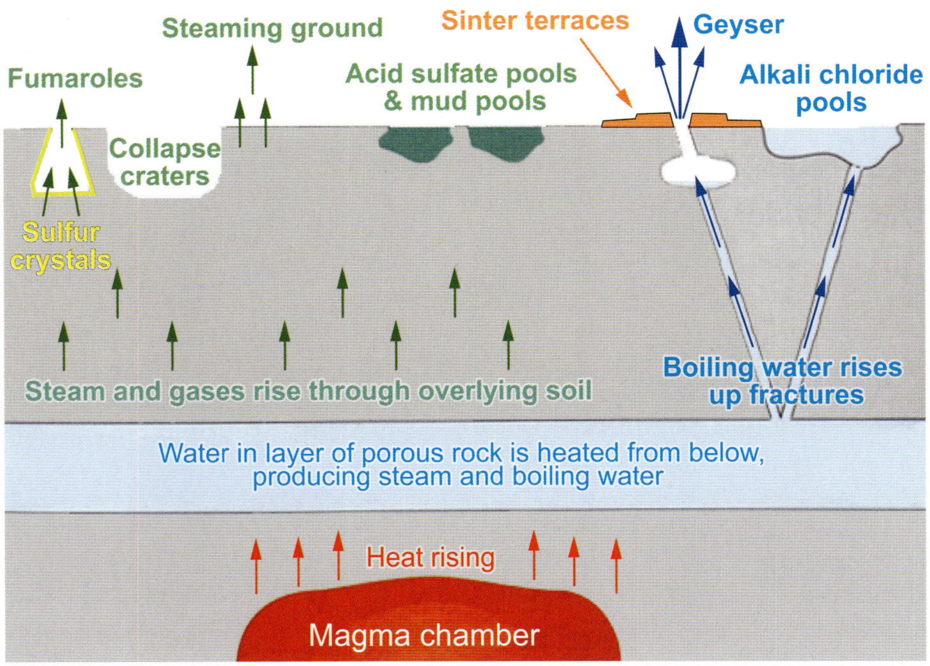

TBP 8.6 Cartoon cross-section showing a reservoir of boiling hot water and volcanic gas trapped in a porous layer at shallow depth. Rapid rise of boiling water through fractures in the rock results in clear pools of boiling alkaline chloride water, geysers and silica sinter deposition. Slow rise of steam and hydrogen sulfide through the overlying soil layers results in fumaroles surrounded by yellow sulfur crystals where the ground is dry, and in acid sulfur pools and bubbling mud pools where the ground is wet.

28.

Pōhutu Geyser, Whakarewarewa, Rotorua

New Zealand's largest and most regular geyser

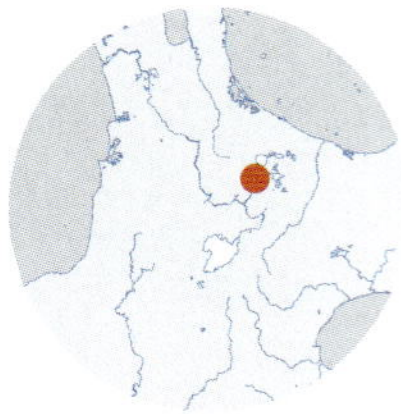

Whakarewarewa is the only major geyser field still active in New Zealand. When Europeans arrived in Aotearoa there were 130 natural geysers in five geothermal fields. One field was destroyed by the Tarawera Eruption, another was drowned by construction of a hydroelectric lake, and in two more the geysers stopped because of extraction of steam for electricity production.

The surface discharge features at Whakarewarewa include 500 hot springs, both alkaline and acidic, bubbling mud pools and fumaroles, but the main attraction is the alignment of seven geysers on Ngā Puia o Whakarewarewa / Geyser Flat. Several of the smaller geysers play almost constantly, but the crème de la crème is Pōhutu Geyser (28.1, 28.2). It generally erupts for several minutes every hour or so, spouting steaming-hot water 20 m into the air but occasionally up to twice that height.

28.1 Pōhutu Geyser is one of New Zealand's major tourist drawcards. As the boiling water erupted from the geysers cools, white silica is precipitated and has built a sizeable sinter mound with cascading waterfalls that are more active after each eruption.

28.2 Pōhutu geyser at Whakarewarewa currently erupts a jet of boiling water and steam 20–30 m into the air. Eruptions occur at intervals about one hour apart. PHOTO: ALASTAIR JAMIESON.

Geysers are a rare breed of boiling hot spring that periodically eject jets of hot water and steam. They generally have a shallow reservoir chamber a few metres below the ground that opens through a narrow pipe to the surface. The water in the reservoir progressively heats up, and when it boils a mix of water and steam jets out through the narrow vent as a geyser. As the water starts to geyser, pressure reduces in the chamber, lowering the boiling temperature, and water violently flashes to steam to further power the geyser jet above. The geyser stops when all the boiling water is used up. The reservoir then refills with water and once again the temperature progressively rises. The length of this recharge period between geyser eruptions depends on the rate of inflow and heating of the water to boiling.

The hot geothermal fluids that feed Whakarewarewa rise up a fractured fault zone along the southern margin of the Rotorua Caldera. From here the superheated underground water flows northwards through a horizontal aquifer of fractured rhyolite rock that passes beneath Whakarewarewa and much of the heart of Rotorua. This aquifer is capped by near-impermeable layers of lake sediment. Rotorua businesses that use the hot water resource obtain it from boreholes sunk down 100 or so metres to the aquifer. The main natural discharges of hot water and steam occur up fractures through the cap layer at Whakarewarewa, Kuirau Park, Government Gardens, Sulphur Bay and Ōhinemutu. In the 1980s the extraction of hot water from the aquifer lowered the level, and activity declined at the last of Whakarewarewa's geysers. They were saved by a government ban on hot water extraction from within 1.5 km of these natural wonders.

Best seen from: When Pōhutu Geyser spurts it can be seen at a distance from the carpark at the end of Fenton St, opposite Rotorua Golf Club.

Take a closer look: Tour Whakarewarewa Thermal Area operated by Te Puia for the best views of the geysers and other features. For lesser, free experiences of hot pools and steam, visit Kuirau Park or Sulphur Bay and Government Gardens, Rotorua.

28.3 Sinter stalactites of Whakarewarewa's Geyser Flat sinter mound have formed by silica precipitating out of the cooling waters as it flows down over the low cliffs to the stream.

28.4 Oblique aerial view over Ngā Puia o Whakarewarewa / Geyser Flat sinter terrace, Whakarewarewa, with several smaller geysers erupting. PHOTO: ALASTAIR JAMIESON.

29.

Waiotapu, Rotorua

Most colourful geothermal area in New Zealand

Waiotapu geothermal area has the most colourful sinter, altered ground and hot pools in New Zealand. Pride of place is the 2000 m^2 Champagne Pool (29.1) which fills a 60-m-deep, 700-year-old explosion crater. This steaming, slightly acidic pool releases large quantities of carbon dioxide which gives a fizz to the water – the source of the name Champagne Pool. The pool is surrounded by a raised rim of grey-white sinter. The inside of the rim is lined with a band of wondrous miniature white coral-like spicular sinter (29.2), which grows at or just above the surface of the water by an interplay of microbial activity and precipitation. Just below water level the silica rim is coated in a bright orange and yellow precipitate of arsenic and antimony sulfide minerals (realgar, orpiment, stibnite), which is also rich in gold and silver. Overflow water from the pool has deposited an extensive sinter terrace known as the Artist's Palette and the Primrose Terrace (29.3).

There are many other small explosion craters and collapse pits throughout Waiotapu geothermal area. Some may be the sites of former geysers and hot springs from a time when the

29.1 The most colourful geothermal feature in New Zealand is the Champagne Pool at Waiotapu. It is a 60-m-diameter alkaline chloride pool surrounded by a lip of sinter, tinged orange by trace amounts of arsenic and antimony. The hot mineral-laden water overflows to the right and has formed the large yellow-tinged sinter terrace known as the Artist's Palette. PHOTO: ALASTAIR JAMIESON.

⊙ 29.2 Spicular sinter formation visible around the edge of the Champagne Pool when the water level is low. These spicule fingers are up to 3 cm high. PHOTO: BRUCE MOUNTAIN, GNS SCIENCE, 131406.

⊙ 29.3 The 1.5-ha Primrose Terrace is the largest actively forming sinter terrace in New Zealand. The terrace sinter is deposited out of runoff from Champagne Pool (steam beyond). The terrace consists of numerous miniature steps, each of which holds a shallow pond behind a low silica lip. PHOTO: LLOYD HOMER, GNS SCIENCE, 4088.

field was far more active. Others are a result of collapse into voids created by acid-steam attack on rock. The great range of black, grey, orange, yellow and green colours in these depressions (29.4, 29.5) are due to sulfur, oxides, hydroxides and sulfides of iron, antimony, mercury and other metals.

Take a closer look: Walk around the commercial Waiotapu Thermal Wonderland to see the Champagne Pool, Artist's Palette, boiling pools, collapse pits and fumaroles. Nearby from a carpark on Waiotapu Loop Rd you can watch rising gas bubbles bursting through the largest thermal mud pool in New Zealand for free.

⊙ 29.4 A collapsed crater, Ipu, at Waiotapu containing black scum on the surface of the mud and water – a kind of crude oil brought to the surface by rising steam and hot water from cooked rotting swamp deposits at shallow depth.

⊙ 29.5 The tourist trail through Waiotapu Thermal Wonderland wends its way through a field of small explosion craters and collapse pits. The yellow-green milky water in Roto Karikitea (left) gets its amazing colour from sulfur and other minerals suspended in the water. PHOTO: ALASTAIR JAMIESON.

30.

Huka Falls, Taupō

New Zealand's largest chute waterfall

Huka Falls are the highest falls on New Zealand's longest river, the Waikato, and also New Zealand's largest chute waterfall. Chute waterfalls have a forced water flow through a narrow channel causing the flow to be rapid, frothy and violent (30.2). The river flows from the outlet of New Zealand's largest fresh-water lake, Lake Taupō, all the way northwest to discharge into the Tasman Sea near Auckland. The falls exist because the river has not been able to erode through a hard band of silicified lake sediments that runs across the valley. These pumice-rich sediments have been cemented by hot fluids flowing from the nearby Wairakei Geothermal Field. As the silica-saturated water passed through the permeable sediment, it cooled slightly and silica crystals grew between the particles gluing them together into the erosion-resistant rock.

30.1 Huka Falls are New Zealand's most visited natural attraction. Although the falls are only 11 m high, it is the sheer volume of water that barrels through the narrow gorge and out into the plunge pool that is so impressive. PHOTO: ALASTAIR JAMIESON.

30.2 Huka Falls on the Waikato River are unusual because of the narrow 100-m-long ravine or cataract that has been carved into the hard sedimentary rocks. The water in the cataract above the falls is a sparkling turquoise colour as a result of the clarity of the water, which reflects light that is intensified by the volume of air bubbles in the surging water.

The lake sediments were deposited in an ancient lake, called Lake Huka. It was a forerunner of Lake Taupō and stretched between the location of present-day Taupō and north as far as Waiotapu. This lake was destroyed by the massive caldera eruption that formed much of the Lake Taupō basin 25,400 years ago. Lake Taupō was formed and rose to a level 140 m higher than present, before a rapid breakout flood established the present outlet route of the Waikato River. Over thousands of years it eroded down to the hard lake sediments creating an early version of Huka Falls.

The second-largest eruption out of Lake Taupō, 1800 years ago, threw up a 40-m-high dam of soft ignimbrite (The Big Picture 7, page 105) across the Lake Taupō outlet and the lake level rose to 34 m above the present. Once again there was a rapid breakout flood when 20 km^3 of water eroded away the natural dam in a single event and exhumed the earlier-formed Huka Falls. Since then erosion has carved out a 100-m-long, 15-m-wide and 10-m-deep cataract into the cemented rocks channelling the Waikato River water down to the falls themselves.

30.3 Continued slow erosion of the lip of Huka Falls has been gradually moving the falls back up the river, creating the elongate amphitheatre around the present-day plunge pool.

Take a closer look: The best views of the falls is from the footbridge over the narrow ravine just above the falls and from lookouts downstream on both sides. Commercial jet-boat rides go up the river to the foot of the falls.

31.

Tongariro River mouth, Tūrangi

New Zealand's best bird's foot delta

31.1 This unusual, bird's-foot-shaped delta has built up at the mouth of the Tongariro River where it flows into Lake Taupō. Note the three 'toes' of sediment that have grown out into the lake at the mouths of the three most recently active channels. PHOTO: ALASTAIR JAMIESON.

31.2 View north across Lake Taupō showing the Tongariro Bird's Foot Delta protruding into its southern end. PHOTO: ALASTAIR JAMIESON.

A delta builds where a river flows into a body of water and loses its power to transport sediment. As a result sand is dumped into the quiet waters at the river's mouth. Some rivers transport gravel as bed load, especially during floods, and this too may help build out a delta. Tongariro Bird's Foot Delta, at the mouth of the Tongariro River where it flows into Lake Taupō, is made of sediment eroded from the western side of the Kaimanawa Range extending south as far as the eastern slopes of Mt Ruapehu.

Lake Taupō is 25,000 years old, and so the alluvial wetlands and delta at the south end of the lake at the mouth of the Tongariro River is younger than this. The lower section of the river is sinuous and has built a delta 2–3 km out into the lake. The lower part of the delta has three toes formed by different branches of the river channel. Upstream there are several cut-off meanders where floods have broken through the levees, shortening the river's route and leaving behind abandoned channels (31.1). Each channel is margined by levee banks made of sand dumped on either side during declining flow speeds after floods. The bird's foot delta owes its existence to the lack of strong erosive forces like ocean waves, strong tidal currents or longshore drift.

The Tongariro River is the centre of trout fly fishing in New Zealand, with many named natural fishing pools along the lower, more sluggish and meandering 20 km of the river's course. Partly for this reason there are tight controls on the amount of water that must be allowed to flow down the river and not be diverted to the Tokaanu Power Station. Some of the fishing ponds are on arms of the delta itself.

Take a closer look: Kayaks can paddle 4 km up the river channel from the lake. Walk along rough fishermen's tracks alongside the delta river channel.

31.3 View up the lower meandering reaches of the Tongariro River showing the three arms of the river discharging dense grey volcanic ash into Lake Taupō. The ash had been eroded off the land surface in the catchment soon after an eruption from Mt Ruapehu's Crater Lake in November 1995. PHOTO: LLOYD HOMER, GNS SCIENCE, 49910.

32.

Red Crater and Ngarotopounamu / Emerald Lakes, Mt Tongariro

Tongariro's youngest volcanic vents

Red Crater and Ngarotopounamu / Emerald Lakes are the youngest volcanic centre on the flattish top of Tongariro Volcano (32.1, 32.2). Mt Tongariro Stratovolcano is about 300,000 years old and is composed of several overlapping cones that were erupted from vents along a northeast–southwest alignment. Its summit and upper slopes are composed of lava flows and scoria sheets. Tongariro is also the location of the relatively young volcanic cones and vents (north to south) of Te Maari, North Crater, Blue Lake, Red Crater and Ngāuruhoe. The so-called Central and South Craters (32.5) are not volcanic vents but depressions that were filled with permanent ice during most of the Last Ice Age (70–12 kyr ago) when lava and scoria ridges built up around them.

Red Crater is a small scoria cone with a wide rift down the middle created by eruptions along a fissure also running northeast–southwest (32.2). The Red Crater fissure vents have been active only in the last 2500 years. Early on, fountaining produced 11 tongues of lava that flowed into Central and South 'Craters' and also east into Oturere Valley. A hollow dike, seen cutting through red scoria in the cliff at the southern end of the fissure (32.3), was part of this elongate fountaining conduit that was drained of lava as its sides cooled and solidified into grey andesite.

Red Crater is currently dormant. It has erupted ash on 16 occasions in the last 360 years, with its last eruptions in 1855 and 1926. The three Emerald Lakes fill small phreatic eruption craters produced by the release of steam heated up by magma at depth beneath the northern end of the fissure. Surrounding the lakes is steaming ground and small fumaroles also indicative of underground geothermal activity. The lake waters are acidic with a pH of 3–5. The most recent volcanic activity on Tongariro Volcano was two large phreatic eruptions accompanied by a small lahar from Te Maari Crater area in 2012, causing the temporary closure of the Tongariro Alpine Crossing track.

Take a closer look: These are on the Tongariro Alpine Crossing (19-km walking track) where many people stop for lunch.

⊙ 32.1 Red Crater is one of the scenic highlights of New Zealand's most popular tourist walk – the Tongariro Alpine Crossing. It is the youngest volcanic centre on Mt Tongariro, with all its small eruptions occurring within the last 2500 years. PHOTO: ALASTAIR JAMIESON.

➔ 32.2 Red Crater (top) and its scoria cone are at the southern end of a northeast–southwest-oriented volcanic rift that cuts across the eastern side of the summit of Mt Tongariro. The rift also includes the three small Emerald Lakes that fill several small explosion craters. PHOTO: ALASTAIR JAMIESON.

➔ 32.3 Seen from the Tongariro Alpine Crossing track, the southern wall of Red Crater shows layers of red scoria that was erupted by fire-fountaining from a fissure vent within the last 2000 years. Part of the now-empty conduit that fed the fountaining lava can be seen in the foreground lined with solidified grey lava. Photo 100 m across.

➲ 32.4 Emerald Lakes are three small, water-filled steam-explosion craters at the northern end of Red Crater fissure. The green colour is due to sulfur entering the lakes from the geothermal system beneath. Hikers (left) provide the scale.

➲ 32.5 The Tongariro Alpine Crossing track passes down over Red Crater scoria cone (right) and through Central Crater (foreground), another valley that was filled with ice during the Last Ice Age and is not an actual crater. The lower northern slopes of the scoria cone are yellowish as a result of geothermal alteration of the ground and sulfur deposition. The dark rock on right is part of a 2000-year-old lava flow that was erupted from Red Crater. Ngāuruhoe and Ruapehu volcanoes are in the background.

➲ 32.6 Red Crater rift and Emerald Lakes in summer when there is no snow left on the summit of Mt Tongariro. PHOTO: KAREN BRITTEN, GNS SCIENCE, 196633.

33.

Mt Ngāuruhoe

Tongariro's tallest and youngest volcanic cone

Ngāuruhoe is the southernmost volcanic cone of the beheaded and composite Tongariro Stratovolcano. It is built out of a mix of lava flows, scoria and ash that was erupted from this one vent over the last 7000 years. The larger Tongariro and Ruapehu Volcanoes have a rugged, irregular shape with large radiating valleys and deep depressions that were occupied by glaciers and ice during the Last Ice Age and greatly influenced where lava flows and ash built up during that period. Ngāuruhoe, on the other hand, is smooth and symmetrical because it was formed in much warmer times with no ice about (33.5).

Ngāuruhoe is built up on top of older Tongariro lava flows and is the most intermittently active of a number of vents around the summit and northern slopes of Tongariro.

33.1 The northwest profile of snow-dusted Mt Ngāuruhoe showing many of the lava flows (black tongues) that came down the slopes during the 1948–49 and 1954–55 eruptions. Ngāuruhoe (2287 m) received international exposure as the stand-in volcano for fictional Mt Doom in the *Lord of the Rings* film trilogy. PHOTO: ALASTAIR JAMIESON.

There have been about 60 eruptions recorded from Ngāuruhoe since 1839. Most of these have been ash eruptions, but occasionally magma has welled up inside the throat of the volcano with fire-fountaining building small summit scoria cones and lava spilling out of the summit crater and down the steep slopes (33.1, 33.4). Ngāuruhoe's crater does not fill with water and so all the eruptions have been dry-style with no lahars (volcanic mud flows), unlike Ruapehu and Tongariro.

There have been three major eruptive periods in the last 100 years – during 1948–49, 1954–55 and 1973–75. All three had ash eruptions, fire-fountaining of scoria, ballistic blocks, lava flows, and block and ash flows generated by ash column collapse or debris spalling off from the front of flows. Fountaining in 1954–55 formed the current summit scoria cone within the main crater (33.2). This eruption climaxed with 10 lava flows spilling down Ngāuruhoe's northwest side. Each flow took just 1–2 days to reach its maximum length. Fountaining, ballistic ejecta, ash eruptions and block and ash flows also occurred during the 1973–75 eruption episode. The last volcanic activity from Ngāuruhoe was two small ash eruptions in 1976–77. Since then we have experienced the longest period of volcanic dormancy (inactivity) in recorded historical times.

Best seen from: The distinctive cone of Ngāuruhoe is easily seen by travellers on the Desert Rd or SH 47 between National Park and Tūrangi.

Take a closer look: Hikers on the Tongariro Alpine Crossing (33.4) walk past the base of Ngāuruhoe, 4–5 km from the Mangatepopo Rd end.

33.3 The symmetrical composite cone of Ngāuruhoe Volcano as seen with winter snow covering. It is the southernmost flank cone of Tongariro Volcano, and has built up in just the last 7000 years. PHOTO: ALASTAIR JAMIESON.

33.4 Dark ribbons of fresh block and ash and lava flows drape the northwest side of Ngāuruhoe Volcano, seen here without snow in summer. These flows were all erupted within the last 100 years. Tongariro Alpine Crossing track is in the foreground. PHOTO: DOUGAL TOWNSEND.

33.2 The summit of Ngāuruhoe showing the most recent scoria cone and crater (right side of summit) that has built up inside the larger summit crater. PHOTO: ALASTAIR JAMIESON.

33.5 Conical Ngāuruhoe (centre) is the youngest and most dominant volcanic feature forming the top of Tongariro Stratovolcano. Blue Lake (left) fills a crater on the east side of the summit and North Crater (right) has a smaller crater nested within it. Mt Ruapehu is in the distance to the south beyond Ngāuruhoe. PHOTO: ALASTAIR JAMIESON.

34.

Crater Lake, Mt Ruapehu

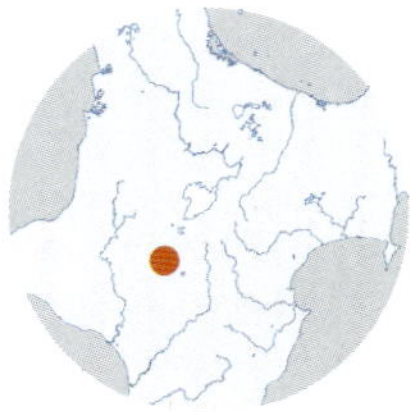

Pulse of the volcano

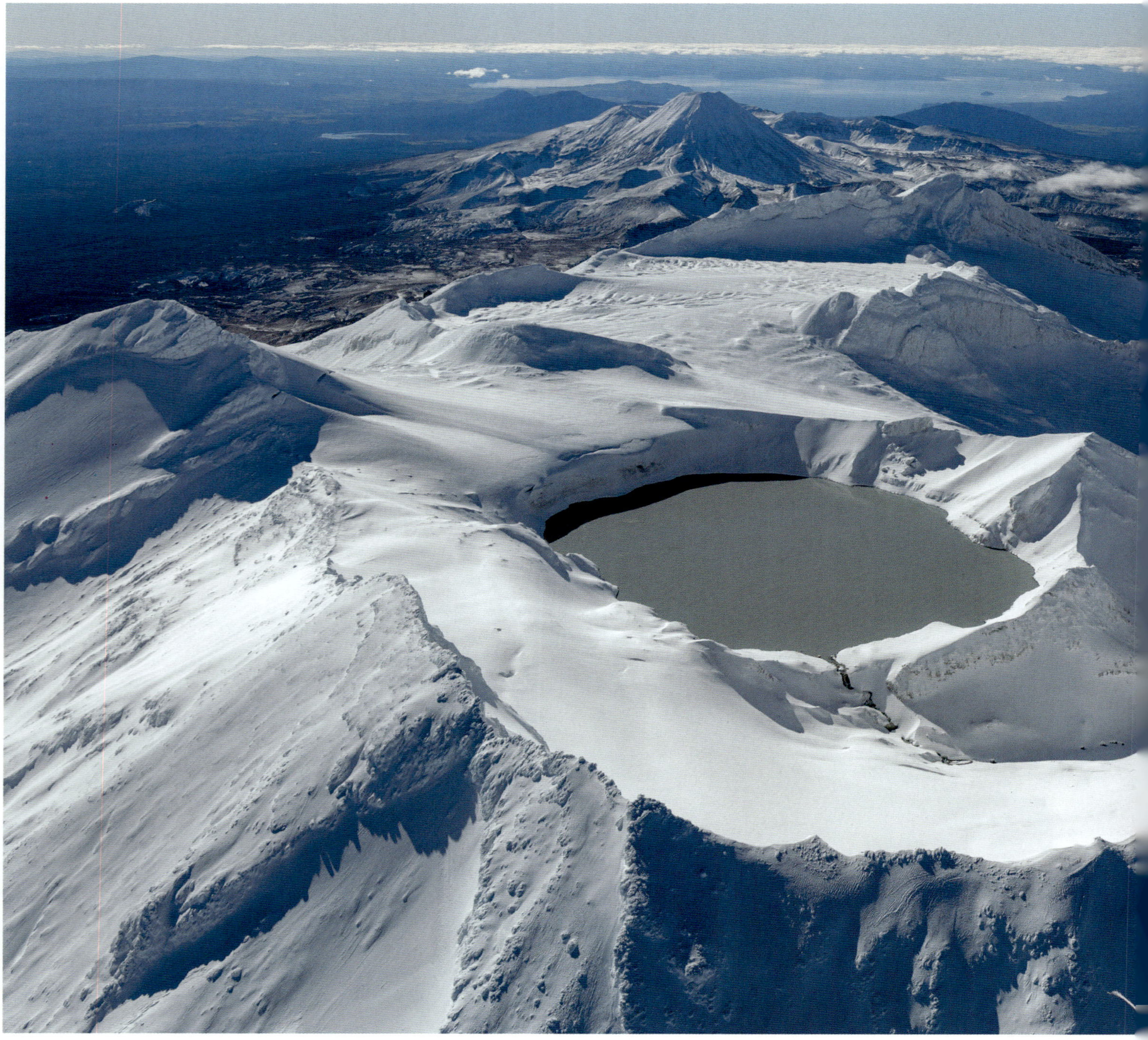

34.1 View northward over the top of Mt Ruapehu with conical Ngāuruhoe and Lake Taupō Volcanoes beyond. The top of Ruapehu has two vent areas - Crater Lake in the foreground and Iwikau Crater, filled with permanent ice, forming the Summit Plateau beyond. Steep-sided Pyramid Peak (right side of Crater Lake) is the highest point on the rim of Pyramid Cone, with the 400-m-wide Crater Lake filling its crater. PHOTO: ALASTAIR JAMIESON.

Crater Lake lies above the periodically active vent on the top of Ruapehu Volcano. Ruapehu is a stratovolcano with a central cone made of layers of andesite lava flows and ejected blocks of solid lava and ash that have been erupted intermittently from several vents over the last 350,000 years. Ruapehu and Tongariro cones are surrounded by a ring plain that stretches to the Desert Rd in the east and the main trunk rail line in the west (34.2). This ring plain has been built by a combination of lahars (volcanic mud flows) and large landslides from the cones, and glacial outwash and alluvium coming down the valleys. Parts of the plain are mantled by thick beds of wind-blown volcanic ash, especially on the downwind northeast side where the ash layers can be seen beside the Desert Rd.

The climate has been colder than the present for 90% of Ruapehu's lifespan. Thus for long periods there was permanent ice around the summit and glaciers extended down valleys to the ring plain (34.5). Many of the rocks on the volcano's surface were erupted under and adjacent to ice during the Last Ice Age period, between 75,000 and 12,000 years ago. Since then most of this ice has melted. This removed some of the support for eroded rocky peaks on the summit and has resulted in two major landslides of unstable rocks (sector collapses) that cascaded down the cone and out onto the ring plain. One such landslide, 10,500 years ago, deposited the hummocky debris avalanche mounds over the ring plain adjacent to the road up to the Chateau. Since then, a small volcanic cone, Iwikau Cone, has erupted in the landslide scar and its crater has filled with permanent ice, now known as the Summit Plateau (34.1).

Another small volcano, Pyramid Cone, has erupted over the last 2500 years, in the summit scar of a second large landslide that swept down the southeastern slopes of the volcano about 5000 years ago. All historic eruptions of Ruapehu have been from Pyramid Cone's vent - Crater Lake (34.1). The largest eruptions have occurred about every 50 years, in 1895, 1945 and 1995–96 (34.3), with 60 smaller eruptions since 1945. In the largest eruptions, most of the crater lake's water was thrown out, mixed with ash and snow and swept down the mountain-side as lahars of dense watery mud carrying along large boulders as they raced down stream valleys. In the two most recent large eruptions (1945, 1995–96), new dams of weak volcanic ash were built across the lake outlet. Crater Lake gradually refilled, and on both occasions (1953, 2007) a major breakout flood resulted as the ash dam failed, releasing a vast quantity of hot acidic water

as a lahar down the Whangaehu Valley. The lahar on Christmas Eve 1953 was not predicted, and swept away the Tangiwai rail bridge moments before the arrival of the Wellington to Auckland Express. The train plunged into the lahar-swollen river killing 151 of the passengers – New Zealand's worst rail disaster.

Best seen from: Ruapehu Volcano is a prominent landscape feature.

Take a closer look: In summer experienced hikers climb to Crater Lake and back from the Top of the Bruce / Iwikau Village (7 hrs). A guided hike is recommended for the less experienced.

↑ 34.2 Snow-capped Mt Ruapehu viewed across its ring plain from the Desert Road.

↓ 34.3 Part of the 1995 eruption sequence from Ruapehu's Crater Lake. The dark colouration blanketing some of the snow is from mud and ash thrown out from the lake. PHOTO: ALASTAIR JAMIESON.

34.4 View northwestward over Ruapehu's Crater Lake, with the volcano's second-highest peak, Paretetaitonga (2751 m), beyond. Crater Lake's overflow at the head of the Whangaehu River is visible in the bottom left of the lake. It was the failure of temporary dams of volcanic ash across this outflow that produced the 1953 (Tangiwai Disaster) and 2007 lahars. PHOTO: ALASTAIR JAMIESON.

34.5 Looking down the impressive U-shaped Wahianoa Valley, on Ruapehu's southwest slopes. It was occupied by a glacier during the Last Ice Age which built up the high lateral moraine ridges on either side (The Big Picture 15, page 283). Since the glacier melted, the summit vents have been on the other side of the summit and thus no lava flows have come down this valley. PHOTO: ALASTAIR JAMIESON.

35.

Mōhakatino coast, North Taranaki

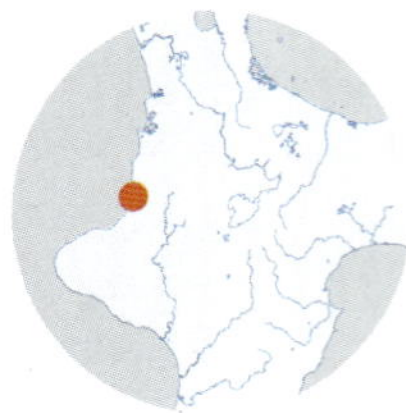

A plethora of sea stacks, caves and arches

35.1 Dozens of sea stacks, caves and arches occur along the Mōhakatino coast, North Taranaki. They are eroded out of mudstone and muddy sandstone layers that were deposited on the seafloor during the late Miocene, about 10 Myr ago.

35.2 Two sea arches have eroded along two parallel joint planes through a small point. Mt Taranaki is visible through the larger arch.

The coastline south of the Mōhakatino River mouth is one of two places in North Taranaki where there are sensational sea caves, arches and stacks eroding out of the relatively soft cliffs. The other, better-known, locality is south of the mouth of the Tongapōrutu River. In both places the cliffs are eroding back quite rapidly into muddy fine sandstone and mudstone that was deposited 11–9 Myr ago (late Miocene) on the seafloor at deep bathyal depths (about 1000 m). These layered strata are tilted gently to the southwest and get progressively younger as you walk south along the coast.

The rocks are cut by steep faults and joint planes, many of which are aligned southwest–northeast. When the tide is high, breaking waves slam into the base of the cliffs, forcibly injecting water along these planes of weakness. Over time caves are eroded into the cliffs, sometimes extending right through a small point to create a sea arch (35.2). Further erosion widens the arch, extending it upwards until the roof collapses in. This creates a sea stack, usually separated from the cliffs by beach sand (35.4). Continued erosion around the base of the sea stacks makes them more precarious until they finally fall over and are removed by the waves. Because of this rapid erosion, few sea stacks along the North Taranaki coast would be more than 100 years old.

The Mōhakatino cliffs and stacks are about 20 m high, and the fresh grey Miocene sedimentary rocks are commonly capped with more weathered orange-coloured sand, volcanic ash and peat (35.7). These younger sediments were deposited on top of a shore platform that had been eroded intertidally by the sea into the underlying strata during the Last Interglacial period, 130–120 kyr ago. Since then, North Taranaki has slowly risen at about 20 cm per 1000 years. As a result, the former intertidal platform now forms an uplifted terrace on top of the cliffs, 15–25 m high above the beach. Its undulating surface is due to old sand dunes deposited on top of the shore platform. State Highway 3 between Awakino and Tongapōrutu runs along this narrow coastal terrace between the modern cliffs on the seaward side and the steep slopes of the uplifted 120-kyr-old cliffs on the landward side.

Take a closer look: From the Mōhakatino River bridge on SH 3, walk 400 m down to the open coast on the south side of the river (only accessible for 2 hrs either side of low tide). Sea caves, arches and stacks are present down the coast for 3 km. Make sure you return before the tide cuts you off.

⊙ 35.3 A 12-m-high, narrow sea stack viewed through an arch on the Mōhakatino coast.

⊙ 35.5 Much of the cliff line south of Mōhakatino River mouth is riddled with short sea caves and tunnels.

⊙ 35.4 Sea stacks, Mōhakatino coast, North Taranaki.

⊙ 35.6 At high tide, the sea laps up against the Mōhakatino cliffs, around the rock stacks and surges in and out of the sea caves, PHOTO: ALASTAIR JAMIESON.

⊙ 35.7 Part of the coast south of the mouth of the Mōhakatino River with three sea stacks and numerous sea caves and tunnels in the cliffs behind. The orange strata forming the upper part of the cliff are beach and dune sands deposited during the Last Interglacial period of high sea level, 130–120 kyr ago. PHOTO: ALASTAIR JAMIESON.

⊙ 35.8 View from the north down the coast from above the Mōhakatino River mouth, showing the low cliffs where the sea stacks and arches occur. Above the cliffs is an uplifted shore platform 20–30 m high that was formed about 125 kyr ago, during the Last Interglacial period of high sea level. The hills rising steeply beyond (left) were the cliffed coastline when the intertidal platform was being eroded by the sea. PHOTO: ALASTAIR JAMIESON.

36.

Taranaki Maunga / Mt Taranaki

The archetypal stratovolcano

36.1 View from the west of the steep upper slopes of Mt Taranaki Stratovolcano (2518 m) – the second-highest mountain in the North Island after Mt Ruapehu (2797 m). Above the bush line (at 1300–1400 m) the young lava flows and small volcanic domes are snow-covered for much of the winter, but this melts almost completely in summer. PHOTO: ALASTAIR JAMIESON.

⊙ 36.2 View from the southwest of Mt Taranaki Stratovolcano with its conical central peak (in forest and snow cover) surrounded by a wide, gently sloping ring plain (in pasture) that extends all the way to the Taranaki coast. The earlier Pouākai peak is to the left. PHOTO: LLOYD HOMER, GNS SCIENCE, 2716.

⊙ 36.3 The profile of Mt Taranaki from the north, with clouds imitating a small eruption. Note Warwick Castle volcanic dome on the east skyline (left).

⊙ 36.4 View over Fanthams Peak (foreground), on the south side of Mt Taranaki, in summer. PHOTO: LLOYD HOMER, GNS SCIENCE, 1403.

The steep-sided cone of Mt Taranaki is the quintessential New Zealand stratovolcano surrounded by a gently sloping ring plain that extends all the way to the coast around Taranaki Peninsula (36.2). It is, however, one of four stratovolcanoes that have erupted along a north-northwest-striking line stretching from New Plymouth to the mountain over the past 1.8 Myr (Quaternary). Paritutū and the Sugarloaf Islands, offshore from New Plymouth, are the minuscule, eroded remnants of the oldest of these volcanoes which erupted about 1.8 Myr ago (36.5). These rocky stacks have survived erosion by the sea because they are the andesite lava plugs that solidified deep within several of its volcanic vents.

The weathered and ash-covered remains of the second volcano form the 684-m-high forested Kaitake Range which erupted prior to 700,000 years ago southwest of New Plymouth. The next youngest was Pouākai which erupted 400,000–300,000 years ago to form the 1400-m-high Pouākai Range, located 8 km north of the summit of Taranaki. Pouākai is a typical stratovolcano cone with evidence of volcanic lahars having swept down its slopes to form older parts of the nearby ring plain.

Taranaki is the youngest and largest of the four volcanoes. It erupted on many separate occasions over the last 200,000 years at least. The volcano consists of a steep-sided central cone made of layers of andesite lava flows, breccias and ash that forms most of the rocky area above the bush line (36.1). Within the National Park and covered by native forest the mountain slopes become smoother and less steep and are mostly composed of coarse volcanic breccia deposits carried down in lahars (volcanic landslides). Outside the circular boundary of the park, the farmed gentle slopes are the ring plain composed of the overlapping deposits of numerous further-travelled volcanic debris avalanche flows and lahars that came down off the growing mountain. The ring plain is 12–20 km wide and extends out further beneath the sea. The last major debris avalanche (Pungarehu, Site 37) removed

➒ 36.5 View over three of the inner Ngā Motu / Sugar Loaf Islands from the top of Paritutū, New Plymouth. These are parts of the more erosion-resistant plumbing beneath Paritutū volcano, the earliest centre of the Taranaki chain to erupt (1.8 Myr old).

much of the cone 25,000 years ago, and the mountain you see today has all built up in the last 10,000 years.

Mt Taranaki has not erupted since the arrival of European colonists, with its last known eruption in about 1790. In the last 1000 years the volcano has erupted at 10–200-year intervals and is currently considered to be dormant (sleeping). The chance of an eruption in the next 50 years is estimated at 50%.

Lava flows that erupted from the main crater on the peak between 1500 and 300 years ago still retain their original uneroded shape, forming many of the radial ridges on the upper part of the mountain (36.7). Lahars and small pyroclastic flows that have flowed westwards from the peak have taken away the western rim of the crater. In the last 1000 years there has been a succession of small lava domes squeezed out of the main vent within the summit crater followed by their rapid collapse while still hot, producing small, hot pyroclastic flows shooting down the mountain valleys. Currently a small dome, formed during the last phase of activity about 1780, is the highest peak of Mt Taranaki (36.7).

The summit is the main vent, but a second vent forms Panitahi / Fanthams Peak, a 1966-m-high basalt scoria cone with andesite lava flows, located south of the summit (36.4, 36.8). Panitahi last erupted about 3300 years ago. Four small hemispherical domes, up to 200 m high (The Dome, Skinner Hill, The Beehives; 36.8), were squeezed out as viscous andesite lava on the surface of the main cone from vertical fractures within the last few thousand years.

Best seen from: In clear weather the iconic cone of Mt Taranaki can be seen from many places all around Taranaki, including from parts of New Plymouth.

Take a closer look: The volcanic cones of Taranaki, Pouākai and Kaitake are all within Egmont National Park, which contains numerous walking and hiking tracks and a number of huts and lodges. For a day visit, take short walks from the visitor centres to viewpoints near the bush line on the slopes of Mt Taranaki at Dawson Falls (Manaia Rd) or North Egmont (Egmont Rd).

➊ 36.6 The succession of three of the four volcanoes that erupted to form Taranaki Peninsula can be seen from the north, above New Plymouth. They are, from oldest to youngest and lowest to highest (right to left): Kaitake, Pouākai and Taranaki Maunga. PHOTO: LLOYD HOMER, GNS SCIENCE, 2023.

⮉ 36.7 Many of the youngest lava flows are clearly visible on the upper northeastern slopes of Mt Taranaki. The summit crater is partly filled with a 240-year-old elliptical dome protruding from the main vent and forming the highest and most central peak on the volcano. The sharp eastern rim of the summit crater can be seen on the left. PHOTO: ALASTAIR JAMIESON.

⮈ 36.8 The Beehives (bottom right) are two 90-m-high hemispherical domes on the southern flanks of Mt Taranaki, 5 km from the summit. They were squeezed out about 3300 years ago as viscous lava from a vertical fracture through the mountain. These domes are now forest-covered within Egmont National Park. The volcanic cone (right) below the summit is that of Panitahi / Fanthams Peak. PHOTO: ALASTAIR JAMIESON.

37.

Pungarehu, Cape Egmont

New Zealand's best debris avalanche mounds

⊙ 37.2 An area of dense, somewhat larger, debris avalanche mounds near their source – the volcanic cone of Mt Taranaki. The cone has been rebuilt since the Pungarehu collapse 25,000 years ago. PHOTO: ALASTAIR JAMIESON.

⊙ 37.1 The Pungarehu debris avalanche mound field consists of thousands of small, 5–30-m-high, rounded hills that are lumps of Mt Taranaki rock that were swept out onto the ring plain when a large portion of the Mt Taranaki volcanic cone collapsed 25,000 years ago. This is the largest and best example of such a phenomenon in New Zealand. PHOTO: ALASTAIR JAMIESON.

The coastal highway around Taranaki Peninsula passes through a huge field of giant mole hills near Pungarehu – the best example in the country of volcanic debris avalanche mounds (37.1, 37.3). The mound field covers at least 250 km^2 and stretches 20 km along the highway each side of Cape Egmont. This area of flat land covered by hundreds of tiny hills is the surface of a large volcanic landslide (debris avalanche) deposit.

The landslide involved the largest known collapse of the cone of Mt Taranaki. In this event more than 7 km^3 of material spilled up to 30 km westward from the collapsing cone. It flowed out across the ring plain and reached at least 8 km off the present coast (which was land at that time). The steep cones of stratovolcanoes are made of lava flows interlayered with volcanic blocks and ash. Periodically the slopes become over-steep and unstable, and parts collapse in a rocky avalanche. Pungarehu debris avalanche, and most of the other 15 cone collapses known to have occurred during the 200-kyr history of the mountain, was triggered by magma intrusion expanding inside the cone possibly associated with an earthquake.

As the avalanche flowed down, large blocks of andesite lava, up to tens of metres across, were fractured and broken as they jostled along in a turbulent and sheared matrix of volcanic ash and blocks. As this passed over the ring plain, the blocks were rafted along in the top of the thinning flow and as it slowed they were left strewn across the surface as hummocky mounds (37.2). Smaller debris avalanche mounds (formerly referred to as lahar mounds) occur in a number of places around Mt Taranaki and were left by other volcanic landslides.

Take a closer look: Close views are obtained along SH 45 between Wārea and Rāhotu or along Kahui and Opourapa side roads.

37.4 Many of the Pungarehu debris avalanche mounds, like the four distant ones in this photo, have smooth surfaces as they have been mantled with volcanic ash erupted from nearby Mt Taranaki since they were emplaced. Some of the mounds, like those in the foreground, have rocks protruding from their surfaces and are composed of larger, more angular and less weathered blocks of andesite lava. These blocks protrude through the mantling ash, some of which has been washed off by rain. PHOTO: ALASTAIR JAMIESON.

37.3 Many of the roads on the ring plain west of Mt Taranaki pass through vast areas of land covered by debris avalanche mounds such as these. The top of the mound on the right has been terraced as part of the defences for a small Māori pā (pre-European defended fort) in pre-European times.

37.5 Some of the mounds nearer the coast were used as small pā (defended forts) in pre-European times. These two mounds have archaeological evidence of this use in the form of ring terraces that have been dug around the top. These would have had defensive wooden stockades around the edge of the terraces. Also clearly present are numerous pits, especially on the left mound. These are the excavated lower parts of small buildings used for storage of harvested crops within the protective defences of the pā. PHOTO: ALASTAIR JAMIESON.

38.

Rangitīkei River, Mangaweka

Quintessential fluvial terraces and abandoned meander channels

Some of the best-developed and most easily seen flights of river terraces and abandoned meander channels in New Zealand occur in the middle stretches of the Rangitīkei River around Mangaweka. River terraces are found in many of the country's valleys, and record episodes of erosion by the river (degradational terraces) or rapid deposition of sediment building up flood plains (aggradational terraces). Aggradational terraces usually develop on both sides of a river and have thick beds (>3 m) of overlying gravel and sand (38.4). Degradational terraces are often cut on only one side of the river and may have a thin deposit of gravel left behind on top, being the last mobile bed load that up until that time had been assisting the erosion. In the Rangitīkei and most other

↓ 38.1 View from Ruahine Rd across the Rangitikei River showing Holocene meanders incised into the 18,000-year-old Last Glacial aggradational river terrace.

38.2 View east over incised meanders, abandoned on left and active on right, of the Rangitīkei River. SH 1 (bottom left) passes along an aggradational river terrace that was formed during the peak of the Last Ice Age (30,000–18,000 years ago). At that time vast quantities of gravel and sand were flushed down the river from erosion of the deforested Kaimanawa and Ruahine Ranges. PHOTO: LLOYD HOMER, GNS SCIENCE, 48195.

New Zealand valleys, aggradational terraces were formed during glacial periods when the mountain ranges in their headwaters (i.e. Kaimanawa and Ruahine Ranges) had a much lower snow line with reduced vegetation cover and erosion was greatly increased. As a corollary, degradational terraces in New Zealand are mostly cut during warm interglacial periods, like the last 12,000 years, when sediment supply diminished as forest cover increased.

Sometimes a river is able to erode a deeper channel due to rapid uplifting of the land or increased capacity to erode due to reduced sediment load. When this happens, former river-level terraces are left high on the valley sides above the new flood level. If there are several episodes of increased river erosion interspersed with periods of stability or sediment deposition, then a flight of river terraces at different elevations may result (38.5).

The Rangitīkei River is the third-longest river in the North Island and flows south then southwest to the sea east of Whanganui. It is an excellent example of a consequent river, which means that its flow direction is a result of the tilted surface of the land. Uplift in the headwaters of the Rangitīkei began around 4 Myr ago (Pliocene). In its middle reaches around Mangaweka, the Rangitīkei River is eroding into relatively soft mudstone and fine sandstone (papa) that

38.3 Vertical view of a section of the meandering middle reaches of the Rangitīkei River at Ōhingaiti, with SH 1 passing across an 18,000-year-old river terrace on the north side. Note the present meandering route of the incised river with abandoned channel loops. PHOTO: GOOGLE EARTH.

was deposited on the floor of the shallow sea in the late Pliocene as the northern reaches were starting to be pushed up out of the sea. A flight of 11 river terraces is recognised between Mangaweka and Marton, having been formed, then progressively uplifted and abandoned over the last 350,000 years.

In the middle reaches around Mangaweka, all the terraces were formed within the last 50,000 years. Several prominent terrace levels on both sides of the valley, 80–160 m above the present riverbed, were formed by aggradation during colder intervals with greatly increased sediment discharge during the Last Ice Age (40,000–15,000 years ago). At this time the river meandered its way across the wide flood plain and, when sediment supply decreased significantly, it began to erode a deep gorge through the alluvial sediments and underlying mudstones, following its most recent meandering route (38.2). Thus in the last 15,000 or so years the Rangitīkei River has eroded out a 300–600-m-wide meandering gorge with 80–130-m-high bluffs on either side. In places, the original meander loops have been cut off by the river taking a shortcut across a narrow neck between loops. This has created a number of dry semicircular amphitheatres a few metres above river level, marking the position of the river channel a few thousand years ago.

Best seen from: The river terraces are best seen as you drive along SH 1 between Mangaweka and 8 km north of Hunterville.

Take a closer look: Closer overlooks are available a short distance from SH 1 along Ruahine Rd (Manuwatū Scenic Route) or Otara Rd.

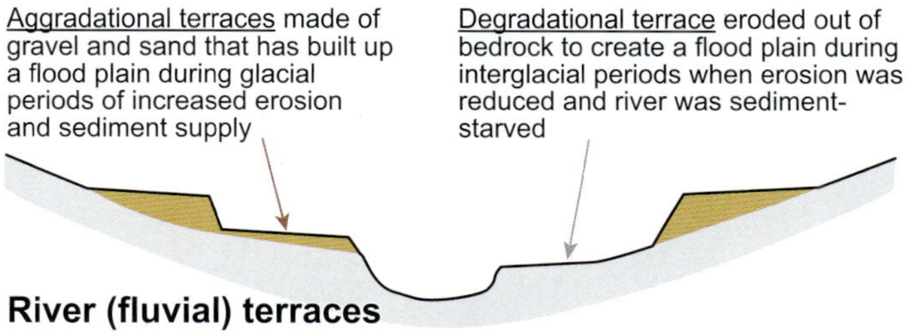

⊖ 38.4 Simplified cross-section through a river valley showing the difference between aggradational and degradational river terraces that are left behind on the valley sides as the river incises deeper into the bedrock during periods of increased erosion, decreased sediment supply and/or land uplift.

⊖ 38.5 View down the Rangitīkei Valley at Mangaweka showing 80-m-high bluffs around the margins of river meanders eroded into the Last Glacial fluvial terrace in the last 12,000 years or so. A higher and older fluvial terrace formed between 50 and 30 kyr ago can be seen on the left. PHOTO: MICHAEL TUREK, AWASTONE, MANGAWEKA.

39.

Kai Iwi Beach, Whanganui

Sedimentary strata deposited during the sea-level cycles of the Ice Ages

⊙ 39.1 The sea cliffs west of Whanganui are internationally famous for their remarkable sequence of young marine strata deposited during high sea levels (warm periods) in the latter part of the Ice Age climate cycles. The flat top of most of the cliffs between Whanganui and Hāwera is the edge of an uplifted coastal terrace that extends several kilometres inland and was eroded as a shore platform 120,000 years ago.

39.2 A series of marine terraces that become progressively higher as they extend 20 km inland from the Whanganui coast, seen here at Kai Iwi Beach. Each formed as an intertidal to shallow subtidal platform during a high interglacial sea level and has been progressively uplifted in the last 450,000 years and cut into by stream erosion. PHOTO: LLOYD HOMER, GNS SCIENCE, 48883.

The sedimentary sequence of rocks exposed in the cliffs of the Whanganui coast and the lower reaches of nearby rivers are of international significance. They are the only place in the world where there are shallow marine sedimentary rocks that span the entire period of the Ice Age cycles of the last 2.6 Myr (Quaternary) and provide a record of the associated cyclic sea-level changes. The existence of such a unique sequence of strata is related to the geological history of subsidence followed by uplift in the Whanganui Basin over the last 4 Myr (The Big Picture 9, page 145).

Between 4 and 0.5 Myr ago, the coastal Whanganui region was part of the subsiding Whanganui Basin, with a subsidence rate of 500–1000 m per Myr. The region of subsidence has now migrated south beneath the western Cook Strait where subsidence is now starting to impinge on the Marlborough Sounds (Site 53). In the last 450,000 years, the tectonic movement has reversed and the Whanganui coastal region has been going up. As a result the sedimentary sequence that was deposited between 2 and 0.5 Myr ago has been uplifted 350 m, so far, and is now being exposed by erosion in the coastal cliffs and terraces west of Whanganui. To the north, up the Whanganui and Rangitīkei Valleys, older strata (4–2.5 Myr old; Pliocene) have been uplifted even higher above sea level and are being exposed by river erosion (Site 38).

During the last 450,000 years there have been five warm periods with sea levels reaching about as high as today, separated by long colder periods when sea level was much lower. During each period of high sea level, marine erosion cut into the soft strata and created a wide shore platform. With the progressively rising land a flight of these uplifted platforms is now preserved as terrace levels in a belt extending

39.3 Okehu Bluff, 2 km west of Kai Iwi Beach, northwest of Whanganui, consists of sediment deposited during four successive interglacial intervals when sea level was high about 1 Myr ago. The sharp change from underlying grey strata to overlying orange-brown strata is an erosion surface that is capped by shallow coastal sediment deposited during the Last Interglacial period (~125,000 years ago) and subsequently uplifted.

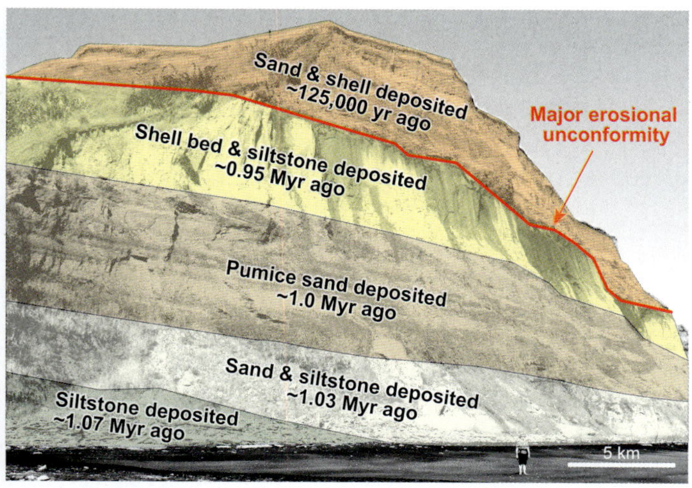

39.4 Explanation of the age and nature of the five packets of shallow marine sedimentary strata forming Okehu Bluff (in 39.3).

➲ 39.5 The cliffs on either side of Kai Iwi Beach settlement (centre) are capped by an extensive terrace that was formed intertidally during the Last Interglacial period (~125,000 years ago) and has subsequently been uplifted. PHOTO: LLOYD HOMER, GNS SCIENCE, 48882.

up to 20 km inland parallel to the coast (39.2). The oldest and highest terrace is furthest inland and they become progressively younger and lower towards the coast.

The strata now eroding in the coastal cliffs have been uplifted slightly more in the northwest near Hāwera and become younger towards the southeast near Whanganui. This sequence records 22 climate cycles deposited 2.0–1.6 Myr and 1.2–0.4 Myr ago. There is an erosional gap in the middle of the cliff sequence, but strata deposited during this period (1.6–1.2 Myr ago) are present in river sections inland. These sedimentary strata in the coastal cliffs were deposited in the sea during warm interglacial periods when sea level was high and this area was underwater. During cold periods, sea level was up to 100 m lower and this area was land and little or no sediment was deposited.

The sedimentary record of each of these climate and sea-level cycles usually consists of several parts (39.6). Each cycle began with rising sea level advancing across the area and eroding away the soil and weathered rock that had developed on land. On top of this erosion surface, beach and shallow marine sand, gravel and broken-up shells accumulated. As sea level rose to its highest level the area became deeper and little or no sand or gravel was transported out to that depth. Often during this period, beds of clams with both shells still joined together in life position were preserved in the accumulating mud. When sea level began to fall again, as global climate cooled, layers of silt were deposited over the shell bed and maybe a thin layer of shallow subtidal and intertidal sand as the coastline withdrew from across the Whanganui area.

Best seen from: The coastal terraces can be appreciated as you drive along SH 3 between Whanganui and Hāwera.

Take a closer look: The coastal cliffs between Hāwera and Whanganui can be viewed from the ends of any of the roads that lead down to the coast. Good examples (east to west) are at Castlecliff, Kai Iwi, Ototoka, Waiinu, Waverley, Waipipi, Manawapou and Waihī beaches.

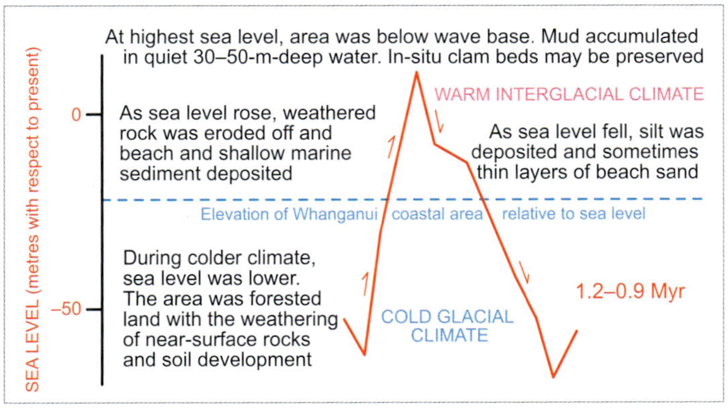

➲ 39.6 Graph showing sea-level rise and fall during an Ice Age climate cycle and a summary of the resulting changes in sediment deposition seen in the Whanganui coastal cliffs near Kai Iwi Beach.

THE BIG PICTURE 9

Whanganui Basin

THE WHANGANUI BASIN is the name geologists give to one of the youngest areas of rapid subsidence and marine sediment accumulation in the New Zealand region. The basin extends from Ohakune in the north to the northern coast of the Marlborough Sounds in the south (TBP 9.1). The northern part of the basin started subsiding earliest (about 5 Myr ago; Pliocene) and the centre of subsidence has progressively migrated southwards to now sit under western Cook Strait, offshore from Levin. The southwards-moving depression was often filled with seawater and accumulated ~3000–4000-m thickness of shallow marine sedimentary strata. As subsidence has moved southwards it has been followed by tectonic uplift, and this has been pushing up the northern part of the basin's sedimentary sequence to create the eroding land that now lies between Ruapehu and the Whanganui Bight (TBP 9.2).

The most accepted hypothesis to explain the southwards-migrating Whanganui Basin is that the deep edge of the subducting Pacific Plate beneath the Australian Plate has been moving southwards over this period of time (TBP 9.1). This edge is 100–200 km deep within the upper mantle, and as it is advancing southwards it is dragging down the overlying portion of the Australian Plate, creating the foundering depression. After a few million years, the edge has migrated further south and the depressed portion of the Australian Plate left behind to the north slowly rebounds and uplifts the sedimentary sequence out of the sea.

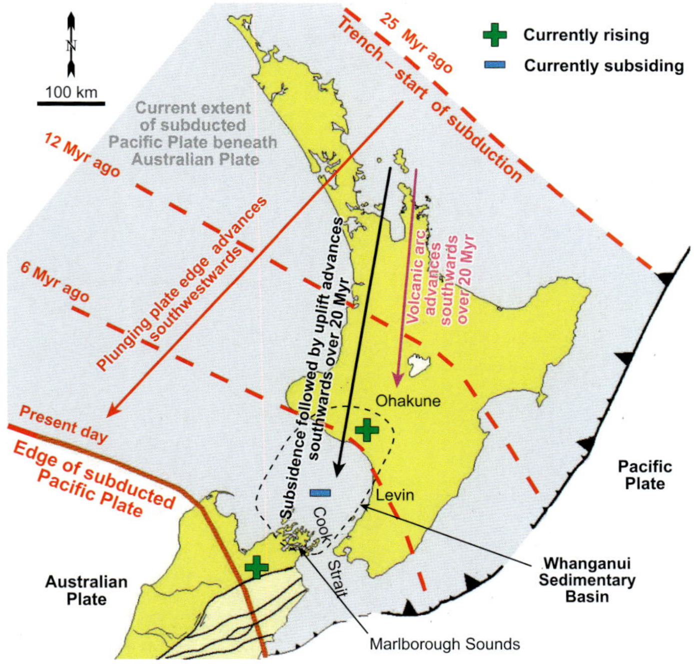

TBP 9.1 Map showing southwards migration through time of: (1) the edge of the subducted Pacific Plate, followed by (2) sedimentary basin subsidence, then (3) uplift, and behind that (4) the southern end of the volcanic arc.

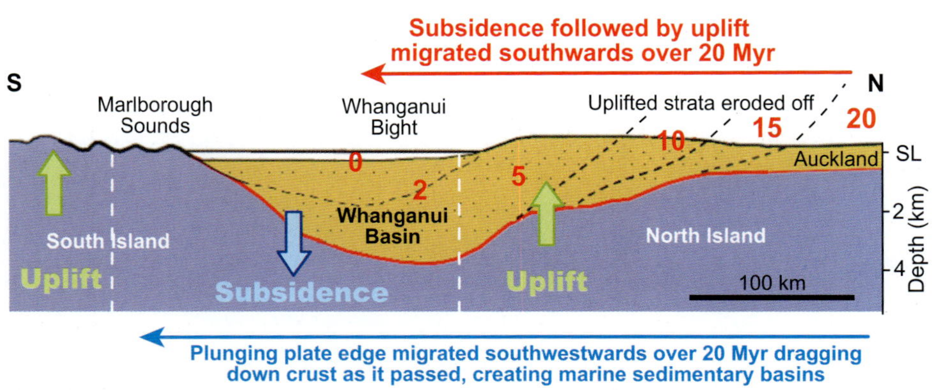

TBP 9.2 North to south cross-section between Auckland and Marlborough, illustrating the current tectonic setting with the northern Whanganui Basin being uplifted and eroded and the southern part subsiding and accumulating a thick sequence of marine sediment. These major vertical movements are inferred to be driven by the drag-down effect of the southwestwards-migrating edge of the subducted Pacific Plate beneath the Australian Plate.

40.

Te Āpiti / Manawatū Gorge

A record of the uplift of the North Island's axial ranges

⊙ 40.1 The Manawatū Gorge is the largest example of an antecedent river gorge in New Zealand. The Manawatū River has retained its west-flowing course and eroded the gorge through the axial ranges (Ruahine to the north and Tararua to the south) as they progressively rose across its course over the last 2 million years. View to the east from above Ashurst with the river flowing towards the photographer. PHOTO: GRAHAM HANCOX, GNS SCIENCE, 201390.

The Manawatū Gorge is the largest and best-known example of an antecedent river in New Zealand and one of the most remarkable in the world. An antecedent river is one that maintains its original course despite changes in the topography through which it flows, usually uplift. The Manawatū Gorge provides the most tangible evidence for the young age of the axial ranges of the North Island. These extend as a linear belt of uplifted basement greywacke mountains extending from the Raukūmara Range in the north through to the Remutaka Range and Cook Strait in the south. The Manawatū Gorge passes through an area of lesser uplift (300–400 m above sea level) between the Ruahine (1000–1100 m high) and Tararua Ranges (500–800 m high). The ridges along the crest of these ranges are relatively smooth as they retain the original planar surface that was eroded to sea level by about 5–4 Myr ago, prior to uplift (40.1).

Soft mudstone and limestone overlie this old erosion surface, stretching right across the top of the uplifted ranges north of the gorge (Manawatū Saddle). These soft rocks were deposited on the floor of the sea between 5 and 2 Myr ago (Pliocene) before this part of the axial ranges was pushed up. North and south of this area, however, forerunners of the Ruahine and Tararua Ranges existed as low-lying fingers of land separating the Tasman Sea from the East Coast fore-arc basin to the east (The Big Picture 10, page 154). The Manawatū seaway between these two fingers was up to 50 km wide but relatively shallow.

From about 2 Myr ago (Pleistocene), compressional plate-tectonic forces started uplifting the axial ranges and the coastal ranges further east. As they rose, the Manawatū River began flowing from east to west through the uplifted seaway and continued eroding down the gorge as the axial ranges continued to rise. Around 1 Myr ago the rate of uplift exceeded the rate of down-cutting and a large shallow lake and swamp became ponded on the upstream (Woodville) side of the gorge, but this has since been drained naturally.

The eastern entrance to the gorge cuts through the North Island shear belt of major faults, which includes the northern end of the active Wellington Fault (Site 52). Here, not only are the axial ranges going up but they are also slipping sideways to the north during each major earthquake rupture on the fault lines.

Best seen from: The old Manawatū Gorge road is permanently closed because of repeated slips. For views of the gorge, drive to either end along the old SH 3.

Take a closer look: For the fit and energetic, walk all (11 km, 4–5 hrs one-way) or part of the Manawatū Gorge Track above the south side of the gorge. Five viewpoints give panoramic views over the gorge and surroundings.

⊙ 40.2 View east over the eastern half of the Manawatū Gorge. The railway line (left) and highway (right, now closed) were constructed through the gorge on either side of the river. PHOTO: LLOYD HOMER, GNS SCIENCE, 52907.

⊙ 40.3 The Manawatū River has eroded a deep gorge through the highly fractured greywacke and argillite rocks of the axial ranges. As a result the steep gorge sides are lined with landslides. Earthworks needed to put a main highway through the gorge undercut a number of older slips, which reactivated over and over during bad weather. By 2017 the cost of clearing and restabilising the slopes forced the New Zealand Transport Agency to permanently close the gorge highway and re-route it over the Manawatū Saddle to the north. PHOTO: GRAHAM HANCOX, GNS SCIENCE, 130571.

41.

Horoera Point, East Cape

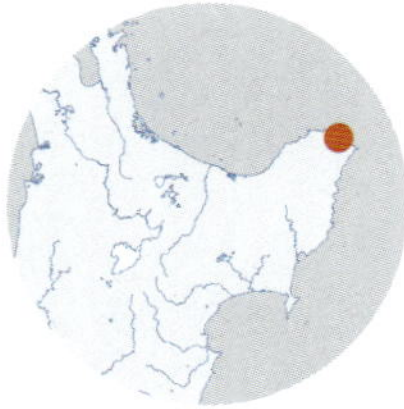

Internationally rare giant doughnut-shaped concretions

41.2 A steeply inclined conical concretion formed by precipitation of dolomite in and around a pipe-like conduit up which methane migrated towards the seafloor.

41.1 Giant, 6-m-diameter, ring-shaped doughnut concretion eroding out of the mudstone rocks forming the foreshore at Horoera Point, near East Cape. The doughnut is tilted at about 30° towards the camera, parallel with the weakly layered strata. It contains a central sediment-filled pipe through which methane-rich fluids flowed towards the seafloor and around which this concretion grew about 8 Myr ago.

Concretions come in many shapes and sizes, with the best known being the large spherical ones, as at Koutu (Site 5) and Moeraki (Site 84). In the foreshore beside the road to East Cape lighthouse are some remarkable ring-doughnut-shaped concretions (41.1). This is one of the few places in the world where they can be seen. These are probably the most exceptional and unusual examples of hydrocarbon-seep-related carbonate concretions anywhere in New Zealand.

These East Cape concretions were formed within layers of muddy sediment beneath the seafloor about 8 Myr ago (late Miocene). Detailed geochemical studies suggest that methane was produced by the cooking of organic matter within older sediments that were deeply buried to about 5–7 km, where temperatures and pressure are considerably enhanced. The methane-rich fluids ascended through pipe-like conduits of fractured rock associated with faulting. It is believed that when the rising fluids encountered an impermeable clay-rich layer within unconsolidated sediment just a few tens of metres below the seafloor, they were trapped there. The methane-rich fluids filling the blocked pipe penetrated outwards into the surrounding sediment layer, forming a ring-doughnut-shaped 'bubble' saturated in methane (hydrate). Periodically the impermeable seal was breached and the fluid within the conduit migrated upwards again and flowed out as a cold methane seep on the seafloor. Each time the pressure was released the trapped methane began to oxidise, resulting in precipitation of finely crystalline dolomite (calcium magnesium carbonate) between the sediment grains, creating this unusual-shaped concretion of hard rock within the softer surrounding mudstones.

The concretions we see on the foreshore today have recently been exposed by erosion of the soft mudstone containing them. They are predominantly of ring-doughnut shape, averaging 4 m across. There are also a few upright conical and pipe-shaped concretions, where most of the dolomite growth occurred within the conduit (41.2). All these concretions retain an internal pipe that is now filled with soft, easily eroded sediment (41.4). The presence of internal pipes characterises concretions here, and in other places, that were formed around conduits that hosted upwardly migrating hydrocarbon-enriched fluids. Thus they differ from the majority of concretions, which grow outwardly around a shell, pebble or rip-up clast within sedimentary strata.

Take a closer look: Seen in the foreshore adjacent to East Cape Rd on the west side of Horoera Point, about halfway between Te Araroa and East Cape lighthouse. Wander down onto the tidal platform, if the water is low enough, for a closer inspection and photo opportunities.

⊙ 41.3 Four doughnut concretions oriented parallel to the right-tilted layering of the strata and several upright conical concretions protrude above the intertidal mudstone shore platform west of Horoera Point. They were formed within, and until recently were completely buried by, the mudstone. Marine erosion has removed the mudstone from around the erosion-resistant hard concretions.

⊙ 41.4 Two upright pipe concretions that grew around the plumbing beneath a seafloor methane seep.

42.

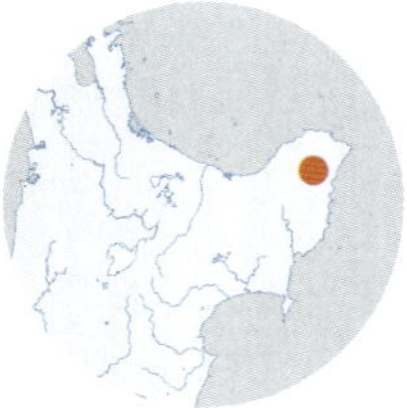

Mts Hikurangi, Aorangi, Wharekia and Taitai, Ruatōria

New Zealand's best examples of displaced sandstone mountains (klippen)

42.1 Three of the Ruatōria sandstone klippen mountains from the east: Hikurangi (far left top), Aorangi (foreground) and Wharekia (middle right). The distant high point, Whanakao, is also part of a large, displaced slab of hard sandstone. PHOTO: LLOYD HOMER, GNS SCIENCE, 44196.

Four high sandstone mountains inland from Ruatōria are New Zealand's most impressive examples of klippen (42.1). Klippen (singular klippe) are giant blocks of rock (hundreds of metres to many kilometres across) that have been moved from their original setting by tectonic forces such as low-angle faults (thrusts). Each klippe is now an isolated erosion-resistant block of sandstone sitting on top of and surrounded by rocks that are distinctly different and usually softer and more eroded.

The four klippen lie in an area west of Pacific Coast Highway 35, between Tapuaeroa and Mata Rivers (42.2) and comprise Taitai (700 m), Wharekia (1106 m), Aorangi (1273 m) and Hikurangi (1752 m). Mt Hikurangi is the highest non-volcanic mountain in the North Island after the four tall volcanoes of Ruapehu (2797 m), Taranaki (2518 m), Ngāuruhoe (2287 m) and Tongariro (1978 m). The four Ruatōria klippen are dominantly made of hard sandstone with lesser breccias and siltstones. These were deposited about 100 Myr ago (early Cretaceous) as sand, gravel and silt on the floor of the sea, 50 km or more north of present East Cape. These sands were deeply buried by younger sediment and as a result were compressed into hard sandstone rocks.

About 23 Myr ago the whole sedimentary pile offshore was pushed up out of the sea by the compressional forces between the Australian and Pacific Plates. These same obliquely convergent forces caused the East Cape area to subside deep beneath the waves. Huge slabs (nappes) of the uplifted rocks slid southwards one after the other into and over the deep East Cape seafloor as the East Coast Allochthon. The four Ruatōria klippen are part of one of these displaced slices that probably broke up into a number of separate blocks as they slowly slid into place. As they moved along, they became surrounded by highly sheared and fragmented siltstone. In the 18–20 Myr since these rocks came to a standstill, they have been pushed up again and much of the softer siltstone has eroded away from around the partly folded, fractured and tilted blocks of sandstone that now form the four high-standing klippen mountains (42.3). Many more, mostly smaller, sandstone klippen now form high hills throughout the East Cape area.

Best seen from: On a clear day the four klippen can be seen from Pacific Coast Hwy 35 just south of the Ruatōria turnoff, or for closer views drive up and back along Tapuaeroa Rd.

Take a closer look: Fit and experienced trampers can climb the track up Mt Hikurangi (1752 m) and back in a full day or stay overnight in the hut near the summit. Safer, less demanding options are commercial guided part-walk and part-4WD or helicopter trips up the maunga with Ngati Porou's Maunga Hikurangi Experiences.

42.2 The four big sandstone klippen viewed from a viewpoint on Pacific Coast Highway several kilometres south of the turnoff to Ruatōria. From the left they are Hikurangi (partly obscured), Aorangi, Wharekia (partly obscured) and Taitai.

42.3 The northwest side of Taitai viewed from Tapuaeroa Valley. The synclinal (downfold) form of the mountain can be seen in the inwards-dipping slopes of the sandstone strata on the left (north) and right (south) sides of the mountain. The central portion of the klippe is also composed of erosion-resistant 100-Myr-old sandstone.

42.4 Two of the sandstone klippen from Tapuaeroa Rd. Taitai (700 m) is on the left and Wharekia (1106 m) on the right.

42.5 View east from above Hikurangi showing the other three high-standing rocky klippen. PHOTO: LLOYD HOMER, GNS SCIENCE, 13796.

43.

Waimata, Gisborne

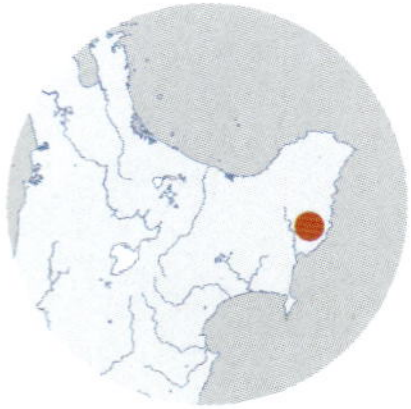

Active mud volcanoes – a rare phenomenon

Waimata mud volcanoes comprise at least four areas within 4 km of each other, inland of Gisborne. At these places mud, water and gas slowly ooze out at the surface, or rarely fountain explosively. In New Zealand, mud volcanoes occur in two settings. One is geothermal (The Big Picture 8, page 108), where they are associated with rising steam heated by shallow magma chambers. The second is where natural gas and salty water escapes to the surface bringing with it a slurry of cold, clay-rich mud and sometimes rock fragments. Most examples of this rare phenomenon are on the North Island's east coast, but examples are also known in Northland and Canterbury.

The cluster of mud volcanoes at Waimata are thought to be erupting from upwards-directed intrusive tongues, or diapirs, of mud from several kilometres depth. These may be rising up through fractured rock in a zone of faults. For the mud to have intruded upwards requires a combination of factors, which include: the presence of smectite clay at depth (probably from East Coast Allochthon sedimentary rocks) which swells and flows in the presence of water; the generation of natural hydrocarbon gases (mostly methane) which rise to the surface with the flowing clay; and some rock compression as a result of the collision between the two tectonic plates. Most of the time the mud and gas rise slowly and ooze out at the surface in a sedate manner. At times, one or more of these vents have become blocked and inactive for many years.

The shape of the mud volcanoes at the surface in the less explosive sites depends on the amount of near-surface groundwater and the resulting viscosity of the exuded mud. With relatively high water content, slurries of mud may flow

43.1 The deposit of mud and rock fragments seen in this photo erupted from a Waimata mud volcano in December 2018. The light colour is from the smectitic swelling clay that flowed up from considerable depths and erupted at the surface over a 5-hour period. It dammed a small valley, creating a pond. Over the coming years the deposit will partly wash away and gradually be covered in grass so it is far less recognisable. PHOTO: MURRY CAVE, JANUARY 2019.

from a vent, building up a gently sloping small shield volcano (43.2). If the mud is thick and viscous, petite steep-sided cones may be built as globs of thick mud are periodically fired out of the vent (43.3). At Waimata, and other places, these small landforms are affected by rainfall. Sustained heavy rain can wash away complete cones, and the presence of more abundant ground and surface water during winter may turn mud volcano vents into mud pools. As the water dries up in summer, the exuding mud may return to building the small mud volcanoes.

Every so often a mud volcano along the East Coast has had a rather violent eruption lasting a few hours to a week or more. Two such eruptions occurred in the upper Waimata Valley in December 2018 and December 2021. These eruptions appear to have been triggered by severe earthquake shaking, which seems to liquefy the mud at depth and cause it to flow upwards in great quantities. In most instances the mud eruption does not happen during nor immediately after the earthquake. Instead, ground-swelling, updoming and cracking have been noticed first and are thought to be the result of the increasing pressure of the rising mud at shallow depth beneath a cap. At the 2018 Waimata eruption site, at least a metre of updoming of the land was noticed after a September 2016 earthquake (magnitude 7.1 Mw) 100 km away near East Cape, but the cap did not burst until two years later.

For five hours this 2018 eruption squirted a huge volume of mud, rock fragments, saltwater and gas up to 100 m into the air and over a farmer's paddock. In that short time, a 1.2-ha area of grass had been buried by a mound of up to 20,000 m^3 of material rising to 5 m high in the middle (43.1, 43.4). This site had not been an active area of mud extrusion for more than a century, but old deposits show that this was not a new vent – it had just been dormant.

Best seen from: These are on private farmland. The site of the December 2018 mud volcano can be seen from Waimata Valley Rd about 20 km into the hills from Ormond.

⊙ 43.2 Several small mud shield volcanoes produced by the slow extrusion of water-rich mud within one of the subdued Waimata mud volcano fields. These tiny landforms have mini-crater lakes at their summits that periodically overflow and carry mud down the slopes in miniature lahars. PHOTO: MURRY CAVE.

⊙ 43.3 Several tiny steep-sided mud volcanoes produced by the accumulation of a heap of globs of thick mud thrown out of the vent by the power of large gas bubbles bursting at the surface of the partly mud-filled crater.

⊙ 43.4 On 15 December 2018 the Waimata mud volcano ejected up to 20,000 m^3 of mud and rock onto private farmland and built up this 5-m-high mound of flowing mud. PHOTO: MURRY CAVE, JANUARY 2019.

⊙ 43.5 The material that spewed out from the 2018 Waimata mud volcano consisted of broken-up rock in a porridge-like matrix of smectite-rich mud. Photo 1 m across. PHOTO: MURRY CAVE.

THE BIG PICTURE 10

Earthquakes and the Hikurangi Subduction Zone

THE EAST SIDE of the North Island from East Cape to Cook Strait overlies the Hikurangi Subduction Zone where the leading edge of the Pacific Plate dives down (is subducted) beneath the edge of the Australian Plate (TBP 10.1). The subduction boundary between the plates is a low-angle (2–5°), west-sloping thrust fault which comes up to the seafloor offshore in the Hikurangi Trench. This trench is parallel to the east coast and marks the actual plate boundary at the surface.

South of Hawke Bay and north of the Chatham Rise the crust of the Pacific Plate (the deeply submerged Hikurangi Plateau) that is being subducted is thicker than normal oceanic crust because it consists of a considerable thickness of ancient igneous rocks of early Cretaceous age (135–115 Myr old). This extra thickness imparts considerable buoyancy to the subduction zone and has resulted in numerous thick slices of younger sedimentary rocks being stripped off the top of the down-going Pacific Plate (Hikurangi Plateau). These younger sedimentary rocks are then shoved along on top of the gently west-sloping plate boundary and effectively added to the underside of the overlying edge of the Australian Plate. These added slices of rock, separated by low-angle thrust faults, have built up a belt (up to 100 km wide and up to 20 km thick) of rock, known as an accretionary wedge (TBP 10.1).

North of Hawke Bay, the accretionary wedge along the coast narrows considerably with its eastern margin currently being trimmed off and carried down the subduction zone with the subducting plate. On land there is a 40-km-wide belt known as the fore-arc basin which separates the accretionary wedge from the uplifted greywacke basement rocks of the North Island's axial ranges (Kaimanawa and Ruahine Ranges). Thick deposits of mud and sand accumulated on the seafloor of this fore-arc basin throughout much of the last 10 or so Myr, although it has been land for most of the last 1 Myr.

In the Hikurangi Subduction Zone, the Australian and Pacific Plates are pushing together obliquely, with convergence rates of 20 mm per year at Cook Strait rising to 60 mm at East Cape. Most (~80%) of this relative convergence occurs as movement on the subduction plate boundary, with the remainder taken up by thrusting movement along low-angle faults that splay off the subduction boundary and cut through the upper (Australian) plate. The margin-parallel component of the oblique convergence occurs as sideways movement (east side southwards relative to the west side) along coast-parallel faults in the upper plate and as slow clockwise rotation of the eastern strip of the North Island. The main zone where this sideways movement is occurring is in the North Island shear belt, which lies along the margin between the uplifted axial ranges to the west and the fore-arc basin to the east.

Detailed monitoring in the last two decades has shown that some of the convergence movement on the subduction

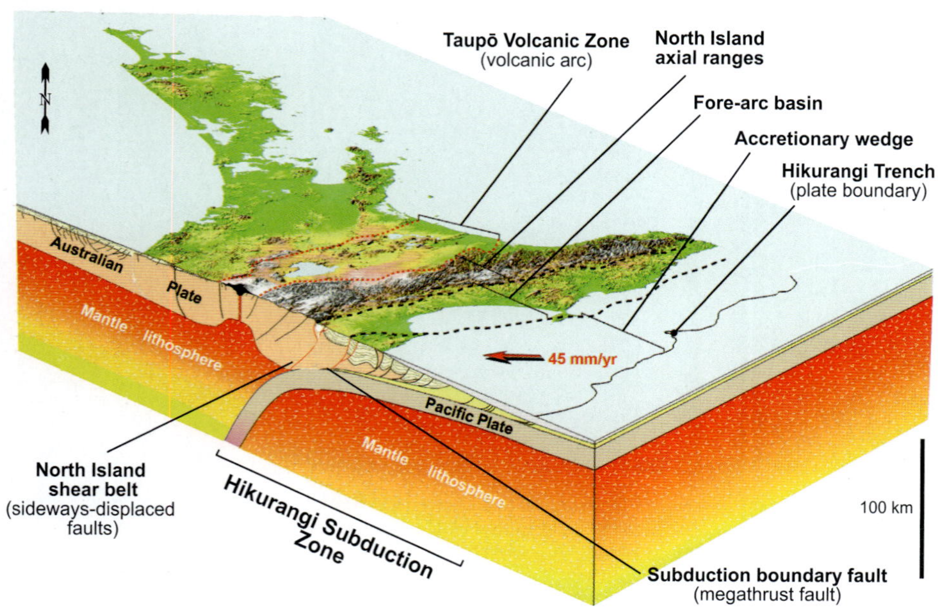

TBP 10.1 Block diagram showing the Hikurangi Subduction Zone with the Pacific Plate being subducted down beneath the edge of the Australian Plate along the low-angle plate boundary megathrust fault. Steeper thrust faults and sideways faults, subparallel to the plate boundary and coast, cut through the upper (Australian) plate. Episodic rupture along any of these faults releases stress that has built up from the convergent movement between the plates and results in earthquake shaking, sometimes damaging. MODIFIED FROM DIAGRAM BY KYLE BLAND.

➲ TBP 10.2 Map of the Hikurangi Subduction Zone which runs through the east side of New Zealand's North Island. The surface trace of active faults is shown in red, and the depth to the subducting-plate boundary megathrust fault is contoured in grey. The background red shading shows the region of the subduction megathrust boundary between the plates that is currently stuck (coupled) and is building up stress. MAP COURTESY OF KATE CLARK.

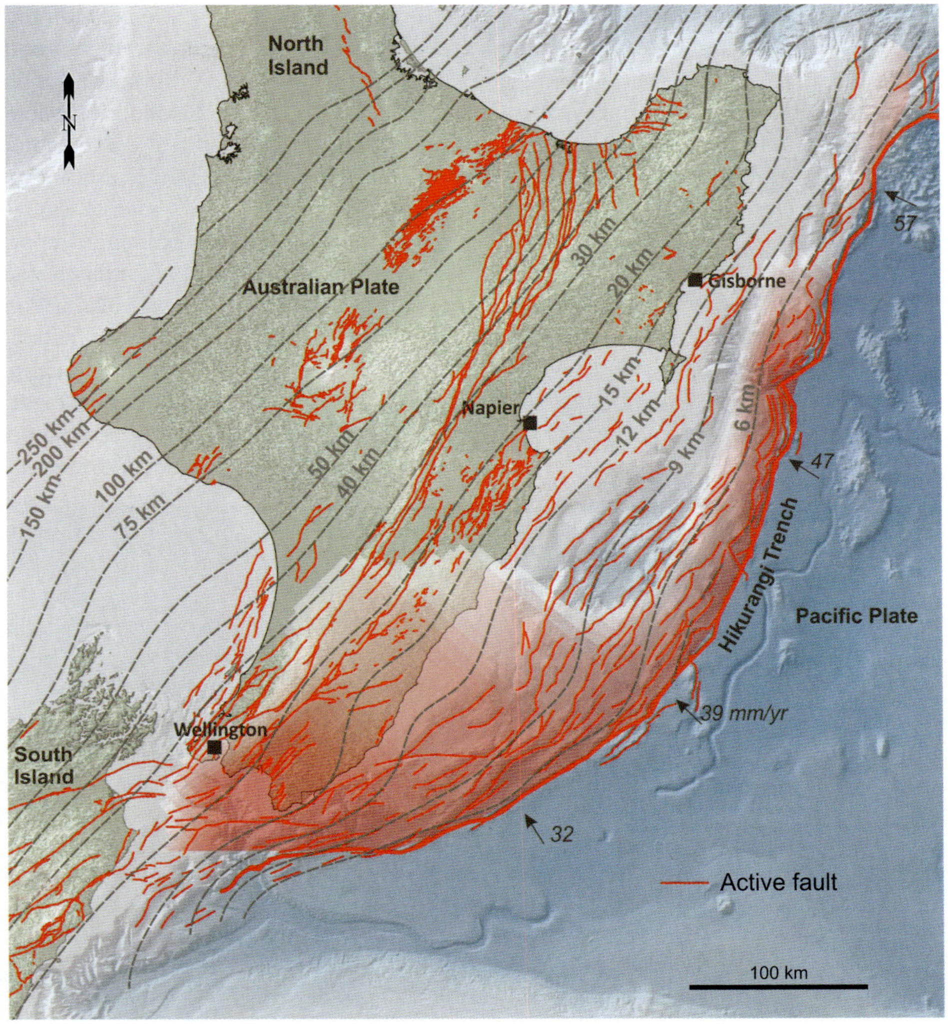

boundary thrust fault occurs in episodic (every 1–2 years) pulses of slow slip (up to 3 cm of movement spread over 1–6 weeks) with no detectable shaking. These patches of slow slip occur on shallow regions of the plate boundary, 2–15 km below the seafloor offshore of the northern two-thirds of the North Island's east coast. At greater depths beneath on-land parts of this northern region, movement on the plate boundary thrust appears to be a slow, steady creep.

To the south, off the Wairarapa coast and extending beneath Manawatū and Wellington, there is no record of slow slip or sliding movement at present on the shallower parts of the plate subduction boundary (less than 30 km depth). Thus, beneath the southern North Island the plate boundary is believed to be locked with stress progressively building (TBP 10.2). This stress will reach a point where the subduction fault will rupture and release the built-up energy in a Great Earthquake, also called a Megathrust Earthquake. All the largest known earthquakes on Earth have been the result of rupturing of subducting-plate boundary thrust faults like the one that forms the Hikurangi Subduction Zone beneath the eastern North Island. There have been no Great Earthquakes in New Zealand in historical times (last ~200 years) but there is a geological record of major earthquake displacements along the coast, sometimes with associated tsunamis, that indicate that they have occurred in the past every 500–900 years or so. The geological record also indicates that ruptures have occurred in the past on the subduction boundary fault beneath the northern portion of the subduction zone, even though at present much of the building stress in that region seems to be released in slow slip events.

The majority of earthquakes that occur along the east side of the North Island, however, are the result of ruptures on the faults (TBP 10.2) that cut through the upper (Australian) plate and have components of both sideways (lateral) and vertical movement. These are all a result of the oblique convergent movement between the two plates in the Hikurangi Subduction Zone. The largest of these in historical times have been the Wairarapa Earthquake (1855, Mw 8.2) and the Hawke's Bay (Napier) Earthquake (1931, Mw 7.8).

44.

Kahutara Point / Table Cape, Māhia

Best-preserved sequence of uplifted marine terraces in New Zealand

Table Cape was named by Captain James Cook for its high, flat terraces. These were wide shore platforms that were eroded intertidally by the sea out of moderately soft sandstones and mudstones. They have been uplifted by a succession of large earthquakes over the last 200,000 years. Each earthquake uplifted the region by 1–3 m.

The tidal platform that existed when each uplift occurred was added to the coastal plain, and waves began eroding the next intertidal platform out of the newly uplifted rocky seabed (44.3). After being uplifted and added to the land, each terrace progressively accumulated a covering of wind-blown dune sand, layers of volcanic ash erupted from the centre of the

44.1 The coast around Table Cape on Māhia Peninsula is fringed by one of the best-preserved sequences anywhere in the world of young marine terraces uplifted by a succession of earthquakes. These four young marine terraces form the low coastal fringe that has been pushed up out of the sea by earthquakes in just the last 3500 years. The higher terraces capping Table Cape (left) and sloping down to the west (middle right) are shore platforms that were formed at sea level between 80,000 and 200,000 years ago and have been progressively uplifted by dozens of major earthquakes. PHOTO: LLOYD HOMER, GNS SCIENCE, 2875.

⊙ 44.2 View east across the modern intertidal shore platform and low grassed fringe of marine terraces of Table Cape that were uplifted by four earthquakes in the last 3500 years. The high, flat-topped plateau of Table Cape is a series of uplifted shore platforms that date back to the Last Interglacial period of high sea level 125,000 years ago.

⊙ 44.3 Narrow section of intertidal shore platform west of Table Cape that has been eroded by the waves out of soft sandstone and mudstone layers. These layers were deposited horizontally on the floor of the sea 10–5 Myr ago (late Miocene) and have subsequently been uplifted and tilted down to the west (left). Every time a large earthquake has uplifted this area, the intertidal rock platform has become a low coastal terrace and a new intertidal platform starts eroding out of the uplifted seafloor rock just offshore.

⊙ 44.4 View west along the north coast of Māhia Peninsula near Māhia. The light-coloured rock is the soft Miocene sandstone and mudstone. The horizontally layered orange rock (left) is the cover beds of sand, loess and volcanic ash that were deposited on top of the intertidal shore platform after it had been uplifted above sea level during earthquakes. The small rounded islet with a narrow shore platform brim is known as The Old Man's Hat.

North Island and, during cold glacial periods, a thick deposit of loess (wind-blown silt) probably blown in from the barren plain of the aggrading Waipaoa River, 40 km to the north, when sea level was much lower (44.4). It would not have taken many decades for forest to colonise and become established on each of the newly uplifted terraces, and soil development contributed to the accumulating cover beds.

The uplift of the Māhia Peninsula is believed to be the result of episodic upward thrust movements on the north-northwest-striking Lachlan Fault located just a few kilometres off Table Cape. The rate of uplift is greatest in the east and decreases westward, resulting in all the terraces on the peninsula being tilted down to the northwest (44.1). Further west, in the vicinity of the sand tombolo that connects Māhia to the mainland, the land subsides slightly during each earthquake rupture event.

The highest and oldest terrace on Māhia Peninsula (not seen at Table Cape) was cut out of the sedimentary rocks during the warm interglacial period, with high sea level, about 200,000 years ago. The high plateau terraces of Table Cape were formed during the high sea level of the Last Interglacial period, 125,000 years ago (44.2). Since then the average rate of earthquake-related uplift has been 1 m per 1000 years and this terrace reaches a height of 130 m above today's sea level. Terraces cut when sea level was moderately high (20–40 m below present), during warm periods 105,000 and 80,000 years ago, have also been uplifted and are recognised on parts of the peninsula. These higher terraces are separated by degraded ancient sea cliffs, 10–30 m high. Any terraces that were eroded intertidally when sea level was more than 40 m lower than now have been eroded away or submerged by rising sea level after the end of the Last Ice Age.

The low fringing coastal plain around Table Cape has all been formed in the last 5000 years when sea level has been within 2 m of its present elevation. This flight of four terraces at Table Cape (44.1) has been dated as being uplifted during earthquakes at about 3500, 1900, 1600 and 250 years ago. Elsewhere there is a still-higher terrace from this period that was uplifted about 4500 years ago. Each of these major shallow earthquakes would have measured Mw 7.5–8, comparable to the Mw 7.7 Napier Earthquake of 1931.

Best seen from: Table Cape, and the uplifted marine terraces are clearly seen from Mahia East Coast Rd as you drive around the north coast of the peninsula and ascend to the plateau-like tops.

Take a closer look: To see the sequence of terraces uplifted in the last 4500 years walk 2.5 km along the beach from the end of Nukutaurua Rd and view them in the paddock at Table Cape itself.

45.

Lake Waikaremoana

New Zealand's largest landslide-dammed lake

There are hundreds of lakes, big and small, in New Zealand that have been formed behind landslide dams across valleys. The largest of all is Lake Waikaremoana, inland Wairoa. The landslide consisted of 2.2 km^3 of sandstone. This sandstone was originally deposited as a thick bed of sand on the floor of the sea about 8 Myr ago (late Miocene). It has even thicker layers of soft mudstone below and above it. Since deposition these rocks have been pushed up out of the sea and tilted down to the southeast. Erosion of these rocks during uplift has created Panekiri (Panekire) and Ngamoko Ranges on either side of the natural dam. They are strike ridges capped by the erosion-resistant sandstone (45.1, 45.6, 45.7).

About 2200 years ago the landslide slid down from the east (Ngamoko Range) and filled a 300-m-deep gorge on the Waikaretāheke River. Erosion had undermined the toe of the dipping sandstone strata. Like many landslides down the east side of the North Island, this one too is believed to have been triggered by intense ground shaking associated with an earthquake rupture on a nearby fault. A gigantic slab of sandstone detached and slid obliquely down the bedding planes, with material being transported 2–4 km downslope on a gentle incline of 6–8° (45.3, 45.4). As it moved along, reaching speeds of 40 metres per second, the front of the sandstone block collapsed and disintegrated into a debris avalanche of blocks which shot ahead of the gliding slab of rock.

It was the broken-up blocks of the debris avalanche that actually filled the gorge and formed the dam (45.2, 45.5). The slab arrived fractionally later, compressing the avalanche debris and bulldozing it up into pressure ridges around its toe. Tongues of debris flowed back up the river valley (now under the lake) and also downstream. The total landslide covers 18 km^2 and the resulting Lake Waikaremoana is 250 m deep. Prior to the 1950s, lake water naturally seeped through many parts of the debris dam, but these seepages were plugged by hydroelectricity engineers to create an overflow for electricity generation.

45.1 Lake Waikaremoana is the largest and deepest lake in New Zealand that has been created by a landslide dam. The landslide, forming the low lip to the north (left) of Panekiri Bluff, is believed to have been triggered by a large earthquake 2200 years ago. PHOTO: LLOYD HOMER, GNS SCIENCE, 14423.

Best seen from: View the landslide blocks and rugged topography from Onepoto Rd and SH 38 which passes through the landslide at the southeast end of Lake Waikaremoana. Walk to Lou's Lookout (45-min. return) for a view over some of the landslide dam and lake.

Take a closer look: Walk the Onepoto Caves Track through the fractured landslide block with numerous small fissures and overhangs between jostled blocks (2 hrs return).

➲ 45.2 SH 38 passes over parts of the blocky debris avalanche landslide that created Lake Waikaremoana's natural dam.

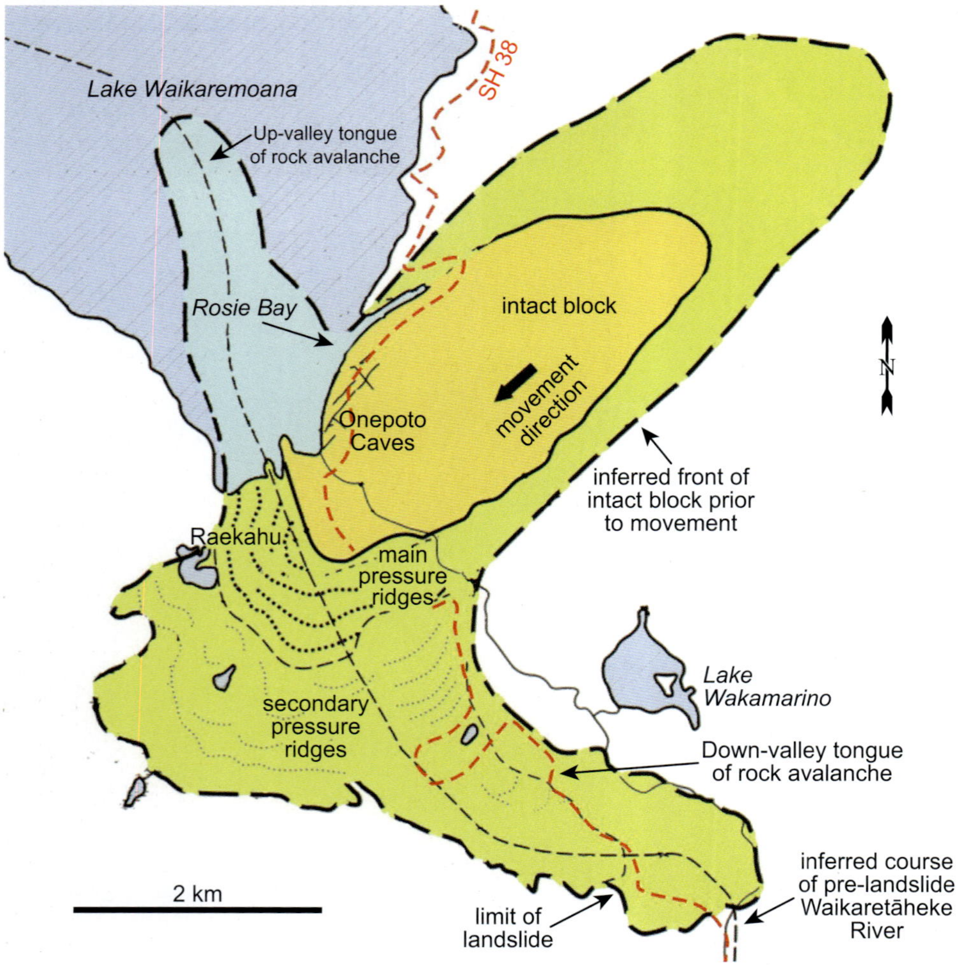

45.3 Map showing the extent of the Lake Waikaremoana landslide, its movement direction, and its division into an intact glide slab with avalanche debris blocks forming the dam. MODIFIED FROM DAVIES ET AL. (2006).

45.4 Cross-section through the Waikaremoana landslide. MODIFIED FROM DAVIES ET AL. (2006).

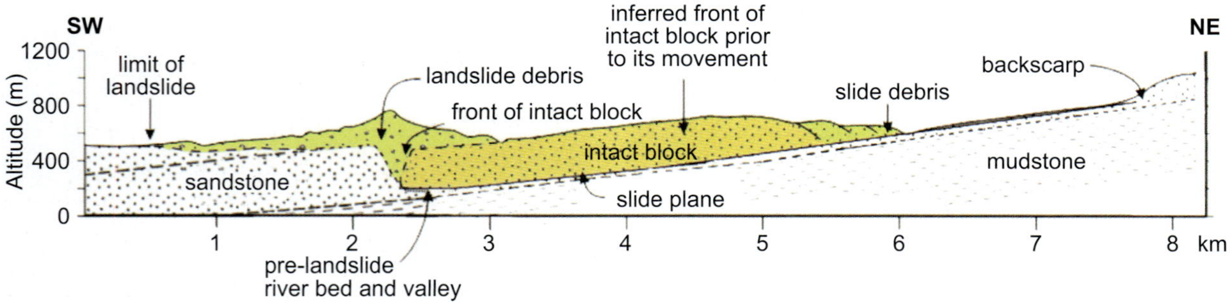

➲ 45.5 Face of the landslide dam that created Lake Waikaremoana.

➲ 45.6 View across Lake Waikaremoana from the landslide dam to Panekiri Bluff of hardened sandstone tilted to the left.

➲ 45.7 The high sandstone of Panekiri Bluff overlooks the south end of Lake Waikaremoana. The landslide dam that created the lake is hidden around the corner behind Panekiri Bluff in this photo. PHOTO: LLOYD HOMER, GNS SCIENCE, 4778.

46.

Ahuriri Lagoon, Napier

Tidal lagoon uplifted out of the sea by the 1931 Hawke's Bay Earthquake

46.1 This is all that remains today of the huge tidal Ahuriri Inlet that was largely uplifted out of the sea during the 1931 Hawke's Bay Earthquake. PHOTO: LLOYD HOMER, GNS SCIENCE, 128364.

One of the two largest historic earthquakes in the North Island resulted in the uplift and draining of the large coastal Ahuriri Lagoon. During the 1931 Hawke's Bay Earthquake the whole of Ahuriri Lagoon was uplifted by 1.5–2 m. Prior to the earthquake, the lagoon consisted of 35 km^2 of intertidal and shallow subtidal mudflats separated from the Pacific Ocean by a narrow gravel spit (46.2). The Tūtaekurī River flowed into the southern end of the lagoon and its flow kept open the tidal mouth of the lagoon, close to Scinde Island.

Study of cores of sediment from the lagoon show that its most recent version was formed about 7500 years ago when sea level rose to its present height and a gravel spit built across between Scinde Island and headlands to the north. The fine gravel of the spit is composed of greywacke rock, eroded from the high inland Ruahine Range and transported down rivers to the coast. The sediment cores record that the uplift during the Hawke's Bay Earthquake was the exception, and that eight of the ten large earthquakes that have occurred in this region in the last 7500 years were accompanied by subsidence of 0.5–1.2 m. This is one of the few places along the East Coast which is currently subsiding, and these subsidence events are interpreted as resulting from Great Earthquake ruptures on the subduction plate boundary, 20 km beneath the lagoon (The Big Picture 10, page 154).

The Hawke's Bay (or Napier) Earthquake on 3 February 1931 was New Zealand's deadliest natural disaster. The Mw 7.7 event resulted in 256 deaths – 161 in Napier, 93 in Hastings and 2 in Wairoa. The earthquake was generated by rupture on the north-northeast-trending Napier Fault with the epicentre at shallow depth, 20 km or so northwest of Napier. Most of the rupture did not break the land surface, but two lines of small fault traces were generated above it between Hastings and Lake Poukawa. Napier Fault dips at 40° down to the west as it cuts through the upper (Australian) plate that runs beneath central Napier. During the earthquake an area stretching 120 km by 15 km and overlying the fault was pushed upwards with a maximum uplift of 2.7 m at Old Mans Bluff (46.8).

The lagoon water drained out within a few hours. Complete draining and reclamation of the flat floor of the former lagoon took many years, with the newly formed land converted into fertile farmland (46.4) and some later used for Napier Airport. In recent decades, Napier suburbs have expanded out over the uplifted southern area that was once swamp or tidal lagoon. The lesson from the sediment cores is that in the next big earthquake the land is likely to go down again.

Best seen from: Look over the uplifted lagoon flats from the west end of Scinde Island or from Rorookuri Summit Track, on an uplifted island at the north end of the lagoon.

Take a closer look: Take a walk around the uplifted inner mouth of Ahuriri Lagoon along the Ahuriri Estuary Walkway.

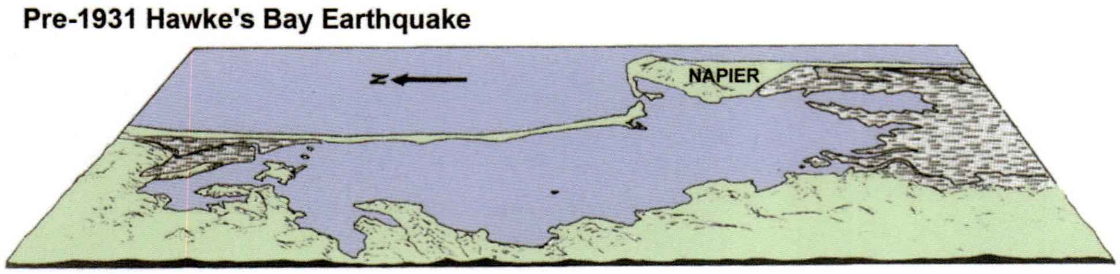

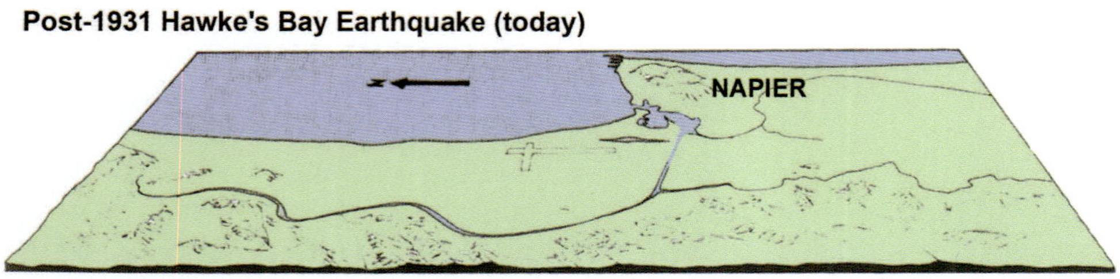

46.2 Sketches of how Ahuriri Lagoon was changed by the uplift during the 1931 Hawke's Bay Earthquake. MODIFIED FROM COX AND HAYWARD (1999).

46.3 Historical photograph taken from the northern slopes of Scinde Island across the mouth and middle section of Ahuriri Lagoon prior to the 1931 earthquake.

46.4 View northwest in 2021 from a similar location to 46.3 prior to the 1931 uplift.

⊙ 46.5 All the flat land in this picture was intertidal mudflats of Ahuriri Lagoon before it was uplifted above high-tide level in the 1931 Hawke's Bay Earthquake. View from the north towards Napier (distant left).

⊙ 46.6 After the Hawke's Bay Earthquake, this jetty on the side of uplifted Ahuriri Lagoon was left high and dry.

⊙ 46.7 View east across the central part of the uplifted Ahuriri Lagoon showing one of a number of small islets that became landlocked after the 1931 earthquake. Napier Port behind.

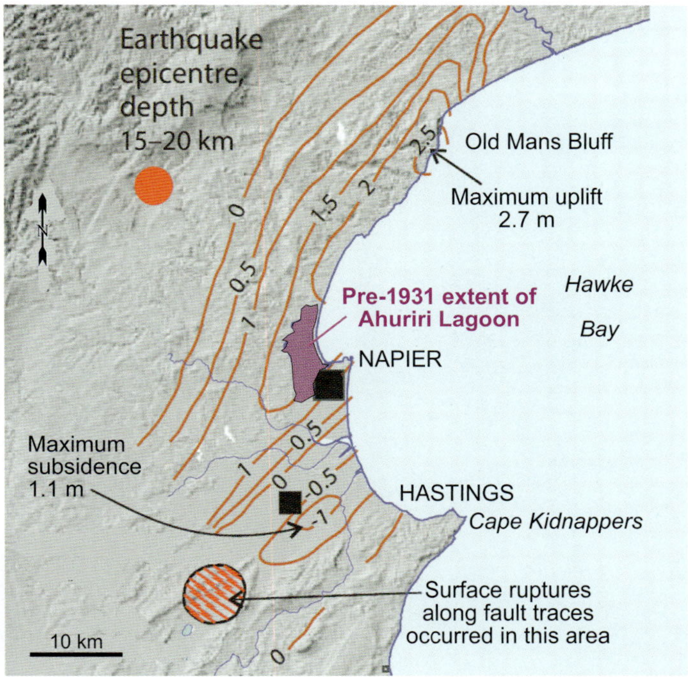

⊙ 46.8 Map showing the elliptical area that was uplifted during the 1931 Hawke's Bay Earthquake. Contours of uplift (and subsidence) are in 0.5 m intervals. MODIFIED FROM BALLANCE (2017).

⊙ 46.9 As you walk along the north side of Ahuriri Estuary Walkway, keep an eye out along the shoreline for areas where beds of shells can be seen embedded in the muddy gravel. These shells lived at or below low-water level prior to 1931.

47.

Te Kauwae-a-Māui / Cape Kidnappers

High cliffs of young sedimentary rocks recording Ice Age sea-level cycles and uplift

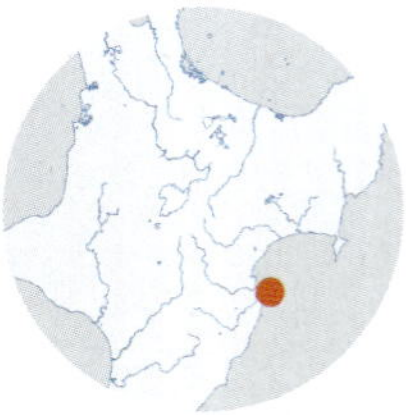

⊙ 47.1 These cliffs, 3 km east of Clifton, are composed of 4–3-Myr-old shell-bearing grey sandstone (lower third of cliff to left). It is overlain by reworked cream volcanic ash and ignimbrite layers interspersed with brown conglomerate and sandstone that was deposited on the ancient Heretaunga coastal plain about 1 Myr ago. There is an erosional time gap (no rocks present) of about 2 Myr between these two groups of rocks.

47.2 Western end of the Cape Kidnappers cliffs section showing sediments deposited during three climate cycles between 800,000 and 500,000 years ago. The two bands of orange-brown conglomerate were deposited during cold intervals when sea level was lower, and the light grey mudstones and sandstones were deposited in estuaries and coastal ponds during warm intervals when sea level was higher.

The Te Kauwae-a-Māui / Cape Kidnappers cliff section is one of the two best places in New Zealand (along with Whanganui coast, Site 39) to see a sedimentary rock record of the high and low sea-level changes that occurred during the climate cycles of the Ice Ages. The cliffs are made of relatively soft sedimentary rocks that are rapidly eroding. The top of the 100–120-m-high cliffs is a relatively flat terrace that was an intertidal shore platform during the Last Interglacial 130–120 kyr ago. This area has been pushed upwards at a rate of about 1 m per 1000 years since that time, forming an upfold (anticline) with the beds in the cliffs tilted at 10° down to the west. As a result we can see exposed in the cliffs the layers of sediments that were deposited here between 5 million and 500,000 years ago when this area was slowly subsiding.

The crumbly rocks forming the cliffs at Clifton are the youngest (deposited about 500,000 years ago) and as you move eastwards you go down through the sequence getting progressively older. For the first 4–5 km, the 400-m-thick layered sequence comprises four repeated cycles of sediment types (47.2). At that time the Clifton–Cape Kidnappers area was near the coast on an actively forming ancestral Heretaunga Plains. The area was subsiding at about 1 m per 1000 years, and the sediment accumulated at about the same rate. The cold–warm cycles were each about 100,000 years long.

During the long cold glacial periods of each cycle, sea level was 30–100 m lower than present and the coast was well out into Hawke Bay. At these times the bush line in the inland axial ranges was much lower than today and erosion was greatly accelerated. Vast quantities of greywacke gravel were carried down the rivers and spread out over the growing plains. The gravel (conglomerate) units in the cliffs were mostly deposited on the braided river plain during these cooler periods. During the shorter warmer interglacial periods the amount of gravel coming down the rivers decreased. The sea was up close to the present level and the rivers deposited fluvial sand near

47.3 The rock sequence that forms the rapidly eroding sea cliffs between Clifton and Cape Kidnappers was pushed upwards by plate-tectonic forces by as much as 120 m over the last 120,000 years. The rocks were uplifted higher in the east than the west, and so the near-original horizontal layers are tilted down to the west. During the many earthquakes that accompanied the uplift events, stresses within the rocks caused them to fracture creating a number of faults that displace the rocks on either side of them, as in this photo.

⊙ 47.4 Cape Kidnappers and the pinnacled stack offshore are eroded out of 5-Myr-old marine mudstone that underlies the colourful, layered sequence seen in the cliffs between here and Clifton. PHOTO: JULIAN THOMSON.

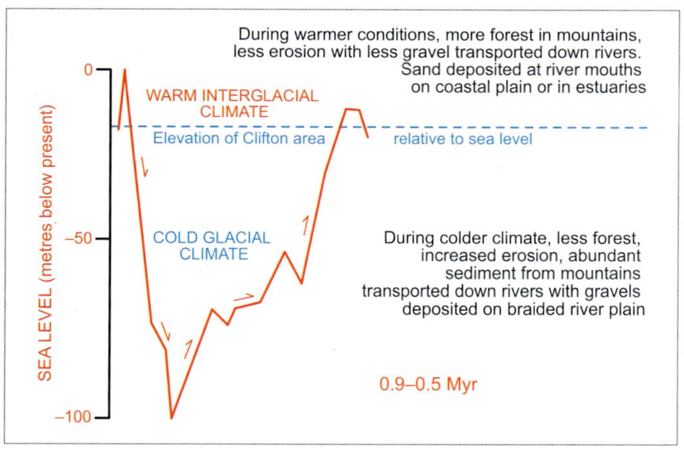

⊙ 47.5 Graph of the changing sea level through an Ice Age climate cycle, and a summary of the resulting changes in sediment deposition seen in the Cape Kidnappers cliffs, east of Clifton.

their mouths. Forest became established on the coastal plain (47.8) and small ponds and swamp developed and accumulated peat and mud, layers of which can be seen in the cliffs. In at least two of these warm periods, the sea advanced over the Clifton area forming tidal lagoons or estuaries. Finer-grained sand and mud were deposited in these, and often contain fossil shells of estuarine clams and snails that can be seen in the cliffs (47.7).

At the base of these four cycles there is 40–50 m thickness of white volcanic ash layers interspersed with yellow-brown sandstone and conglomerate (47.1). These were deposited on the ancient Heretaunga Plain around 1 Myr ago. The ash layers are full of fine pumice fragments. They were erupted from a large, now buried, caldera volcano near Mangakino about 150 km away near the centre of the North Island. Some of the ash was erupted high in the air and blown here, while some arrived as hot, ground-hugging ignimbrite flows of gas, pumice and crystals (The Big Picture 7, page 105).

Still further east, in the cliffs and underlying the white ash layers, is a bedded sequence of sandstone containing fossil scallop and oyster shells that was deposited on the shallow seafloor about 4–3 Myr ago (Pliocene). Hardened layers in these rocks form Black Reef that sticks out into Hawke Bay (47.6). These overlie still older (about 5 Myr), light-grey mudstone that was deposited in even deeper water. Along with the overlying rocks it has only recently (last 150,000 years) been uplifted and eroded to form Cape Kidnappers

47.6 The small gannet colony at Black Reef (6 km east of Clifton) has established on the partly eroded remains of a former rocky shore platform that was uplifted from the intertidal zone by 4 m during a huge earthquake about 2200 years ago.

itself (47.4). The four youngest cycles of sediment in the Cape Kidnappers cliffs mainly consist of conglomerate and sandstone that was deposited on land on the Heretaunga coastal plains during high and low sea levels, whereas the cycles of sediment seen in the Whanganui coastal cliffs (Site 39) consist almost entirely of sandstone and mudstone deposited in shallow marine environments during high sea level (warm interglacial intervals).

Best seen from: Distant views of the Cape Kidnappers cliffs can be obtained along the foreshore south of Napier.

Take a closer look: Traverse the base of the cliffs at low tide by walking (8 km each way, at least 4 hrs) or take a commercial tractor-trailer ride. A large rock-fall in 2019 highlighted the hazards of approaching too close to the base of the cliffs.

47.7 Shells of cockles fossilised in 600,000-year-old mudstone that accumulated in an estuary, 1 km east of the Clifton.

47.8 Fossil stump of a tree that grew in river plain muds near the coast about 700 kyr ago, 2 km east of Clifton.

48.

Te Mata Peak, Havelock North

Limestone ridge pushed up high (cuesta)

Te Mata Peak ridge is the best-known cuesta in New Zealand. A cuesta, or strike ridge, is a ridge with a gentle dip slope on one side and a steep slope or bluffs on the other. The ridge is generally made of a layer of tilted hard rock lying between layers of softer rock that have been eroded away.

In southern Hawke's Bay, just south of Havelock North, the dissected country consists of a series of elongate ridges oriented approximately north–south parallel to the plate boundary offshore. These ridges are the tops of upthrust slivers of mostly young sedimentary rock within the compressed accretionary wedge (The Big Picture 10, page 154). The best-known and most easily accessed of these ridges is Te Mata Peak (48.1), 3–5 km south of Havelock North.

The Te Mata Peak ridge consists of hard limestone that accumulated as an elongate shell bank in shallow water (0–50 m depth) about 3.5 Myr ago (late Pliocene). This shell bank was one of a number that built up on the shallow submarine crests of these upthrusting ridges. The shell banks were at least 40 km offshore from the east coast of the North Island at that time and were separated from it by a deep subsiding basin (East Coast fore-arc basin). Any suspended mud that reached this area, from erosion of the distant North Island coast, was swept away by strong currents, leaving behind a relatively pure deposit of shells.

The thick shell-bank deposits were subsequently buried by further sediment. Percolating groundwater started to dissolve some of the shells and new calcite crystallised out of the carbonate-saturated fluids in the gaps between the shells, gluing them together into the relatively hard limestone of today. Subsequently, further uplift has pushed this limestone

48.1 View north from Te Mata Peak along the line of limestone bluffs on the ridge's east side.

48.2 Bluffs of Te Mata Peak cuesta from Tukituki Valley floor.

high up out of the sea and tilted it slightly (10°) down to the west. Softer sedimentary rocks around it have been eroded down so that the ridge now stands high with its crest 300–400 m above sea level. The tilt of the limestone layer produces the gently sloping ridge crests (dip slopes) on the down-tilted west side. On the east side, the Tukituki River has eroded down through fractured and softer rocks in the sheared boundary between two of the upthrust slices. This erosion has slowly eaten its way westwards, creating the line of 100-m-high limestone bluffs of Te Mata Peak ridge (48.2, 48.4).

Best seen from: The spectacular high cliffs can also be viewed from below while driving along Waimarama Rd near Craggy Range Winery.

Take a closer look: For panoramic 360° views go to the lookout at the Summit carpark, at the end of Te Mata Peak Rd (5 km). There are numerous scenic walking and biking trails in Te Mata Park.

48.3 Numerous well-preserved fossil scallop shells and oysters, such as these, can be seen in the limestone forming the crest of Te Mata Peak. The limestone is predominantly composed of the broken-apart plates of barnacles, broken clam shells and bryozoans. Some beds are packed with large oyster shells and lampshells.

48.4 Limestone bluffs of Te Mata Peak cuesta.

48.5 The southern end of Te Mata Peak cuesta is broken into several blocks.

49.

Rangiwhakaoma / Castlepoint

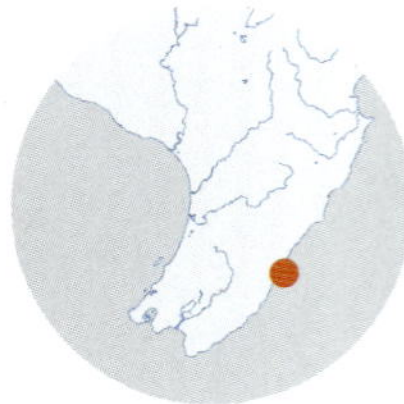

Fortress-like promontory and fossil-rich limestone

Rangiwhakaoma / Castlepoint is a 162-m-high fortress-like promontory with a 1.5-km-long arcuate rocky peninsula extending north parallel to the Wairarapa coast (49.1, 49.4). The castle and peninsula owe their existence to the cemented limestone made out of fossil shells which is far more resistant to erosion than the mudstone that has been eroded away on either side. This limestone originated as shell and sand that accumulated on the seafloor just 2.4 Myr ago (early Pleistocene), making it one of the youngest limestones in New Zealand and about 1 Myr younger than the limestone forming Te Mata Peak (Site 48).

Recently geologists have recognised that this limestone is quite different from most others on the North Island's east coast. It contains a diverse mixture of whole and broken-up fossils – primarily of clams, barnacles and bryozoa (49.5, 49.6). Study of these shows that they lived on the seafloor in a wide range of water depths from close to shore at no more than 50 m depth down to depths in excess of 400 m (bathyal).

The layering in the limestone shows evidence of considerable disruption, squeezing and folding, and in places includes large blocks of limestone or siltstone, up to 4 m or more in size. These sediments were clearly emplaced by seafloor landslides carrying shells, sediment and blocks of rocks from shallow depths downslope to considerable depths.

Offshore from Hawke's Bay and the Wairarapa today there is a narrow continental shelf that slopes gently out to about 200 m water depth. Beyond the edge of the shelf is the continental slope which is much steeper and is interrupted by a number of slope-parallel, elongate depressions (sedimentary basins 5–50 km long) that trap sediment. Here the seafloor is also scarred by many submarine canyons that carry sediment in landslides and slurries from the shelf down to the ocean depths. It is now inferred that the limestone at Castlepoint is the uplifted and eroded sediment that once filled one of these submarine canyons that has now been uplifted.

49.1 View south along the sliver of mostly west-tilted limestone from Castlepoint lighthouse peninsula to the 162-m-high castle in the distance. The limestone was deposited by slurries of shell and sand flowing down a submarine canyon. The light-grey rock, to the right of the lighthouse, is a large block of mudstone that was plucked from the canyon wall and carried down the canyon in an undersea landslide.

49.2 View north from the upper slopes of the castle showing the shore-parallel finger of hard limestone that follows the arcuate strike of the bedding out to the lighthouse. A hidden fault runs from the photographer northwards under the lagoon and sand tombolo and separates the young limestone from much older mudstone that underlies the hills on the left.

49.5 In some places the unbroken shells of fossil clams and snails can be seen in the Castlepoint limestone. These are two shells of an offshore scallop, *Mesopeplum*. Photo 15 cm across.

49.6 Close examination of the limestone is usually rewarded by the recognition of the fossil shells of a number of different marine organisms, like this barnacle. Photo 10 cm across.

49.3 View south along the west-dipping layers (bedding planes) of shelly limestone that form the rocky reef between the Pacific Ocean and the lagoon at Castlepoint. The small flat area on top of the castle is the much-eroded remnant of an uplifted marine terrace that was eroded at sea level several hundred thousand years ago and has subsequently been uplifted to 160 m elevation.

49.4 The arcuate line of the erosion-resistant limestone that forms the lighthouse peninsula and castle to the south is clearly visible. PHOTO: LLOYD HOMER, GNS SCIENCE, 4979.

Periodically, loose shell and sand on the shelf around the upper margin of the canyon was mobilised by strong earthquake shaking and slid as a dense watery slurry down into the canyon. Sometimes these slides undercut the compacted mudstone of the canyon wall and blocks of this also slid down the canyon. Several of these blocks of soft grey mudstone can be seen within the limestone near the foot of the Castlepoint lighthouse today. The grey mudstone in the lower part of the sea cliffs below the lighthouse is inferred to be part of the canyon wall. A fault zone separates the limestone sliver of coastal Castlepoint today from the 10-Myr-old mudstones that the adjacent mainland is made of.

Following the filling of the canyon with sediment it would have been buried and preserved by further sediment accumulation on top of it. Groundwater passing through the shelly sediment partly altered the composition of the aragonitic and calcareous shells, and resulted in the growth of sparry calcite crystals in the voids that glued the shells and sand particles together into the hard cemented limestone. Eventually, plate-tectonic forces pushed this area up from the depths and tilted the limestone to the west, as seen in the sloping bedding planes of the rocky peninsula (49.1, 49.3). Episodic uplift during large earthquakes continues through to the present day and sea erosion has exhumed the hard limestone sliver and created the magnificent coastal landforms.

Take a closer look: From the end of Jetty Rd, walk across the high-tide beach and take the path up to and around the lighthouse. The fossils are protected, so please only take photos. For the more energetic, climb the track to the top of the Castle for breathtaking views (1-hr return).

50.

Pūtangirua Pinnacles, Cape Palliser

New Zealand's best example of badlands erosion

⊙ 50.1 Walking up the gully among Pūtangirua's Pinnacles you can appreciate the scale of the badlands erosion of this 7-Myr-old alluvial fan deposit.

⊙ 50.2 Aerial view over the eroding gully where most of the Pūtangirua Pinnacles occur. PHOTO: LLOYD HOMER, GNS SCIENCE, 22689.

⊙ 50.3 Pūtangirua Pinnacles, in southern Wairarapa, are the most impressive example of badlands erosion in New Zealand. This is the view from the lookout over part of the eroding gully.

➔ 50.4 The cemented layers within the alluvial fan conglomerate are gently tilted to the east in these Pūtangirua Pinnacles.

Pūtangirua Pinnacles, in southern Wairarapa, are the best example of badlands erosion in New Zealand. Badlands are areas of intense erosion characterised by strongly dissected and gullied landscapes with sparse vegetation. At Pūtangirua, heavy rain during storms has washed out the softer silty matrix of the conglomerate rock and eventually created pinnacles or 'hoodoos' (50.3, 50.5). The conglomerate originated as an alluvial gravel fan deposited at the mouth of a steep stream about 7 Myr ago (late Miocene). The gravel consists of greywacke clasts up to boulder size in a silty matrix. The vague layering and highly variable size and degree of rounding of the cobbles and pebbles are characteristic of alluvial fans.

Soon after it accumulated, this subsiding area was drowned by the sea and the gravels buried by the accumulation of mud and sand on the seafloor. The gravels were compacted into conglomerate rock and percolating groundwater deposited calcite between the grains in some places, creating thin cemented layers. Rapid uplift (about 1 m per 1000 years) in the recent past has created the elevation for erosion to strip off the overlying marine sedimentary strata and start cutting down deeply into the fan conglomerate.

The gullies erode out during episodes of heavy rainfall. When a large boulder or hard cemented layer is uncovered they protect the conglomerate directly beneath from further erosion. Eventually a whole series of pinnacles are formed, each with a hard cap on top of them (50.3). In some places the pillars are stepped as there are erosion-resistant caps of boulders or cemented layers at different levels.

Take a closer look: From Pūtangirua Pinnacles carpark, take the Pinnacles Track to the spectacular overlook or walk up the Pūtangirua Stream bed and left into the first major tributary valley, 1.2 km upstream (2–3 hrs return).

➔ 50.5 A 20-m-high pinnacle eroded out of weakly layered conglomerate.

51.

Cape Turakirae, Wellington south coast

Series of raised storm-beach ridges record major earthquakes

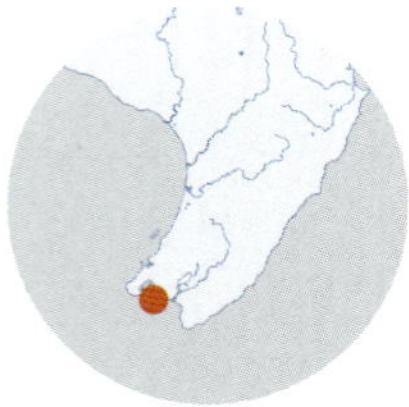

Turakirae Head, on Wellington's storm-battered south coast, is a classic example of an uplifting coastline with one of the finest series of raised storm-beach ridges in the world. The uplifted shore platform around the base of the steep hills (51.3) holds a record of four major earthquakes over the last 7000 years that pushed the land up a maximum of 25 m.

When viewed from the air, four ribbons of light-grey ribbons run along the platform parallel to the coast (51.1, 51.4). Each was once a gravel beach thrown up above high tide by surging waves during large southerly storms. The gravel is mostly rounded cobbles and pebbles of greywacke (51.2) that have come down the Ōrongorongo River at the southern end of the Remutaka Range and been moved around the coast in storms.

The lowest ribbon is the modern one that has been formed since the huge Wairarapa Earthquake in 1855 (Mw 8.2) uplifted

51.1 View west over the uplifted shore platform between Turakirae Head (foreground) and Baring Head (middle distance). Four distinct ribbons of light-grey gravel storm beach run along the platform parallel to the coast. The lowest is the modern beach ridge, and the other three have been uplifted during major earthquakes 165 (AD1855), 2200 and 5000 years ago. A series of uplifted, marine-cut shore platforms, similar to the one in the foreground, can be seen on the ridge crests on the far side of the bay and have been progressively uplifted in major earthquakes going back at least 300,000 years. PHOTO: LLOYD HOMER, GNS SCIENCE, 1752.

⊙ 51.2 This area at Turakirae Head was the subtidal seafloor prior to it being uplifted to become land during the 1855 Wairarapa Earthquake. The gravel ridge (lower right) was the active storm beach at that time, whereas the gravel ridge beyond the people is the modern storm beach that has formed in the last 165 years. Note that gravel beaches are relatively smooth and dry and do not have soil development nor much vegetation. Between the beach ridges the land is more rugged and rocky. Peaty soil is accumulating in the damp hollow between the ridges and vegetation is already well established despite its young age.

⊙ 51.3 Looking east across the bay at the mouth of the Ōrongorongo River to the southern end of the Remutaka Range ending in Turakirae Head. The whole range has been uplifted out of the sea in less than 1 Myr with the low uplifted shore platform, barely visible above sea level, showing how much uplift (maximum 25 m) there has been in the last 7000 years.

the Remutaka Range and all of Wellington. During that earthquake the West Wairarapa Fault, which runs along the Wairarapa side of the Remutaka and Ruahine Ranges, ruptured with as much as 12 m of sideways movement (west side to the south). At the same time the Remutaka Range was buckled upwards ever so slightly as part of the growth of the north–south-trending Remutaka Anticline (upfold). The maximum amount of uplift in 1855 (6.5 m) occurred at the axis of the anticline, just east of Turakirae Head, with the amount of uplift progressively decreasing to 1.5 m at the mouth of the Wainuiomata River 5 km to the west.

The other three higher-level storm-beach ridges record earlier but similar major earthquakes caused by simultaneous rupture of the fault and upwards flexure of the anticline. These earlier earthquakes have been dated as occurring about 2200, 5000 and 6800 years ago with maximum uplifts along the coast east of Turakirae Head of 9 m, 6 m and 4 m, respectively. The highest and oldest beach ridge is only visible in the northeast part of the platform and is not seen in the accompanying photos. During each earthquake it is not just the coast that is uplifted but all of the land west of the fault. This can be seen by a series of uplifted marine terraces on the ridge crest between the Ōrongorongo Valley and Wellington Harbour. Each marine terrace was once a shore platform that was eroded by the Cook Strait storms during a warm interglacial period when sea level was close to its present elevation. A stepped sequence of at least three marine terraces can be recognised on the ridge rising up to at least 350 m elevation (51.1), with the highest probably dating back to at least 300,000 years ago.

Take a closer look: From the carpark at the end of Coast Rd, walk (or cycle) along West Coast Track into Turakirae Head Scientific Reserve (2 hrs return on foot).

⊙ 51.4 View northeast along the Wairarapa side of Turakirae Head showing the modern beach ridge and three older ones uplifted by earthquakes. The highest and oldest ridge is only recognised at the north end of this section of coast. PHOTO: LLOYD HOMER, GNS SCIENCE, 1273.

52.

Wellington Fault

Fault line that formed Wellington Harbour and the Hutt Valley

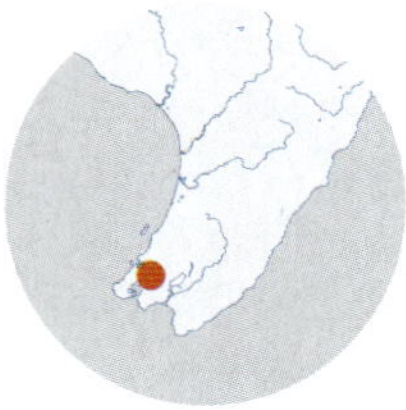

↑ 52.1 The Wellington Fault scarp forms the western side of the Hutt Valley (distance) and Wellington Harbour, and extends south through the hilly suburbs of western Wellington city. The eroded flat surface on top of the basement greywacke rocks has been uplifted 200–350 m on the west side (left) and downthrown up to 600 m beneath the inner harbour (middle) and 300 m beneath Petone (distant). PHOTO: LLOYD HOMER, GNS SCIENCE, 22993.

The Wellington Fault has one of the most prominent scarps in the country. It is also part of the longest fault line in the North Island and can be traced all the way from Cook Strait to the Bay of Plenty. Wellington Harbour and the Hutt Valley owe their existence to the episodic displacement across the fault going back about half a million years.

Wellington Fault is one of the major sideways-moving (strikeslip) faults that run parallel to the plate boundary through the North Island. It is believed to extend all the way down to the subduction interface thrust between the two plates, 25–30 km beneath Wellington. The most recent rupture on the Wellington Fault was between 1640 and 1840, prior to European colonisation. The time between the three most recent ruptures has averaged 1100 years and so a major earthquake on this fault is not expected any time soon. There are, however, a number of other large parallel faults through the Wellington region, such as the Wairarapa and Ōhāriu faults, which generally rupture independently and together result in more frequent locally generated major earthquakes for the southern end of the North Island.

⊙ 52.3 In Tōtara Park, Upper Hutt, the Wellington Fault trace runs obliquely through California Park along the top of the fault and records the displacement of only the last few earthquake ruptures. Prior to that, the Hutt River had eroded the main fault scarp back to the foot of the hills (top right). PHOTO: LLOYD HOMER, GNS SCIENCE, 1803.

⊙ 52.2 Looking south down the Lower Hutt Valley from above Avalon with the straight Wellington Fault scarp rising steeply on the west (right) side (in the shade). The valley has been tilted down to the west (right) by about 1000 rupture events on the fault over the last 500,000 years, which have also moved the west side up to 4 km to the south. Gravel and sand transported down the Hutt River have largely filled this elongate fault-angle depression creating the flat alluvial valley floor. PHOTO: LLOYD HOMER, GNS SCIENCE, 21876.

52.4 View west across the alluvium-filled flat floor of Petone at the southern end of the Hutt Valley. The prominent Wellington Fault scarp forms the uplifted far side of the valley and the inner Wellington Harbour. Matiu / Somes Island in the middle of the harbour is on the crest of a narrow upfaulted ridge of greywacke.

During each of the last few displacement events, the west side of the Wellington Fault has moved northwards by about 5 m with a lesser and more variable amount of vertical movement across it. The fault is steep and its trace across the countryside is fairly straight except for several prominent bends where subsidiary north-trending faults join it. Interaction between these bends and the dominantly sideways displacement across the fault has created several down-folded basins on the eastern side. These basins are mostly filled with young alluvium and underlie the Upper Hutt–Trentham and Lower Hutt–inner Wellington Harbour areas. The basins are separated by an upfolded and faulted area of basement greywacke that now forms Taitā Gorge. Another uplifted area of greywacke east of the fault forms Miramar Peninsula and the shallow entrance to Wellington Harbour.

Total uplift of the west side of the Wellington Fault has been relatively uniform (200–350 m), creating the prominent scarp along the west side of Wellington Harbour and the Hutt Valley (52.1, 52.2). Downthrow on the east side has been highly varied, however, with the Upper Hutt basin subsiding to 150 m below sea level and the inner-harbour end of the Lower Hutt–Wellington Harbour basin subsiding 600 m below sea level. As these basins progressively subsided they filled with alluvial gravels and sands brought down the Hutt River. Some of these gravels form the groundwater aquifer beneath the Lower Hutt Valley that currently supplies much of the city's freshwater. During the warm interglacial high-sea-level intervals the sea invaded the southern basin, forming successive Wellington harbours which at times advanced northwards at least as far as central Lower Hutt.

Vertical displacements across the Wellington Fault have down-tilted the eastern block to the west, with greatest subsidence next to the fault and the greywacke basement tilted up forming the Hutt Valley eastern foothills. This westward tilt caused the Hutt River to flow down the west side of the valley where it has eroded the fault scarp back by up to 200 m. In a few places out on the valley floor, such as at Tōtara Park, there is a low fault trace produced by the last few displacements on the Wellington Fault (52.3).

Best seen from: This fault scarp can be seen from many parts of the Hutt Valley and from the Cook Strait ferries on Wellington Harbour. Other panoramic views are from lookouts on Wainuiomata Road saddle, the top of Mt Victoria, Hataitai or Wrights Hill, Karori.

⬆ 52.5 View south down the Hutt Valley with the Wellington Fault on the west (right) side. The flat, alluvium-filled floor of the down-folded Upper Hutt basin is in the foreground with the upfolded greywacke ridge (forested) cut by the Taitā Gorge beyond. PHOTO: LLOYD HOMER, GNS SCIENCE, 1837.

⬇ 52.6 View from the south above Cook Strait across Wellington Harbour. The Wellington Fault scarp forms the straight western (left) boundary of the harbour and runs north, defining the west side of the Hutt Valley. You can also follow the fault line along the valleys south to the Cook Strait coast (foreground left). PHOTO: LLOYD HOMER, GNS SCIENCE, 1949.

The highest parts of the South Island's Southern Alps with Aoraki / Mt Cook (right) and Horokoau / Mt Tasman (left) on the skyline, and La Perouse on the main divide (foreground). PHOTO: ALASTAIR JAMIESON.

South Island

53.

Marlborough Sounds

Drowned dendritic valley systems

⊙ 53.1 The Marlborough Sounds are the best example in New Zealand of an area of intensely dissected dendritic valleys that have been flooded by rising sea level after the end of the Last Ice Age. The valleys were eroded 10–5 Myr ago out of an uplifted range of greywacke and schist. The sounds area has been slowly subsiding in the last 1.5 Myr. View north down Tennyson Inlet to the outer Pelorus Sound. PHOTO: LLOYD HOMER, GNS SCIENCE, 11539.

53.2 Anakoha Bay in the outer Pelorus Sound is a long, narrow, V-shaped embayment formed by the drowning of a deeply-incised stream valley. 1200-m-high Mt Stokes, the highest point in the Marlborough Sounds, is partly shrouded in cloud at the head of the bay.

53.3 Cape Jackson is the 150–300-m-high snake-like crest of an almost completely submerged ridge that extends 6 km out into Cook Strait. The mouth of Queen Charlotte Sound is on the left. PHOTO: ALASTAIR JAMIESON.

Ironically the drowned Marlborough Sounds landscape owes its origins primarily to uplift that occurred here for most of the last 5–10 million years. That uplift has now migrated south and is uplifting the mountain ranges east of Nelson. During the period of uplift, erosion carved deep valleys into the uplifted schist and greywacke rocks draining into the subsiding marine basin near Whanganui to the north. These elongate valleys were mostly eroded along fault zones of broken-up rock that are aligned in a northeast–southwest direction. Many smaller subsidiary valleys are aligned north–south, which may indicate a secondary set of fault lines with more easily eroded broken rock. The depth of the eroded valleys was determined by the height of sea level during each cold glacial period and so the floors of the main northeast-flowing valleys were probably 50–100 m below present by the middle of the Ice Ages (1.5 Myr ago). Each time sea level rose during the intervening warm interglacials, the sea invaded the valleys and created a network of sounds not too dissimilar to what we see today.

The tectonic setting of the Marlborough Sounds changed about 1.5 Myr ago, and instead of being pushed up they started to subside. This is because the zone of subsidence that had been advancing south through the western North Island over the last 10 Myr or more had now reached the Sounds (The Big Picture 9, page 145). At the present time the rate of subsidence in the Sounds is ~0.5 m per 1000 years

⬆ 53.4 Te Hoiere / Maud Island (368 m), in Pelorus Sound, is a partly submerged former hill with spurs stretching out in four directions. PHOTO: ALASTAIR JAMIESON.

⬇ 53.5 Melville Cove (foreground) in Port Gore on the north coast of the Marlborough Sounds.

◀ 53.6 At Te Aumiti / French Pass, the rising sea level has flooded over a saddle that joined D'Urville Island (foreground) to the mainland. The slightly eroded rocky top of the saddle now forms a reef traversed by extreme tidal currents. PHOTO: ALASTAIR JAMIESON.

▼ 53.7 Long Island in the middle of the Cook Strait entrance to Queen Charlotte Sound is the crest of a narrow ridge that was partly drowned by rising sea level after the end of the Last Ice Age about 10 kyr ago. Kokomohia Islands (in the foreground) are part of the same ridge. The waters surrounding these islands became the South Island's first marine reserve in 1993. PHOTO: ALASTAIR JAMIESON.

and has averaged about 0.3 m per 1000 years over the last 1.5 Myr. The eroded floors of the drowned valleys have now subsided to ~500 m below sea level near their entrances on Cook Strait, and 250–300 m below sea level further south, off Picton. The floor of the sounds themselves are not that deep, however, as they have been filling up with sediment brought down from the surrounding hills and especially the Pelorus and Rai Valleys, but also by mud carried in from Cook Strait in suspension.

Between 75 and 12 kyr ago, when the climate was colder and sea level 50 m or more lower than present, the Marlborough Sounds were flat-floored valleys with thick alluvium beneath. Rising seawater started to invade the valleys from 12 kyr ago onwards, reaching its present extent by about 7500 years ago. The Marlborough Sounds would have existed for periods of 10–30 kyr on at least 10 occasions in the last 1 Myr, separated by even longer intervals when they were forest- and shrub-filled valleys. Today the numerous dendritic stream valleys are a series of short embayments along the sides of each of the main sounds (53.1). Several of the lower ridges between valleys in the northern part of the Sounds are deeply submerged and now remain as narrow finger-like peninsulas and islands projecting into Cook Strait (53.3, 53.7).

Best seen from: The partly submerged hills and valleys of the Marlborough Sounds can be appreciated from many of the roads that meander their way among them or from the Cook Strait ferry. Good views include those from Karaka Point and Shakespeare Bay lookouts near Picton, Cullen Point Lookout near Havelock, or the road to French Pass.

54.

Wairau Bar and Waikārapi Lagoons, Blenheim

Narrow conveyor belt that transports gravel across mouth of Wairau Valley

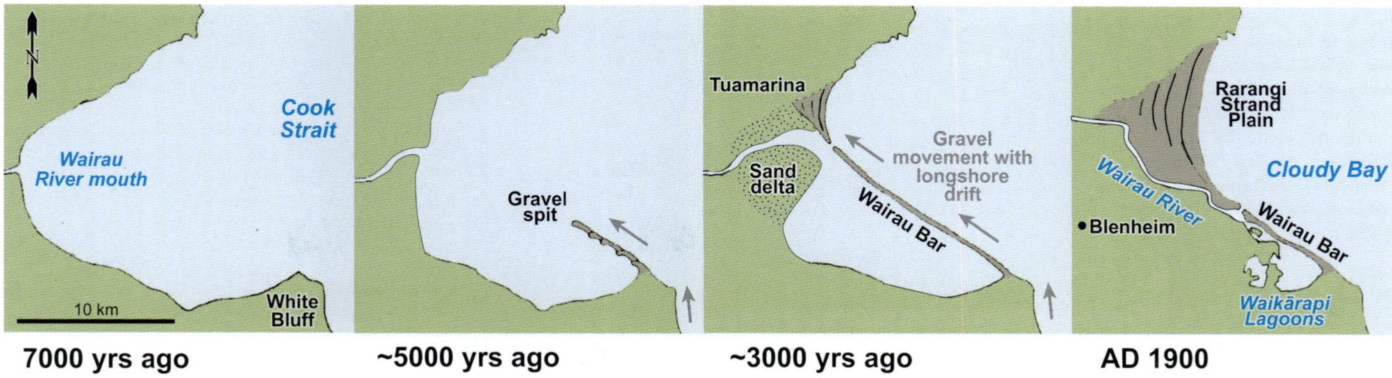

⊖ 54.1 View northwards over the Waikārapi Lagoons, at the mouth of the Wairau River Valley. This unusual bay head delta with its meandering stream course bordered on each side by slightly raised silty sand and salt marsh is the only feature of its kind in New Zealand. Several of the less-meandering cross-cutting channels were dug by pre-European Māori in the 18th century as part of a network of ponds and channels for husbandry and harvesting of fish (especially eels) and moulting ducks. PHOTO: ALASTAIR JAMIESON.

⊕ 54.2 Maps showing inferred evolution of coastal landforms in Cloudy Bay at the mouth of the Wairau Valley over the last 7500 years. MODIFIED FROM PICKRILL (1976).

The Wairau Bar and Waikārapi Lagoons are part of a complex coastal landform system at the mouth of the Wairau Valley, near Blenheim. These coastal landforms have evolved and changed in response to sediment being transported in by the large Wairau River and gravel being swept northwards up the coast by longshore drift. Deposition of these sediments in the valley mouth has been influenced by tectonic subsidence along two major fault lines and fluctuations in sea level over the last 10,000 years (Holocene).

The Wairau Valley is eroded along one of New Zealand's largest and more active faults – the Wairau Fault, which is the northern extension of the Alpine Fault (The Big Picture 12, page 248). The southeast side of the valley mouth follows the Vernon Fault which is subparallel to the Wairau Fault. There has been considerable uplift on the south side of the Vernon Fault, forming the Vernon Range, over the last million years or so. The Wairau Valley mouth is currently subsiding at an average rate of 1–3 mm per year.

The Wairau Valley has eroded along the fractured rocks of the Wairau Fault, and during the lower sea level of the Last Glacial the valley floor was 30–40 m below present in the vicinity of the present coastline. As sea level rose after the Last Glacial, it migrated up the valley to form a deep sheltered embayment extending 6–8 km inland, almost as far as Blenheim, by 7500 years ago (54.2). The rising sea also began eroding into the gravel scree and weakly cemented conglomerate cliffs of White Bluff (54.4). These cobbles and pebbles were moved northwards along the coast by longshore drift driven by southerly swells, and began building a narrow gravel spit obliquely northwestwards across the mouth of the Wairau Valley.

By about 4000–5000 years ago the straight, narrow gravel spit (ancient Wairau Bar) had reached the north side of the valley near Tuamarina (54.2). From then on, the spit acted like a conveyor belt with longshore drift moving cobbles and pebbles along the spit in the swash zone during periods of large southerly swells (54.3). The gravel crosses the mouth of the Wairau River in cycles of alternate spit growth across in front of the mouth during dryer summer months followed by

➲ 54.3 The Wairau Bar is a narrow gravel spit that has grown from the south (foreground) across the mouth of the southern part of the Wairau Valley over the last 5000 years. The bar encloses the semi-tidal Waikārapi Lagoons (left) and supplies gravel via longshore drift to the Rarangi Strand Plain in the north (54.5). PHOTO: ALASTAIR JAMIESON.

breaching in the next flood with some of the spit gravels then being added to the northern bank (54.5). From south to north the cobbles and pebbles become smaller, being somewhat reduced in size by abrasion as they travelled up and down the beach on their journey north.

On the north side of the river mouth, the gravel mixed with sand coming down the river and progressively built out a large area of flat coastal land seaward of SH 1 and extending from Rarangi in the north to the Wairau River mouth. This is the Rarangi Strand Plain, which consists of a succession of beach ridges that were thrown up during storms and are made of pea-sized gravel along with sand. As the strand plain grew seaward and south, it progressively pushed the natural mouth of the Wairau River to where it now lies near the middle of Cloudy Bay (54.6).

While the strand plain grew north of the river mouth, shelly mud and fine sand accumulated in the sheltered lagoon inside the gravel spit to the south. By about 4000–5000 years ago the lagoon had filled up with sediment to become extensive tidal flats. Sand and silt carried down the Wairau River spilled southwards into the ancient Waikārapi Lagoons, assisted by inflowing tidal currents. Over time an extensive low-lying flood-tide delta formed, with numerous moving channels and fine gravel and sand islands (54.5).

Some of the earliest Polynesians to colonise Aotearoa set up camp on one of these delta islands just over 700 years ago (AD1285–1300). The Wairau Bar archaeological site on the southern side of the natural mouth of the Wairau River is now a nationally important historical site because it contains the oldest, well-dated archaeological evidence of early Polynesian settlement. This extensive site contains numerous artifacts and the remains of the early settlers' diverse sea- and land-derived diet, including the butchered remains of a number of extinct larger birds, such as moa, Haast's eagle and New Zealand swan and raven.

Sand and silt carried by flood waters down the small streams on the southern and western sides of the Waikārapi Lagoons built unusual bay head deltas out into the shallow lagoon. These had a meandering stream channel down the middle and slightly raised, sandy salt marsh built up on either side (54.1). By the time the earliest European maps of the lagoon were produced (1840s) many of these deltas had coalesced and also joined with the flood-tide delta from the north, reducing the size of the lagoons. Since then there has been ongoing subsidence, some accompanying mid-19th-century earthquake shaking and liquefaction, and the lagoons have grown larger. About 30 cm of sea-level rise in the last century or so is also responsible for increased erosion of the delta islands and their salt marshes so that the lagoons are now considerably larger than they were 170 years ago.

Best seen from: The best views are from the air, to or from Blenheim Airport.

Take a closer look: Walk the Wairau Lagoons Walkway, accessible from the end of Hardings Rd. Visit the natural mouth of the Wairau River and Lagoons at the end of Wairau Bar Rd.

➋ 54.4 The eroding conglomerate of White Bluff supplies the greywacke cobbles and pebbles that are transported along the Wairau Bar during large southerly swells. PHOTO: ALASTAIR JAMIESON.

➋ 54.5 The Wairau Bar gravel spit is a conveyor belt that moves cobbles and pebbles northwards (towards the camera) along it in the swash zone during periods of large southerly swells. The gravel crosses the mouth of the Wairau River in cycles of alternate spit growth across the mouth (as starting to develop here) during dryer summer months, followed by breaching of the spit in the next flood with some of the spit gravels being added to the northern bank and Rarangi Strand Plain (54.6). PHOTO: ALASTAIR JAMIESON.

➋ 54.6 Rarangi Strand Plain is a 35 km^2 area composed of a succession of fine gravel and coarse sand beach ridges that has built out seawards on the north side of Cloudy Bay over the last 5000 or so years. As the plain grew, it progressively pushed the Wairau River mouth to the south. In 1963 the Wairau Diversion of the river course (barely visible in the distance) was excavated through the strand plain to allow major flood waters to drain more directly and quickly out to sea. PHOTO: LLOYD HOMER, GNS SCIENCE, 1899.

55.

Sawcut Gorge, Ward

New Zealand's best example of a slot canyon

55.1 Visitors pass upstream through Sawcut Gorge – New Zealand's most mind-blowing slot canyon. It has eroded and dissolved its way down through hard limestone. In the middle of the day in summer, sunlight filters down into Sawcut Gorge, bouncing shades of yellow light off its curving sides while illuminating patches of the shingle floor.

Sawcut Gorge is one of the most remarkable small 'slot canyons' in the world. It occurs on Isolation Creek, a tributary of the Waimā River in southeast Marlborough. Prior to the 2016 Kaikōura Earthquake (The Big Picture 13, page 261) there was public access to it through private land, but unfortunately the valley-bottom track was destroyed by rock landslides brought down by the shaking.

Sawcut Gorge is the narrowest part of a much longer section of steep-sided gorge, mostly eroded into limestone. Walking up Isolation Creek you are suddenly confronted by a 50-m-high wall of rock that looks impenetrable, but getting closer you can see a narrow slot like a sawcut passing down right through the wall with the stream flowing out at the bottom (55.2). This cut is 2–5 m wide, 50 m long and at least 50 m high. It is possible, except in flood, to walk through it by wading ankle- to knee-deep in the cold stream water (55.1). Sawcut Gorge has cut down through a hard, fine-grained limestone, described in more detail under Site 56.

At first glance, it is puzzling to understand how the slot was formed, but there are clues that help. A tantalising explanation might be that the canyon was formed as a cave and the roof has collapsed in on it, as is the case for many limestone gorges. However, in this instance the width of the cut is similar all the way up and all the walls are water-smoothed with no indication of more rugged surfaces that would be left by rocky roof collapse. It seems clear that the cut has been dissolved and eroded by Isolation Creek starting at the top and progressively cutting down just like a sawcut, although a little more irregularly (55.3). Another clue to Sawcut Gorge's formation is the presence of a vertical joint through the limestone that is parallel to the gorge and only 2 m away (55.2). Thus the slot has dissolved and eroded its way down along a second vertical joint, widening the slot to fit the stream flow as it went. There are probably faults upstream and downstream of the sawcut, along which softer rocks had been displaced to be level with the gorge at either end. Water erosion has removed much of the softer rocks and widened out the gorge above and below its entrances.

Take a closer look: The gorge track has been closed since the 2016 Kaikōura Earthquake which brought down huge landslides.

⊙ 55.2 The downstream side of Sawcut Gorge showing the narrow slot that the stream has dissolved and eroded along a joint plane through the hard limestone. A parallel joint can be seen 2 m to the right of the sawcut. Person in entrance to the gorge for scale.

⊙ 55.3 Looking skyward from the stream up through Sawcut Gorge slot canyon. At times barely 2 m wide, cut through cemented limestone, this cleft soars up for some 50 m to the narrow strip of sky and greenery above. PHOTO: PETER CROSSLEY.

56.

Mead Stream, Chalk Range, Clarence Valley

Southern Hemisphere record of asteroid impact that killed off the dinosaurs

The gorge of Mead Stream, a tributary of the Clarence River, contains the most complete sedimentary record in New Zealand of a significant period of geological time, between 80 and 40 Myr ago. The sedimentary strata are exposed almost continuously in the walls of the gorge that cuts through the Chalk Range, between the Seaward and Landward Kaikōura Ranges (56.2, 56.5). One reason why this sequence is so long and complete is because the sediments were deposited very slowly on the floor of the deep ocean well away from any land that would have contributed large volumes of eroded sand and mud to the accumulating pile. Thus the thickness of strata for such a long period is quite small (~500 m thick). A second reason why this long sequence of time can be seen in a short gorge is that the originally horizontal layers have been tilted up to ~45° and thus the stream cuts right through the section at right-angles to the lie of the strata (strike) (56.1).

The Chalk Range is a remarkable example of a strike ridge with nearly equally steep (45°) slopes on either side (56.2, 56.3). It runs northeast–southwest for 40 km, parallel to the Clarence River and Kaikōura Ranges. The range is made of erosion-resistant cherty limestone that has been uplifted and tilted to the west. Slightly softer layers above and below this section have been eroded down, leaving the elongate symmetrical ridge upstanding.

The cherty limestone (56.5) consists of the skeletons of microscopic plankton that fell to the seafloor after they died. They mostly consist of the skeletons of four groups of microscopic plankton. Two are algal phytoplankton – diatoms with siliceous skeletons and calcareous nannoplankton; and two are single-celled protists – radiolaria with siliceous skeletons and calcareous planktic foraminifera. Paleontologists have undertaken detailed study of these microfossils. Their

56.1 The Chalk Range is an amazing white ribbon (strike ridge) of erosion-resistant limestone that stretches for 40 km between and parallel to the Seaward and Inland Kaikōura Ranges. Numerous tributary streams of the Clarence River have eroded their way through the hard limestone strata. One of these, Mead Stream, is in the foreground. PHOTO: ALASTAIR JAMIESON.

⊙ 56.2 Mead Stream has cut a deep gorge through the symmetrical slopes of the Chalk Range in inland Marlborough. The sequence of limestone strata exposed at stream level in the bottom of the gorge is one of the longest and most complete in New Zealand and spans the interval between 80 and 40 Myr ago (Cretaceous to Eocene). PHOTO: CHRIS HOLLIS.

evolutionary sequence through the strata is used to date them, and the chemistry of their skeletons records the history of the ocean's climate and the impact that major climate changes had on the plankton. Similar sequences of strata have been studied in drill cores from the deep ocean floor, but Mead Stream is one of the best places on land anywhere around the world for these studies.

One of the most dramatic events in the history of life on Earth was the extinction of the dinosaurs as a result of the impact of a huge asteroid 10–15 km in diameter that hit the Gulf of Mexico 66 Myr ago. The record of this event is preserved in the strata at Mead Stream (56.4) and a number of nearby places in Marlborough. Not only the dinosaurs were wiped out but also many other land and shallow marine groups, such as ammonite and belemnite squid. The Mead Stream section records the devastation this event caused to the microscopic plankton with 90% of the species becoming extinct and just a few surviving through. The asteroid's impact is believed to have generated a global dust cloud that shut out sunlight, causing freezing conditions and shutting down photosynthesis for a decade or more. Some of the distinctive (iridium-bearing) dust generated from this impact is recorded in the Mead Stream strata.

The chemistry of the microfossils in Mead Stream also records several times between 56 and 40 Myr ago (Eocene) when the Earth's temperature was the hottest it has been in the last 100 Myr. The hottest of these warm periods (at the end of the Paleocene, 56 Myr ago) came on very quickly and lasted only a few hundred thousand years, and is associated with an increase in greenhouse gases.

Take a closer look: This feature is in private land but can be visited on tourist ecotours.

56.3 View northeast along the Chalk Range from above Swale Stream. For most of its length the range is a single strike ridge of hard limestone that has been uplifted and tilted steeply to the west (left). The Blue Mountain Range is in the distance. PHOTO: ALASTAIR JAMIESON.

56.4 In the middle of this photo (next to the orange cobble) there is a thin, excavated, clay layer that records the second-largest extinction event to have ever occurred on Earth. The extinction of the dinosaurs, large marine reptiles and many other groups of plants and animals brought to a close the Age of Dinosaurs, 66 Myr ago. It resulted from a large asteroid strike in the Gulf of Mexico. The deep-water sedimentary strata here at Mead Stream do not contain fossils of any of these larger organisms but they do record the vast die-off that occurred in four major groups of ocean plankton. PHOTO: KYLE BLAND, GNS SCIENCE, 237198.

56.5 A 20-m-thick sequence of thin-bedded limestone – part of the 500-m-thick section exposed in the Mead Stream gorge that was deposited on the floor of the ancient Pacific Ocean. These layers are mostly composed of the fossil skeletons of microscopic plankton that record the history of the climate in this part of the world between 80 and 40 Myr ago. PHOTO: JAMES CRAMPTON, GNS SCIENCE, 237154.

57.

Lake Rotoiti

Accessible Alpine (Wairau) Fault trace

One of the easiest places to see the Alpine (Wairau) Fault trace is at the north end of Lake Rotoiti in Nelson Lakes National Park (57.1). Here the northeast–southwest-oriented fault is parallel to the direction of oblique compression between the Pacific and Australian tectonic plates, and as a result almost all the displacement across it is sideways, with the south side moving westward with respect to the north side. A small amount of vertical displacement across the fault over the last 7–5 Myr has uplifted the greywacke rocks of the Spenser Mountains and St Arnaud Range 1000 m higher than the hills on the north side of the fault (57.3).

The fault trace at Nelson Lakes is a straight-line continuation of the Alpine Fault that runs along most of the length of the Wairau Valley, reaching the coast of Cook Strait at Blenheim. The Wairau Valley section of the fault and its continuation southwest through Nelson Lakes is also known as the Wairau Fault. Some geoscientists prefer to say that the Wairau Fault used to be the northern sector of the Alpine Fault up until about 2 Myr ago, and that this northern section has now migrated further south to the parallel Hope Fault. This is because in recent times the Hope Fault has become more active than the Wairau Fault, and the Hope is now considered to be the location of the plate boundary. If the Alpine Fault is synonymous with the transcurrent (sideways-moving) portion of the plate boundary, then today the Hope Fault should be considered the northern end of the Alpine Fault. If, however, the Alpine Fault is defined as the major rupture along which there has been 480 km of sideways displacement of rocks over the last 25 Myr (The Big Picture 12, page 248), then the Wairau Fault is the Alpine Fault's older northern section.

Athough the Hope Fault strand of the Alpine Fault is now more active, the Wairau Fault strand is not inactive. River terraces formed in the last 11,000 years (Holocene) in the Wairau Valley have been displaced by several ruptures on the Wairau Fault. Around St Arnaud the trace of the Alpine (Wairau) Fault can be followed as a distinct uplifted scarp from south of Tophouse Saddle down to the shore of Kerr Bay on Lake Rotoiti,

57.1 View south over Lake Rotoiti in the Nelson Lakes National Park. Cutting diagonally across the photo is the trace of part of one of New Zealand's most defining geological features – the Alpine Fault (Wairau Fault strand). The trace (dashed) runs through the Kerr Bay campground (left), then passes under the lake and across Brunner Peninsula, under the lake again, then up to the Mt Robert carpark and into the distance along Speargrass Valley (right). PHOTO: ALASTAIR JAMIESON.

where it passes right through the middle of the Department of Conservation's Kerr Bay campground (57.4). The fault line passes under the north end of the lake and obliquely through Brunner Peninsula (57.5). From Lake Rotoiti the fault rises up and through the saddle north of Mt Robert, along the straight portion of the Speargrass Valley (57.1, 57.3) and then diagonally across the southern end of Lake Rotoroa. In many places, like the Speargrass and Wairau Valleys, the creeks and rivers have eroded out along the sheared and crushed rocks in the wide displacement zone of the Alpine Fault.

Best seen from: The best place to view the trace of the Alpine Fault is from the lookout 100 m north of the carpark at the end of Mt Robert Rd.

Take a closer look: Lake Rotoiti's Brunner Peninsula Nature Walk passes over the fault scarp as a 3-m-high step on the west side of Brunner Peninsula where the shoreline has been displaced with the south side moved westward (1-hr return on loop tracks).

57.2 View southwest across Lake Rotoiti's West Bay to Mt Robert (1421 m). The Alpine Fault passes through the saddle to the north (right) of Mt Robert and along the Speargrass Valley to the north of the uplifted Robert Ridge on the skyline.

57.3 View southeast across the Alpine Fault with the uplifted Robert Ridge on the far (south) side and the lower flat surface (Waipounamu Erosion Surface, The Big Picture 19, page 315) with tarns on the north. The Alpine Fault runs down the middle of Speargrass Valley at the foot of the 1400–1700-m high ridge. PHOTO: ALASTAIR JAMIESON.

57.4 View northeast across the northern end of Lake Rotoiti showing the arrowed trace of part of the Alpine Fault (Wairau Fault strand) from Kerr Bay (foreground) to Tophouse Saddle (distance). PHOTO: ALASTAIR JAMIESON.

57.5 The trace of the Alpine Fault can be seen crossing Brunner Peninsula at the northern end of Lake Rotoiti. Note that the south side (foreground) has been uplifted slightly and moved sideways to the southwest (left) by displacement along the fault since the landform was made during the Last Ice Age (about 15,000 years ago). PHOTO: ALASTAIR JAMIESON.

58.

Te Taero a Kereopa – Te Tahuna a Tama-i-ea / Nelson Boulder Bank

The world's longest boulder bank

Nelson Boulder Bank is New Zealand's best-known linear boulder spit and an icon of Nelson city. At 13 km, it is the longest boulder bank of its kind in the world. It is composed of rounded boulders, cobbles and pebbles all derived by erosion from Mackay Bluff and Cable Bay, just up the coast to the north (58.1, 58.2). They are all made of granodiorite (58.8), a plutonic rock that was a magma chamber probably beneath an active volcano about 140 Myr ago. The molten magma slowly cooled and crystallised into the solid granodiorite rock composed of interlocking crystals of mostly feldspar, hornblende and quartz.

The present boulder bank is inferred to be just the most recent of many similar banks that built up along this side of Tasman Bay during warm interglacial periods of high sea level during the Ice Ages. The modern version grew on top of the weathered and down-faulted remains of the youngest of the earlier banks that would have grown across this bay around 125,000 years ago (Last Interglacial). Today's Nelson Boulder Bank started forming about 9000 years ago when rising sea level, after the end of the Last Ice Age, started eroding into huge scree deposits that would have accumulated along the foot of Mackay Bluff over the preceding 100,000 years, when sea level was a lot lower.

Over the next few thousand years this eroding gravel was progressively transported south to form an elongate bank slightly more seaward than its present location. The growth of the bank was due to southwards-moving longshore drift. Most storm waves approach this coast at an oblique angle from the northwest and carry cobbles and pebbles slightly

➲ 58.1 Nelson Boulder Bank snakes its way along the east coast of Tasman Bay, creating shelter for the tidal Nelson Haven and Port of Nelson. Haulashore Island (bottom left) is the natural tip of the boulder bank, but its connection was snipped when 'The Cut' was put in to provide better access for the port in the early 1900s. Mackay Bluff, the source of Nelson Boulder Bank's gravel, is in the distance. PHOTO: ALASTAIR JAMIESON.

➲ 58.2 Map of Nelson Boulder Bank, which has grown across this embayment in less than 9000 years and created the sheltered tidal inlet of Nelson Haven in behind. MODIFIED FROM BALLANCE (2017).

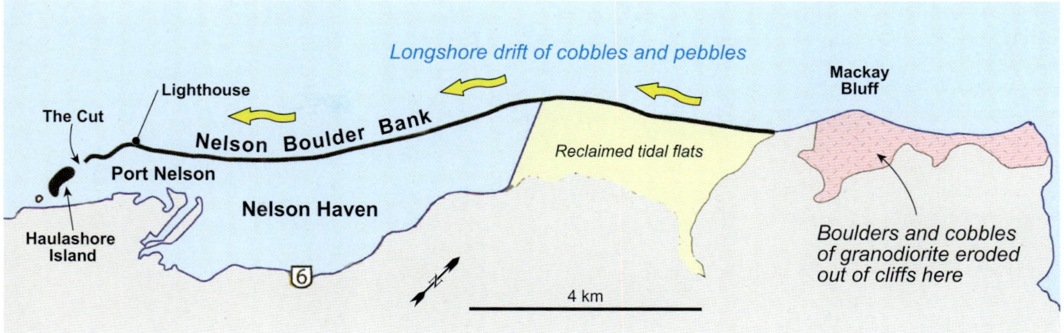

58.3 View from the old lighthouse down on a section of Nelson Boulder Bank, clearly showing the seaward longitudinal storm-beach ridges (left side) and the sequence of oblique ridge 'hooks' – each of which was produced by storms sweeping gravel in towards Nelson Haven (right) around the advancing southern tip of the high storm ridge at the time. PHOTO: MIKE JOHNSTON.

58.4 View from the northeast along the younger half of Nelson Boulder Bank towards Nelson City. Landward of the boulder bank (left), thick shelly mud has accumulated on the sheltered floor of Nelson Haven since the first phase of the present bank was built between 9000 and 6000 years ago. PHOTO: ALASTAIR JAMIESON.

58.5 Profile of part of the cobble and pebble storm-beach ridge that runs along the length of Nelson Boulder Bank near the end of Boulder Bank Drive. This location is at the northeast head of Nelson Haven, near the end of Boulder Bank Drive.

58.6 View west over the entrances to Nelson Haven with Haulashore Island and the southern tip of Nelson Boulder Bank (right). The Cut (right) provides the deep-water entrance for shipping into the Port of Nelson. Prior to its dredging in 1903–06, the south end of the boulder barrier was the left end of Haulashore Island. The prominent hook on the north side of The Cut (right) is due to longshore drift and storms carrying material, including that dredged from The Cut, through the artificial entrance and into Nelson Haven.

58.7 The historic lighthouse on Nelson Boulder Bank, commissioned in 1862, was the second permanent lighthouse in New Zealand. At that time this was the southern end of the storm ridge on the crest of the boulder bank. The lighthouse was decommissioned in 1982. It can be visited by walking down the boulder spit or by taking The Ferry from Nelson port area.

southwards as they move them up the beach. This first phase of construction was fuelled by the large initial supply of gravel from the eroding scree that rapidly built a spit as far south as Haulashore Island by about 6000 years ago. Sea level continued to rise to about 1.5 m above present level and the supply of gravel from the north dwindled. While sea level was at this height (~6000–3000 years ago), most of the boulder bank was probably intertidal, with a higher storm-beach portion closer to the Bluff at the northern end, which today is ~2.5 m higher than further south.

Sea level started to fall 3000–2000 years ago and by about 1000 years ago had fallen to approximately the present level. With this fall in sea level, wave base was lowered and some of the gravel in the intertidal and shallow subtidal gravel bank, especially in the north, became remobilised. These reactivated cobbles and pebbles were thrown up above high-water mark in storms, building the narrow, crestal storm ridge (58.5). This grew southwards and slightly landwards on top of the earlier intertidal bank. The cobbles and pebbles in this storm ridge become smaller and more rounded the further they have been transported from their original source at Mackay Bluff.

As this second phase of the bank extended southwards the southern tip was periodically bent in towards the land by storm waves creating a hooked high-tide ridge. Over time, as the spit tip moved further southwards, a whole series of these hooked oblique ridges built up on the landward side of Nelson Boulder Bank (58.3). In the last 1000 years or so, further longitudinal storm-beach ridges have been added on the seaward side of the bank. The crestal ridge now snakes its way along the full length of the boulder bank, and until 'The Cut' was put in was advancing southward at rates of 7–30 m per year.

Although it is named Nelson Boulder Bank, true boulders (with a diameter of 25.6 cm or more) are less common than cobbles and pebbles. The largest boulders present are no more than 1 m in diameter, with most less than 40 cm, and they become smaller and less abundant the further south you are away from Mackay Bluff. A few local residents have monitored some of the boulders on the bank over a number of years and recorded their actual episodic movement southwards, sometimes in leaps of up to 70 m in just one large storm.

Pre-European Māori used the boulder bank as a source for food, such as birds' eggs, seabirds and seafood. The shores of Nelson Haven were chosen for the colonial settlement of Nelson because of the sheltered port provided by the boulder bank. For the first 60 or so years after establishment, the entrance to the Port of Nelson was via 'the Narrows', south of the end of the bank at Haulashore Island. As moving sand shallowed this hazardous entrance, the decision was made to make The Cut (1903–06) through the boulder bank to provide better access for the harbour (58.1, 58.6). Today Nelson Boulder Bank is a cherished and protected part of Nelson and New Zealand's natural heritage.

Best seen from: For panoramic views over Nelson Boulder Bank and Nelson Haven, drive up one of the roads into the hilly suburbs of Marybank, Atawhai, Brooklands or Stepneyville.

Take a closer look: Visit Glenduan, at the end of Glen Rd; walk/jog the length of the spit from the end of Boulder Bank Dr (15 km return, easier and faster at low tide); take The Ferry from Wakefield Quay to Haulashore Island or the lighthouse (15 mins one-way).

58.8 Fine and coarsely crystalline granodiorite cobbles and pebbles on the southern end of Nelson Boulder Bank. The entire bank is built up from rocks eroded from Mackay Bluff and transported southwest by longshore drift during storms.

59.

Tokangawhā / Split Apple Rock, Kaiteriteri

Iconic granite core stone split in half

59.1 Tokangawhā / Split Apple Rock is a 5-m-diameter core stone of fresh granite that has been eroded out of the surrounding more weathered granite and later has naturally split into two halves along a joint plane.

Tokangawhā / Split Apple Rock is an eye-catching sphere of granite on the west coast of Tasman Bay, north of Kaiteriteri, that has broken into two halves and resembles a giant split apple (59.1). It is the best example in New Zealand of a large core stone that has split in half along a joint plane. The rock is made of granite that forms all the coast of Abel Tasman National Park and south as far as Motueka. This granite body (pluton) was formed by the slow cooling and crystallisation at depth (5–10 km below the surface) of a large molten magma chamber (The Big Picture 11, page 207). This magma chamber probably lay beneath a large rhyolite volcano, gradually collecting, melting and stewing magma, some of which periodically rose to the surface and erupted, around 110 Myr ago (Cretaceous).

Over time, underground stresses produced cracks in the solid granite. Many of the cracks occur in several sets of parallel fractures, known as joints, which cut the rock into a network of large subrectangular blocks. In the millions of years since the granite solidified, this region has been uplifted in several phases and the overlying rocks progressively eroded off. As the granite was approaching the surface, oxygen-rich groundwater began circulating through the joints starting a process of weathering that progressively ate its way into the rock. As erosion wore away the weathered rock and soil on top, it also removed the weathered zones along the joints, leaving behind the fresh unweathered, rounded core stones in the middle of the blocks. The 5-m-diameter Split Apple Rock and the rocks around its base are the remains of fresh core stones from inside the granite.

Split Apple Rock must have had a tight planar joint running through it that the circulating groundwaters did not penetrate. Once this huge rock was exposed to the sun, air and sea it must have been subjected to cycles of heating and cooling, wetting and drying that resulted in slight expansion and contraction of the rock. With these slight movements within the rock, it has eventually fractured along the existing tight joint plane and split open.

Best seen from: Take the track down to the beach from the end of Moonraker Way off Tokongawa Dr between Kaiteriteri and Mārahau, or hire a kayak or take a ferry ride from Kaiteriteri.

59.2 View southwest to Split Apple Rock. PHOTO: ALASTAIR JAMIESON.

59.3 Cottage Loaf Rock is an equally large core stone of granite to Split Apple Rock. This huge intact boulder lies 18 km further north off the coast of Abel Tasman National Park. PHOTO: ALASTAIR JAMIESON.

THE BIG PICTURE 11

The Median Batholith

A BATHOLITH IS a giant body of intrusive plutonic rocks (hundreds of kilometres in size) composed of many individual smaller intrusive bodies called plutons (up to 10 or so km across). The Median Batholith is a giant linear belt of plutonic rocks that stretches from Stewart Island through Fiordland to north Westland and northwest Nelson, separated in the middle by displacement on the Alpine Fault (TBP 11.1). The belt was formed within the continental crust of the supercontinent of Gondwana parallel to the long-lived convergent plate boundary along its southeastern margin. On land this belt now underlies 10,000 km^2.

Most of the Median Batholith is made of granitic rocks (TBP 11.2) that were intruded as a number of separate plutons between 360 and 105 Myr ago (Carboniferous, Jurassic and Cretaceous) into the ancient Paleozoic rocks that now outcrop in western New Zealand. Molten magma was formed partly by partial melting of the mantle above the plate boundary subduction zone, and partly by melting of the base of the existing continental crust owing to crustal thickening and heating as several basement terranes were accreted on to the margin of Gondwana. The buoyant molten magma rose up through the overlying crust towards the surface, stalling at 5–25 km depth, where these large plutons cooled and solidified into coarsely crystalline granitic plutons. As the magma cooled slowly there was time for the crystals of quartz and feldspar and other minor minerals to grow to 1 cm or so across as they knitted together to form the rock.

It is likely that many large rhyolite volcanoes erupted above the Median Batholith, fed from the shallow intruded magma chambers, but it is hard to test this hypothesis as most volcanic rocks that may have been erupted have long since eroded away. Since intrusion of the plutons ceased, many parts of the Median Batholith that now form land have been slowly uplifted and 5–10 km thickness or more of overlying rocks has been stripped off them by erosion.

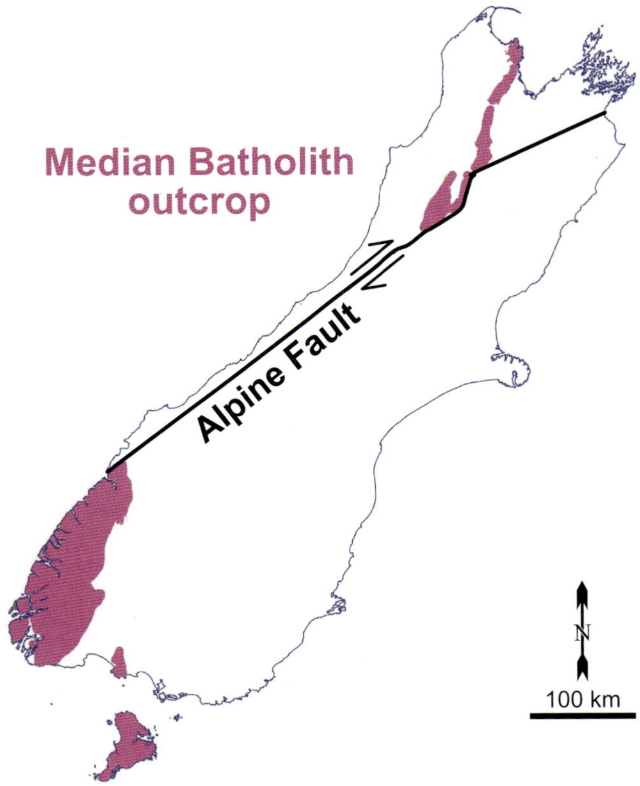

TBP 11.1 Map of the extent of the Median Batholith in New Zealand (it does not occur in the North Island).

TBP 11.2 The Median Batholith is mostly composed of solidified granite plutons. This granite mostly consists of large interlocking crystals of pink potassium feldspar, white plagioclase feldspar, colourless quartz and black biotite. Photo 25 cm across.

60.

Abel Tasman coast

Golden sand beaches from eroding granite

The coast of Abel Tasman National Park in northwest Nelson is famous for its magnificent golden sand beaches backed by native forest (60.3). The sands owe their colour to the iron present in the local granite rock from which they have been eroded. Iron in the near-surface granite oxidises to orange-yellow ferric oxide, which stains many of the feldspar and quartz crystals which form the Golden Bay beach sands. Most of the National Park and almost all of its coastline is composed of biotite granite of the Separation Point plutons which form part of the huge Median Batholith (The Big Picture 11, page 207). These plutons cooled and solidified between 125 and 105 Myr ago (early Cretaceous). Since then they have been uplifted many kilometres and the overlying rocks stripped off by erosion.

The Arthur and Pikikiruna Ranges that form the mountains west of Tasman Bay and the Waimea Plains have been uplifted from sea level to their present height in just the last 2 Myr. During all this time rain and seawater have been working away to erode down the uplifted granite. Numerous streams were eroded below present sea level during each cold Ice Age period. When the sea rose to its present level (7500 years ago) after the end of the Last Ice Age, it flooded up some of these stream valleys, creating estuaries and small inlets such as those at Wainui Bay, Tōtaranui, Awaroa Bay, Bark Bay, Torrent Bay, Mārahau Bay and Otuwhero Inlet (60.4).

In the last 7500 years sand has been thrown up as small sand spits across the entrances to many of these bays. The spits are usually anchored to the upstream side of the bay entrance and grow across the bay with longshore-drift movement of the sand. Thus local longshore drift appears to have been from north to south and then west around the northern half of the Abel Tasman National Park coast, and from north to south, south of Torrent Bay.

➲ 60.1 The magical golden sands around the coast of Abel Tasman National Park are derived from erosion of the local granite and lend their name to Golden Bay. This view is east across the intertidal sands at the entrance to Wainui Bay. Sand spits have built part-way across the entrance from each side. PHOTO: COLIN MONTEATH, ALAMY.

The granite is mostly composed of large crystals of creamy and orange feldspar and translucent-white quartz, and these are the main components of the sand on the beaches and in the sand spits. Over time the feldspar sand grains will wear away or weather to clays, whereas the quartz grains are more inert and remain around a lot longer. In some places around the park the beaches are more orange, such as on the northern beaches and at Kaiteriteri (60.1, 60.2). This may be because of more rapid marine erosion releasing the oxidised iron-stained feldspar grains to the beaches. In other places, such as at Torrent, Bark and Awaroa Bays, the sand can be much whiter (60.4). This is possibly because the iron oxide and feldspar crystals have weathered away, leaving the sand rich in inert white quartz. In this area there is little active coastal erosion and the white sand may be older and recycled from earlier times.

Best seen from: The most accessible golden sand beaches are Kaiteriteri, Mārahau, Wainui Bay and Tōtaranui.

Take a closer look: Walk part or all of the Abel Tasman Great Walk (1–3 days) and visit a number of idyllic sandy beaches surrounded by native forest.

⊙ 60.2 The golden sands at the north end of Tōtaranui Beach with low granite rock cliffs behind. The colour of the sand is often more intense when wet.

⊙ 60.3 The sweeping orange arc of the beach at The Anchorage with Torrent Bay beyond is the most popular place for day-walkers to be dropped by boat charter for the return walk along Abel Tasman Track to Mārahau. PHOTO: ALASTAIR JAMIESON.

↑ 60.4 The gold and white beaches, sand spits and sand bars at the entrance to Awaroa Inlet, Abel Tasman National Park, contrast with the natural turquoise waters on the outgoing tide. Some of the beaches are much whiter than others because of the greater abundance of quartz at the expense of orange, iron-stained feldspar grains. PHOTO: ALASTAIR JAMIESON.

↓ 60.5 The sand on Abel Tasman beaches is all derived by erosion from the 125–105-Myr-old granite rock that forms the coastal cliffs and inland hills. Geologists call the rock Separation Point granite, named from its type locality here at Separation Point. Secluded Anatakapau and Anapai Bays on the northern loop of Abel Tasman Track lie beyond. PHOTO: ALASTAIR JAMIESON.

61.

Tākaka Hill

Best fluted marble karst in New Zealand

⊙ 61.1 The largest area of fluted marble in New Zealand occurs along the crest of Tākaka Hill in northwest Nelson. It formed over tens of thousands of years by solution of the rock by slightly acidified rain runoff dribbling down the sides of marble blocks beneath a forest canopy.

⊙ 61.2 Much of the top of Tākaka Hill, near Ngārua Cave, is covered in thousands of blocks of intensely fluted marble like this. PHOTO: ALASTAIR JAMIESON.

⊙ 61.3 The near-horizontal top of this marble outcrop near Harwoods Hole, Tākaka Hill, has been dissolved into a network of fine runnels that resemble a miniature mountain landscape. The ridges between the solution grooves are razor-sharp and will cut bare feet. Photo 60 cm across.

⊙ 61.4 Aerial view down to Harwoods Hole – New Zealand's deepest vertical shaft at 176 m deep. It is a doline that has collapsed into a large cave system dissolved in marble. PHOTO: ALASTAIR JAMIESON.

⊙ 61.5 The finely fluted marble around Harwoods Hole, Tākaka Hill, was formed through solution of the surface of the ancient marble rock by weakly acidic runoff. This runoff acquired its acidity from the humus that collected around the base of clumps of vegetation growing on top of the marble. The production of these runnels took thousands of years. PHOTO: ALASTAIR JAMIESON.

The best example of surface fluting (karst) developed on marble in New Zealand is on Tākaka Hill between Tasman and Golden Bays in northwest Nelson (61.1, 61.5). Almost all true marble in New Zealand occurs in a 90-km-long belt forming parts of the eastern ranges of the region. There are three main areas of marble karst: from north to south, Tākaka Hill and the west Pikikiruna Range, Mt Arthur and Mt Owen. The latter two areas were extensively eroded by ice and glaciers during the Ice Ages, and surface solution karst features (67.6) have not had time to become well developed there since the end of the Last Ice Age, about 15,000 years ago.

Most of the marble in the Mt Owen and Mt Arthur areas is above the bush line, which currently is about 1200 m elevation. The marble on Tākaka Hill lies at 400–1000 m above sea level, well below the present bush line, and its surface karst (The Big Picture 4, page 83) presumably dissolved under a forest canopy. Slightly acidified rainwater percolated down through cracks in the marble dissolving out grikes and underground cave networks. Water dribbling down the exposed sloping sides of marble blocks beneath the forest dissolved out vertical fluting or developed more intricate patterns like fish skeletons or miniature landscapes on the more horizontal surfaces (61.2, 61.3).

The marble in northwest Nelson is metamorphosed limestone that accumulated on the seafloor as shelly sediment during the Ordovician, 480–450 Myr ago. During metamorphism the limestone was deeply buried, with the resulting high temperatures and pressures causing considerable recrystallisation. This resulted in a dense intergrowth of calcite crystals that increased the rock's cohesiveness and toughness and reduced its porosity to almost zero. This marble comes in various shades of grey and white, and occasionally nearly black or pink-tinged. Marble of pure calcite is pristine white, whereas the grey colour is derived from microscopic flakes of graphite formed by metamorphism of the organic remains of organisms buried in the original limestone. Rusty-brown staining in some Nelson Marble (e.g. in Parliament's Beehive building in Wellington) is caused by iron oxide impurities.

Following metamorphism, the rocks have been uplifted and eroded down. The most recent phase of uplift was within the last 1–2 million years when Arthur and Pikikiruna Ranges, including Tākaka Hill, were pushed up from being a coastal plain.

Best seen from: Visible on north side of the main road as it crosses the top of Tākaka Hill near Ngārua Cave and the Canaan Rd turnoff.

Take a closer look: Walk among the karst to Hawkes Lookout on the top of Tākaka Hill (20-min. return), or visit Harwoods Hole karst (2 hrs return from end of Canaan Rd).

62.

Rāwhiti Cave, Tākaka

Rare tufa stalactites

The large, cavernous entrance to Rāwhiti Cave on the steep slopes of the lower Tākaka Valley in northwest Nelson is adorned with the best and most numerous examples of tufa stalactites in New Zealand (62.1, 62.2). The cave itself was dissolved out of Arthur Marble as much as 1 Myr ago when an underground stream probably flowed through it. Since then the Dry River has eroded across its front, leaving the cave high and dry 300 m above the valley floor. Debris from cliff and roof collapses has formed a 15-m-high barricade across the front of the 40 m by 20 m cavern opening (62.4).

A wooden stairway descends to a viewing platform in the middle of the cavernous opening, where you can appreciate the festoons of thousands of narrow, vertical, calc-tufa stalactites that cover the back wall (62.3). Calc-tufa (not to be confused with tuff, made of volcanic ash) is porous and friable because it is composed of calcium carbonate that has been deposited over and through moss and algae that grows on the surface of these stalactites. Inside caves the stalactites are dense and crystalline because no vegetation can grow in the dark. Here in the cavern entrance the moss and algae can photosynthesise and grow in the percolating light. The water has flowed down through fractures in the marble, dissolving and carrying along the calcium carbonate in solution which is then deposited on the vegetation growing on the stalactites as the water evaporates (62.2).

Stalactites in caves are usually vertical, reflecting the influence of gravity, but many of the large tufa stalactites that hang down from the roof of Rāwhiti Cave's entrance nearer

62.1 Rāwhiti Cave, near Tākaka, has New Zealand's best and most numerous examples of calc-tufa stalactites. Many of the stalactites hanging down from the lip of the cavernous ceiling have developed sloping, tusk-shaped orientations, as a result of calcite precipitation over mosses which grow more vigorously on the outer, sunny sides of the stalactites.

62.2 Numerous unworldly, calc-tufa stalactites adorn the ceiling of the vast entrance cavern to Rāwhiti Cave. Moss and algae grow in the sunlight on the porous stalactites which grow by precipitation of calcium carbonate over and through the vegetation, producing the unusual shapes.

the sunlight are curved outwards like an elephant's tusks (62.1). This is because the moss and algae grow more vigorously on the side with more light, and the tufa stalactites grow out in that direction as the water flows over the growths. In some places the tufa stalactites near the entrance have bulbous patches of moss growth that have trapped the dribbling water on top of them, resulting in the growth of saucer-like rings at various intervals along their lengths (62.2).

Take a closer look: Visit via a strenuous walk up a steep hillside Department of Conservation track (1 hr each way) from the carpark at the end of Packard Rd, Tākaka Valley.

62.4 Looking back out of Rāwhiti Cave from the viewing platform (62.3), the top of the domed entrance can be seen to be adorned with tufa stalactites. PHOTO: ALASTAIR JAMIESON.

62.3 The back wall of the domed cavern in Rāwhiti Cave entrance is covered in thousands of narrow tufa stalactites hanging down in rows beneath elongate cracks in the marble ceiling – conduits for the percolating water. PHOTO: ALASTAIR JAMIESON.

62.5 In the early history of this ancient cave, the domed ceiling collapsed many times and built up a mound of debris across the entrance. On top of some of this mound a number of robust stalagmites, up to 2 m high, have grown up but are currently dormant. PHOTO: ALASTAIR JAMIESON.

63.

Cobb Valley, northwest Nelson

Classic glacier-carved valley and New Zealand's oldest fossils

The Cobb Valley in northwest Nelson is a classic glacially carved valley with smooth valley sides and a cirque with small cirque lakes at its head (The Big Picture 15, page 283). A succession of glaciers would have occupied the valley during each cold glacial period of the Ice Ages for at least the last 1 million years. These glaciers occupied, smoothed off and straightened an originally small, sinuous stream valley. In warmer interglacial periods, every 100,000 or so years, the glacier melted and the valley returned to a forested stream course with the bush line well up the valley sides, as it is today. Moraine debris dumped high on the valley sides shows that during the last few cold periods, at least, the glacier almost completely filled the Cobb and adjacent valleys (63.3). Scattered moraine boulders extend as far as the Cobb hydro dam and mark the terminus of the glacier.

The flat valley floor above the end of the road is alluvial sediment that remote sensing shows fills a former lake that occupied the deeply scoured-out valley floor after the last glacier retreated about 12,000 years ago. Just upstream from the head of the hydro lake, the valley floor is quite uneven with numerous rounded knobs of bedrock protruding through the matakoura / matagouri scrub and tussocks. These are hard rock layers that have been overridden and scraped down by the rocks frozen in the base of the glacier (63.5).

Cobb Valley is a major drawcard for New Zealand geology buffs, partly because of the high diversity of rocks and minerals found there, but also for the location of New Zealand's oldest rocks and fossils (510 Myr old, middle Cambrian). The fossils are mostly trilobite fragments that occur in a small lens of grey marble forming a roche moutonnée, known as Trilobite Rock, near the end of the Cobb Road (63.6, 63.7). Other ancient fossils that occur in the valley include the extinct group known as graptolites, plus ancient sponges, lampshells and molluscs. The sedimentary strata in the Cobb Valley that contain these fossils were deposited on the seafloor off the coast of the ancient supercontinent of Gondwana between 510 and 440 Myr ago (middle Cambrian to Ordovician).

Best seen from: The roadside lookout on the ridge crest between Tākaka and Cobb Valleys provides panoramic views.

Take a closer look: Drive into the glacial valley at the head of the lake to see small roches moutonnées and Trilobite Rock. Tramping tracks take you to different rock types and landforms.

◐ 63.1 The Cobb Valley in northwest Nelson is a superb example of a simple, U-shaped, glacially carved valley typical of a classic mountain valley glacier (The Big Picture 15, page 283). This lower portion of the glacier's valley has been dammed to create a storage lake for hydroelectric power generation. The crests of many of the mountain ridges around Cobb Valley are smooth and gently sloping. They are the eroded remnants of the Waipounamu Erosion Surface (The Big Picture 18, page 309) that has been uplifted along a number of large faults in the last few million years to form the northwest Nelson mountainous country. PHOTO: ALASTAIR JAMIESON.

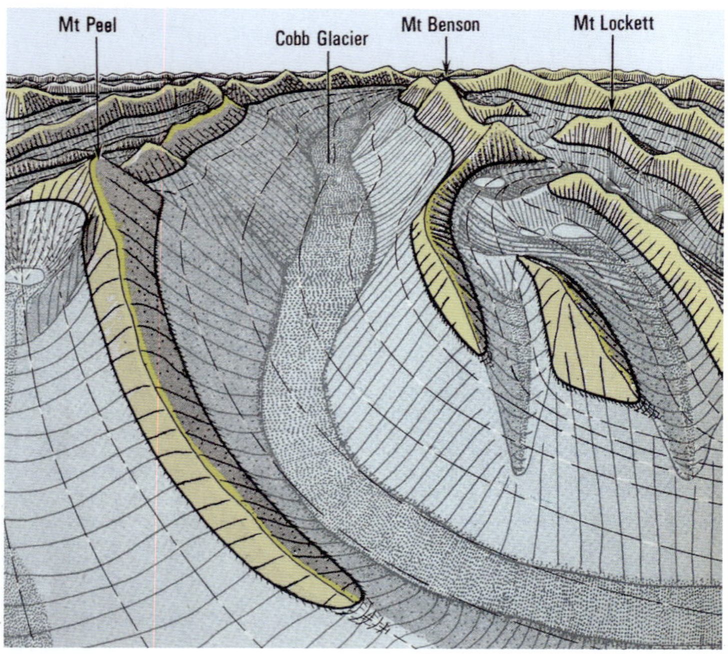

⮉ 63.2 View southeast across Lake Cobb in a moraine-dammed cirque at the head of Cobb Valley. The straight, glacially carved upper Cobb Valley extends 10 km to the Cobb Reservoir, on the bend in the distance. PHOTO: ALASTAIR JAMIESON.

⮈ 63.3 During the Ice Ages of the last 2.6 Myr, the climate has see-sawed back and forth and the Cobb and adjacent valleys have alternated between forested river valleys in warm periods, such as now, and glacier-filled slow-moving ice during the peaks of the cold glacial periods, as depicted in this reconstruction. Only the very tops of the highest ridges would have protruded as rocky peaks above the extensive ice cap of mountainous northwest Nelson at these times. MODIFIED FROM COOPER (1984).

➲ 63.4 View northwest up the length of the upper Cobb Valley. Scour by rocks in the glacier ice straightened the valley and dug the relatively smooth U-shaped profile.

➲ 63.5 A number of low rounded knobs of hard bedrock protrude from the valley floor near the head of the hydro lake in the Cobb Valley. These are small roches moutonnées (a French term meaning rock sheep), typically with a steep downstream slope and a gentler one upstream. These hard layers of rock have been overridden by the Cobb Glacier and smoothed and rounded down by the rocks carried along within the frozen base of the glacier under the enormous pressure of the thick overlying pile of ice. These ice-borne rocks scratched and gouged small grooves in the surface of the bedrocks (as in the photo) as they were dragged over them. PHOTO: ROGER COOPER.

➲ 63.6 The Cobb Valley is a magnet for geologists because of the presence of Trilobite Rock (shown here) which contains New Zealand's oldest rocks and fossils. Trilobite Rock is a 30 m by 10 m roche moutonnée that sticks out above the valley floor between Trilobite Hut and the head of the hydro lake.

➲ 63.7 Trilobite Rock is made of the piled-up broken skeletons of extinct trilobites. The remains of trilobites and other fossils it contains are hard to see because of the recrystallisation of the rock during deep burial and metamorphism to hard marble. Seen here are the head and tail sections of two trilobites found in the rock. Trilobite Rock is protected from hammering or collecting as it lies within Kahurangi National Park. Photo 8 cm across.

64.

Te Waikoropūpū, Tākaka

Largest and clearest freshwater springs in New Zealand

⊙ 64.1 Te Waikoropūpū Springs are the largest karst springs in New Zealand and among the very largest cold-water springs in the Southern Hemisphere. The water is among the purest and clearest in the world. The purple-red colour in fast-flowing areas of the pool downstream is from dense beds of the New Zealand endemic aquatic plant, water milfoil. PHOTO: ALASTAIR JAMIESON.

⊙ 64.2 The main Te Waikoropūpū pool is fed by the Main Spring (top right) and Dancing Sands Spring (top centre, obscured by trees). The aqua-blue spring areas have little vegetation, whereas downstream the red colourations in late summer are from the aquatic plants that thrive in the cool, clean water. PHOTO: ALASTAIR JAMIESON.

⊙ 64.3 A large king salmon hatchery is located adjacent to Te Waikoropūpū Springs because of the constant source of cool, clear water. PHOTO: ALASTAIR JAMIESON.

Te Waikoropūpū Springs is the largest spring in New Zealand (discharging 14,000 litres per second) and has some of the clearest (optically pure) water in the world (64.7). There are two groups of springs in the reserve – the main springs (including Dancing Sands Spring) that rise through the sand on the bottom of the large pool (64.2); and upstream a smaller group of 12 springs known as Fish Creek Springs. The emerging waters are cool with an almost unvarying temperature of 11.7°C.

All of Te Waikoropūpū's water originates from rainfall, but the time it takes from source to the springs varies between 1 and 20 years, with an average time of ~8 years. This mostly depends on the distance it has to travel. The springs are the main discharge point for the Arthur Marble Aquifer, although some water also discharges from freshwater springs offshore in Golden Bay. This aquifer is highly unusual for New Zealand as the water flows through an underground maze of interconnected passages that have been dissolved over several million years along cracks that run through the marble.

Some of the aquifer water comes from rainfall on the marble crest of Tākaka Hill and north along the ridge to Canaan where it directly flows into the subterranean passages. From there it flows northwest down deep beneath the Tākaka Valley, where the marble is capped and confined by impervious coal-measure claystones that prevent the water escaping to the

➋ 64.4 There are over 200 sinkholes like these in the Waingaro Valley near Hāmama and adjacent Tākaka Valley. They are highly unusual sinkholes, in that the ground has collapsed into solution voids in the underlying marble, but the marble itself is buried by younger river gravels which line the sinkholes. Rainwater flowing into some of these sinkholes probably enters the underground aquifer that feeds Te Waikoropūpū Springs, but it is only a very small proportion of the total flow. PHOTO: ALASTAIR JAMIESON.

surface until it reaches Te Waikoropūpū Springs (64.6). Here the marble and capping claystones are folded back up towards the surface and a window has been eroded in the cap rock, exposing the marble.

Where the water flows down beneath the claystones on the east side of the Tākaka Valley the elevation is ~60 m above sea level and the springs are ~20 m above sea level. So, although the water in the confined marble aquifer flows down to depths in excess of 180 m below sea level under Tākaka Valley, gravity builds up pressure within the aquifer and it freely flows up again as artesian water at Te Waikoropūpū. As the water flows up from depth it drags up a little bit of saltwater that has infiltrated through the rocks from the sea, giving a slight salinity to the spring water (64.6).

A second major source of water for the aquifer is from the Tākaka River itself. Between Upper Tākaka and East Tākaka river water flows down through the loose valley gravels and into the marble aquifer, upstream of the claystone cap rock. During dry summers, almost all of the Tākaka River's water disappears through the gravels in this section. The average time for rainwater to travel to the Fish Creek Springs is only ~3–4 years, and for this reason it is thought that the source is much closer, possibly mostly from the middle sections of the Waingaro River where water can flow directly through the riverbed into the marble aquifer (64.4).

Take a closer look: Take the Te Waikoropūpū Springs Walk to a viewing platform over the springs (30-min. loop).

➊ 64.5 The surging waters of the Main Spring can be appreciated from a viewing platform. A loop walking track takes visitors past the smaller Dancing Sands Spring and the clearwater of Fish Creek sourced from the smaller Fish Creek Springs.

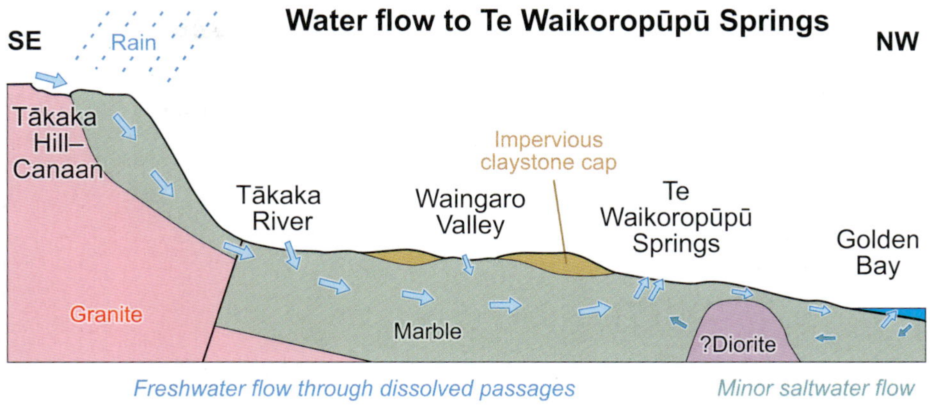

64.6 Southeast to northwest cross-section showing where much of the water that gushes out at Te Waikoropūpū Springs flows underground through subterranean pathways in marble from its source as rainfall on the Tākaka Hill–Canaan area.

64.7 Underwater view of freshwater plants growing in the main springs pool. The visual clarity of the pure spring water is an almost unbelievable 75 m. PHOTO: LLOYD HOMER, GNS SCIENCE, 74976.

65.

Onetahua / Farewell Spit

One of the longest sand spits in the world

Onetahua / Farewell Spit is New Zealand's longest sand spit, a Ramsar Wetland of International Importance and a Flyway Reserve Network Site for migratory shorebirds. The present sand spit has built out from the northern tip of the South Island in just the last 10,000 or so years as sea level rose towards its present height after the end of the Last Ice Age. It undoubtedly built on top of the weathered and eroded remains of a number of older sand spits that would have grown in the same place during earlier interglacial times of higher sea level. Since the last interval of high sea level, there has been a 110,000-year period when the climate was variably cooler and sea level was 30 or more metres lower than present. At the peak cold of the Last Ice Age, just 18,000 years ago, sea level was 130–120 m lower and a forested plain extended from Nelson to Taranaki–Manawatū providing a land bridge between the North and South Islands.

The sand that forms Farewell Spit is composed of about 40% fine rock fragments and subequal amounts (30% each) of quartz and feldspar. The sand has been transported by longshore drift up the west coast of the South Island, and as it rounds the northwest tip the strength of the drift (driven by westerly swells) drops and the sand has been progressively deposited, forming the spit in the lee of Cape Farewell (65.2, 65.5). The amount of sand deposited on the spit each year would fill three Wellington 'cake-tin' stadiums (3.3 million m^3). The source of the sand is from erosion of the high country that drains to the west coast, especially the actively rising schist of the Southern Alps (The Big Picture 12, page 248). Gravel and sand carried down to the coast by rivers is moved northwards along the beaches and in shallow offshore regions. The gravel becomes smaller on its journey north, and only sand passes around the rocky shoreline north of the Heaphy River mouth to reach the spit.

Farewell Spit consists of three landform belts parallel to its length – mobile beach and dunes, fixed dunes, and sand flats (65.1, 65.3). The exposed seaward side consists of a wide, gently sloping beach that passes up above high tide into

65.1 Onetahua / Farewell Spit is New Zealand's longest sand spit. It is 26 km long and extends a further 6 km underwater. The current sand spit is only about 10,000 years old. PHOTO: ALASTAIR JAMIESON.

extensive areas of mobile sand dunes (65.4, 65.5). Sand that is thrown up above high-tide level by waves dries in the sun and is blown south and east to form the dunes. The southern half of the spit is largely composed of older fixed dunes stabilised by vegetation that has grown upon them. A few small freshwater lakes are scattered in hollows between these dunes (65.3).

The third belt is the extensive intertidal sand flats that extend 5–10 km out into Golden Bay off the lee side of the spit (65.2). It is the tasty invertebrates that live in these extensive sand flats that attract vast flocks of wading shorebirds before they undertake their long migratory flights to the Northern Hemisphere to breed each year. Bivalve cockles live in their millions in the surface sand of this huge intertidal flat. Their dead shells are moved along by small breaking waves and tidal currents, and collect in extensive shell banks inside the tip of Farewell Spit, where they provide a relatively stable base for further sand spit growth.

Best seen from: View from Pillar Point (1.5-hr return walk), and from above the Farewell Spit Visitor Centre (20-min. return walk).

Take a closer look: Farewell Spit is a Nature Reserve with free public walking access to both sides of the spit (1–4 hrs return walking options) from Triangle Flat Carpark (end of Freeman Access Rd). Take a 4-wheel-drive bus tour all the way along the spit to the lighthouse.

65.2 The narrow, arcuate sweep of Farewell Spit has built out off the northern tip of the South Island from sand eroded off the growing Southern Alps and transported northwards by currents. When the tide recedes, New Zealand's largest single area of intertidal sand flats (nearly 80 km^2) is exposed in the lee of Farewell Spit. This is a rich source of food for migratory shorebirds, but also the most frequent site in the country for marine mammal strandings. PHOTO: NASA.

65.3 Farewell Spit consists of three distinct zones parallel to the spit's arcuate axis. The outer, northernmost belt (left) consists of the actively moving white sand on the beach and young sand dunes. Older, vegetated dunes occur along the southern side together with numerous small, dune-dammed lakes and ponds. The third, easternmost, belt is the vast intertidal sand flat that has accumulated inside Golden Bay in the lee of the spit. PHOTO: ALASTAIR JAMIESON.

65.4 The mobile sand dunes of Farewell Spit are constantly advancing along their eastern fronts and eroding behind. This wind erosion often exposes intricate patterns of former sand-dune growth, in this instance with a rippled surface that had been buried by younger sand accumulation. The wind also sorts the sand based on grain size and density, creating layers that may be composed of white-coloured quartz, black heavy minerals or yellow-grey mixtures with abundant fine rock fragments and feldspar grains. Photo 2 m across.

➋ 65.5 Mobile, easterly migrating sand dunes, up to 20 m high, form a belt behind the beach along the exposed side of Farewell Spit.

➋ 65.6 View from the east along the landward half of Farewell Spit at full tide. The regularly spaced blue patches on the spit are depressions between the active east-moving sand dunes. Seawater flows into these depressions on spring high tides, forming shallow saltwater lagoons that may be dammed by moving sand and eventually become dune lakes. PHOTO: ALASTAIR JAMIESON.

66.

Wharariki Beach, northwest Nelson

Sea arches eroded along joints in conglomerate cliffs

66.1 Magical Wharariki Beach, northwest Nelson, showing one of the larger sea arches that has been eroded through the 60-Myr-old conglomerate rock forming the Archway Islands. PHOTO: ALASTAIR JAMIESON.

Wharariki Beach is one of the most memorable natural beach landscapes in New Zealand. It nestles in a bay near the northern tip of the South Island's west coast with not a sign of any houses, roads or other human development other than the grazed paddocks of Pūponga Farm Park. The 1.5-km-long white sand beach has four rocky islands, the Archway Islands, in front of its northern half (66.1). In behind the islands a large area of mobile sand dunes has developed. Rocky cliffs bookend the beach with more cliffs and small intertidal rocky stacks towards the south end.

The cliffs are made of layered conglomerate and sandstone (66.6) that was deposited in a subsiding rift valley during the Paleocene (66–55 Myr ago). This Pākawau Rift began foundering between two northeast–southwest-oriented faults about 100 Myr ago, and for the next 45 Myr it slowly subsided as the region was pulled apart by the same parallel extensional forces that were opening the Tasman Sea at that time. On either side there was uplifted land composed of eroding sedimentary rocks and granite. The hills were drained by streams that joined together to form a wide, braided river plain that occupied the rift valley (66.2). Gravel and sand brought down from the hills accumulated on the plain and kept the rift valley topped up with sediment as it subsided by about 1500 m. Within the sequence there are small pockets of carbonaceous mud and wood that accumulated in small ponds in cut-off river meanders (66.4).

The rift stopped subsiding about 55 Myr ago, but in the most recent 7 Myr the same bounding faults have reactivated and reversed their throw direction as the region moved into a regime of northwest–southeast compression (66.2) as a result of slight changes in the direction of oblique collision between the plates. Consequently, the hardened conglomerate and sandstone that had filled the Pākawau Rift have now been pushed up, tilted to the west and are themselves being eroded by streams and the sea. The bounding fault on the east side, called the Wakamarama Fault, runs south down the Aorere Valley from Collingwood with the uplifted Wakamarama Range on its west side.

While the conglomerate and sandstone were deeply buried for millions of years, they were subjected to a wide variety of stresses that produced several sets of steep cracks

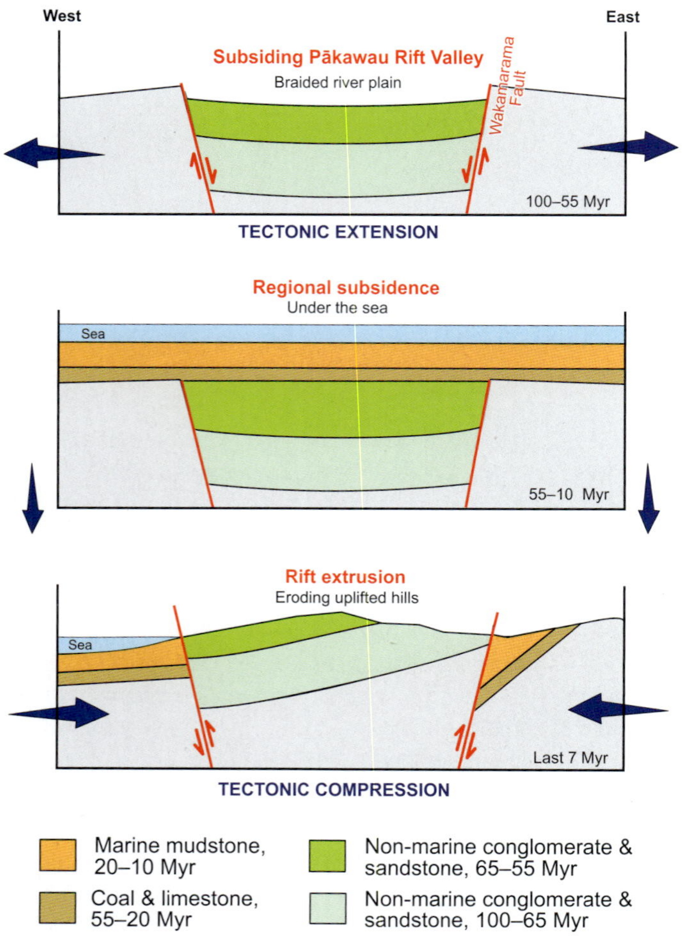

(joints) cutting through them. The most prominent set is aligned northwest–southeast. Every time sea level has been high during the warmer periods of the Ice Ages the sea has bombarded the uplifted rocks and progressively eaten them back, creating bays, islands and cliffs. The most powerful erosive force is the breaking waves, which slowly blast their way along these cracks carving out sea caves (66.5, 66.6). When these cave roofs are undercut and collapse they create guts and rocky stacks and islets. In places the caves have been eroded right through the islands and promontories, creating the sea arches (66.1, 66.3) which Wharariki is famous for. Some of the sandstone layers are not as resistant to erosion as the conglomerate and in a few places these layers have been eroded out to form sea caves or windows.

Take a closer look: To experience this fabulous beach and sea arches, take the walking track from the end of Wharariki Rd (20-min. one-way).

66.2 West to east cross-sections through western northwest Nelson showing (top) tectonic extension resulting in subsidence of the Pākawau Rift in which the conglomerate and sandstone of Wharariki accumulated. This was followed by a period of regional subsidence through to about 10 Myr ago. In the last 7 Myr there was a switch of the local tectonic regime to compression which reversed the throw across the bounding faults, pushing the area up and tilting it to the west. MODIFIED FROM BALLANCE (2017).

66.3 Two of the Archway Islands viewed from Wharariki Beach. PHOTO: IMAGEA ORG, WIKIMEDIA COMMONS (CC-BY-SA-3.0).

66.4 The conglomerate that forms the cliffs and islands around Wharariki Beach consists of rounded pebbles of a variety of hard older rocks. They were eroded off the surrounding uplifted hills between 66 and 55 Myr ago and deposited in the subsiding Pākawau Rift. The upper part of the photo is of a layer of quartz sandstone with black carbonised fossil branches that have been compressed by the weight of overlying sediment that once buried them. Photo 50 cm across.

66.5 A series of sea caves eroded along joints through the conglomerate and sandstone rocks at the northern end of Wharariki Beach. PHOTO: ALASTAIR JAMIESON.

66.6 For several kilometres south of Wharariki Beach, the indented coastline is a maze of sea caves, arches and guts eroded into the conglomerate and sandstone rocks. They underlie a west-tilted, flat-topped plateau of 300-Myr-old limestone (in the distance). PHOTO: ALASTAIR JAMIESON.

67.

Mt Owen, northwest Nelson

Lengthy cave networks inside marble mountains

67.1 Some of the densest and most awesome speleothems inside the upper dry levels of the Mt Owen marble massif are found in Bohemia Cave. Here a mix of stalactites, straws and helictites decorate the ceiling, with older slender stalagmite columns on the cave floor. PHOTO: PETER CROSSLEY.

⊖ 67.2 A diverse group of unusual speleothems in Bohemia Cave. PHOTO: PETER CROSSLEY.

⊕ 67.3 Among the most delicate and rare kinds of speleothem in some marble caves are helectites. These are fragile stalactites that appear to defy gravity by growing and branching in all directions as calcite is deposited by capillary forces on different parts of the water droplets. Photo 30 cm across. Bohemia Cave, Mt Owen. PHOTO: PETER CROSSLEY.

Mt Owen, together with Mt Arthur, is the best example of glaciated surface and subterranean karst in the Southern Hemisphere. This is because of the extensive underground karst passageways that have formed within the huge thickness (500–1500 m) of marble that lies beneath these mountain massifs. Mt Owen contains the longest cave network (Bulmer Cave, 72 km) in New Zealand, and Mt Arthur contains the deepest cave system (Stormy Pot–Nettlebed Cave, 1174 m deep) in the Southern Hemisphere. The subterranean drainage in the Mt Arthur marble block comes together and exits to the east near the foot of the mountain at the Pearse Resurgence, which eventually flows into the Motueka River. This cave system has dissolved out passageways to considerable depth, with divers having descended down a New Zealand record of 229 m below the surface to almost sea level without yet reaching the bottom.

Mt Owen (1875 m) and Mt Arthur (1795 m) dominate the western skyline viewed from Nelson city and are often capped white with snow in winter. Both mountain massifs were ice-covered during each of the glacial periods of the Ice Ages. Small mountain glaciers carved out a number of smooth, amphitheatre-like basins called cirques, polishing and scraping the hard marble. Unlike other areas, small cirque lakes are rare, as the amphitheatre depressions seldom hold water as it flows down through sinkholes (dolines) into the cave passageways beneath. Even when covered in ice, much of the meltwater drained out through the cave networks.

Most of the cave networks high up inside the marble mountains are now dry passageways often decorated with dense stalactite and stalagmite formations (67.1, 67.4). Many of these caves retain evidence that they were dissolved out by flowing streams when groundwater was at this level. This must have been 2–1 Myr ago when the northwest Nelson mountain ranges were much lower and only just starting to be pushed upwards by tectonic forces. As the mountains gradually rose, groundwater levels fell and the subterranean streams dissolved out new cave passages along joint planes at lower and lower elevations. This explains the immense height difference between the upper levels of the cave networks today and their resurgence levels many hundreds of metres lower.

Take a closer look: Located in a remote area of Kahurangi National Park requiring considerable physical effort and caving expertise to visit. To see inside an old marble cave visit Ngārua Cave, top of Tākaka Hill.

⬆ 67.4 Some of the marble caves of Mt Owen and Mt Arthur massifs have enormous chambers, and because the high dry levels are more than a million years old they often contain dense speleothem decorations like these in Bohemia Cave, Mt Owen. PHOTO: PETER CROSSLEY.

⬅ 67.5 The main entrance to Bulmer Cave, New Zealand's longest cave, is high in the subalpine zone on Mt Owen. PHOTO: PETER CROSSLEY.

↑ 67.6 New Zealand's longest cave network has formed inside the marble massif of Mt Owen. On the surface the fresh grey marble has been smoothed off by glacial ice during the Last Ice Age and much is still unvegetated above the bush line. PHOTO: ALASTAIR JAMIESON.

← 67.7 These small conventional stalactites are densely decorated with coralloid overgrowths and interspersed with worm-like helictite strands. Bohemia Cave, Mt Owen. PHOTO: PETER CROSSLEY.

68.

Longford Syncline, Murchison

New Zealand's deepest and most deformed sedimentary basin

The Longford Syncline (68.1) is the most prominent part of New Zealand's deepest and most intensely deformed Tertiary sedimentary basin – the Murchison Basin. The basin contains ~9000 m of sedimentary rocks (68.2), with up to 3000 m more having been eroded off, all of which were deposited on the floor of a rapidly subsiding depression between 35 and 15 Myr ago. The sedimentary sequence records the progressive submergence and deepening of this basin, starting with coastal plain deposits overlain by mudstone that was deposited on the seafloor in deeper and deeper water, reaching 1000 or more metres depth by 20 Myr ago. After this, sediment input increased and the basin began filling up with sediment eroded off the surrounding land. It shallowed up to estuarine and then river valley alluvium by about 18 Myr ago. The centre of the basin continued subsiding and accumulated a further 5000 m or so thickness of river gravels and leaf-bearing pond sediments.

The history of the Murchison Basin and subsequent formation of the Longford Syncline was driven by changes in the direction of oblique collision between the tectonic plates in this part of New Zealand. Most of the subsidence and later folding was accommodated at depth by movement on existing major faults and boundaries between basement terranes (sutures) in the underlying older basement rocks. Early subsidence of the Murchison Basin was due to extension with tectonic forces pulling the area apart, resulting in slow subsidence along reactivated north-northeast-trending normal faults in the basement rocks on either side of the depression.

About 22 Myr ago the regime changed, with onset of oblique compression along the plate margin causing the centre of the basin to subside much more rapidly right through to about 7 Myr ago. Then there was another change in the direction of collision between the plates which brought about

68.1 The magnificent Longford Syncline (downfold) can be seen in the strata forming the tops of the hills in the foreground. This is part of the deepest and most intensely deformed sedimentary basin in New Zealand – the Murchison Basin. Today the Buller River has eroded deep into the syncline and flows south down its axis towards Murchison in the middle distance. PHOTO: ALASTAIR JAMIESON.

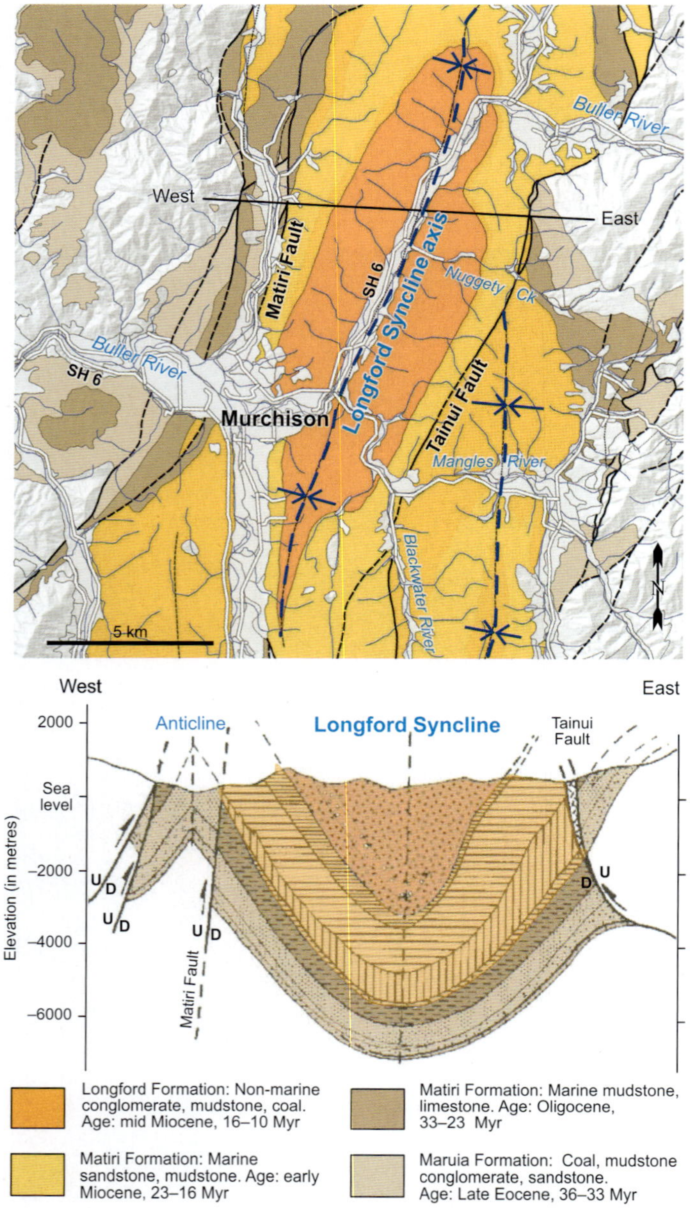

the most intense period of deformation, with compression and reversal of faults within the basement rocks resulting in the folding of the Longford Syncline (68.1) and other associated synclines and faulted anticlines nearby. Squeezing that formed these folds also resulted in 2000–3000 m of uplift with most of the uplifted rocks eroded off from the top.

The Longford Syncline is boat-shaped, steeply folded on each side, especially in the south (68.2). The axis of the fold at each end plunges down towards the deep centre of the syncline, about 8 km north of Murchison. The enormous subsidence within the Murchison Basin resulted in deep burial of organic-rich sediments sufficient to cook and generate hydrocarbons. As a result there are several hydrocarbon gas and oil seeps coming up from the depths of the basin today (68.5), although no oil or gas reservoirs have been found at depth by petroleum exploration and drilling.

Best seen from: Mostly not visible without a charter flight. Steeply dipping strata on either side of the syncline are visible in the hillside beside SH 6, about 3 km east of Murchison (68.4).

Take a closer look: Non-marine sedimentary strata and plant fossils (68.3) that accumulated in the Murchison Basin can be seen 200 m along the closed Nuggety Creek Rd above the Buller River.

68.2 Map and west-to-east cross-section showing the extent and structure of the Longford Syncline and the bounding northeast-trending faults. BASED ON MAPPING BY SUGGATE (1984) AND LIHOU (1993).

68.3 Beautifully preserved 15-Myr-old (middle Miocene) fossil leaves from within the Longford Syncline strata exposed in a cutting of the Nuggety Creek Rd above the Buller River. These leaves are from an extinct large-leaved southern beech tree, *Nothofagus*.

⊙ 68.4 These near-vertical strata have been folded up from their original horizontal orientation by compressive forces between the tectonic plates over the last 7 Myr. These strata were deposited in an alluvial valley about 15 Myr ago and now form part of the steep western limb of the Longford Syncline. They can be seen forming a hillside beside SH 6, about 3 km east of Murchison. PHOTO: ALASTAIR JAMIESON.

⊙ 68.5 Natural gas, generated deep in the Murchison Basin, seeps out at the surface where it is burnt off by a flame that never extinguishes.

69.

Devil's Dining Table, Matiri Range

Exemplary, uplifted limestone mesa

The Matiri Range has by far the best examples in the country of large, uplifted mesas. A mesa is a flat-topped plateau surrounded by vertical bluffs. The Matiri Range contains two such mesas. The Devil's Dining Table (One Hundred Acre Plateau) in the north is the most dramatic and remote (69.1). The 12-km-long Thousand Acre Plateau is slightly lower and much larger, and not fully encircled by bluffs at its southern and northern ends.

Both mesas are capped by hard recrystallised limestone that started life as an extensive shell bank in the shallow sea that covered much of New Zealand 30 Myr ago during the Oligocene. The area subsequently foundered and the shell bank was compressed and recrystallised into limestone beneath the weight of more than 2000-m thickness of mud that accumulated on top of it over the next 20 Myr (Miocene). In the last 7 Myr the whole region has been pushed up out of the sea, in places folded

69.1 The Devil's Dining Table in the Matiri Range, north of Murchison, is New Zealand's most impressive mesa – a bluff-surrounded plateau. The 5-km-long by 0.3–1-km-wide plateau is capped by uplifted flat-lying limestone at an elevation of 1100–1300 m. Thousand Acre Plateau is in the middle distance to the left. The large semicircular embayments in the cliffs on either side are cirques formed at the head of glaciers during the Last Ice Age (The Big Picture 15, page 283). Both mesas are capped by hard, flat-lying limestone that has prevented them from being eroded completely away. PHOTO: ALASTAIR JAMIESON.

69.2 View north along the length of the Devil's Dining Table in the Matiri Range. The Haystack (right distance) is composed of remnant mudstone that overlies the mesa's limestone cap rock and has not been stripped away completely by erosion. PHOTO: ALASTAIR JAMIESON.

(e.g. Longford Syncline) or tilted, but here at Matiri the layers are almost flat-lying and that is why the plateau exists.

As the area was being pushed upwards the soft mudstone was mostly eroded off, but in places the down-cutting has been held up by the hard, erosion-resistant limestone layer. Water and ice managed to dissolve and erode valleys through some of the limestone, leaving the uneroded parts as these magnificent, mostly tussock-covered plateaux, surrounded by vertical limestone bluffs (69.1, 69.2). As in other places, slightly acidic rainwater has percolated down through cracks and dissolved out subterranean channels and passages in the limestone. Dissolved and collapsed dolines pockmark parts of the plateaux (69.3) and lead down into a network of caves that run along the near-horizontal surface contact with the underlying insoluble conglomerate and mudstone. After heavy rain, these cave passages may fill with runoff water from the plateau. Some of this water jets out of a cave exit as a dramatic waterfall from part-way up the bluffs into a tributary of the Matiri Valley (69.4).

Best seen from: Not visible without a charter flight from Murchison.

Take a closer look: Experienced multi-day trampers can walk in to Thousand Acre Plateau (1 day each way) and Devil's Dining Table (2 extra days return).

69.3 The surfaces of tussock-covered Thousand Acre Plateau (shown here) and Devil's Dining Table mesas are pockmarked with thousands of small sinkholes and trenches. Many of these are in rows, presumably aligned along joint planes through the limestone that have been partly dissolved out by weakly acidic rainwater forming a regular network of small caves beneath. Photo about 100 m across. PHOTO: ALASTAIR JAMIESON.

69.4 The Hole in the Wall Waterfall spurts out of the bluffs on the east side of Thousand Acre Plateau. It only flows after heavy rain, some of which flows down through the limestone cap and into a cave network within the mesa. PHOTO: MURCHISON HELITOURS.

70.

Ōpārara Arch, Karamea

Largest natural arch in Australasia

⊙ 70.2 Looking upstream through the 43-m-high Ōpārara Arch. The lower 10 m or so of the walls is composed of granite, with the upper walls and arch made of limestone. PHOTO: EGON EBERLE.

⊙ 70.1 Ōpārara Arch is the largest limestone arch in Australasia. It actually consists of two sections of arch with a 20-m skylighted gap between them. This is the narrower downstream section seen from the walking track. PHOTO: EGON EBERLE.

⊙ 70.3 View from beneath the higher and narrower section of Ōpārara Arch. PHOTO: EGON EBERLE.

⊙ 70.4 Moria Gate Arch is a more easily photographed but smaller version (19 m high, 43 m wide) of Ōpārara Arch. PHOTO: EGON EBERLE.

⊙ 70.5 View through Moria Gate Arch. Calc-tufa stalactites and roots dangle down from the ceiling. PHOTO: EGON EBERLE.

Ōpārara Arch is the largest natural arch in Australasia at 43 m high, up to 79 m wide and 220 m long (70.1–70.3). It is one of three limestone arches that started off as solution caves and have been subsequently eroded much wider and higher by the Ōpārara River which runs through them. The river flows southwards for 20 km along the contact between the limestone sitting on granite before swinging west out to the coast.

The limestone accumulated 30 Myr ago as a thick shell bank on top of an uplifted and eroded surface of 360-Myr-old granite, part of the Median Batholith (The Big Picture 11, page 207). The shells were buried on the seafloor by thick mud deposits. Then groundwater, percolating through the shell bed, recrystallised their calcite, forming a relatively hard limestone rock. The sequence has subsequently been uplifted and tilted down to the east at an angle of 5–15° over the last 7 Myr.

As the Ōpārara River valley was forming, rainwater filtered through cracks in the limestone and dissolved out a cave network along the contact between the soluble limestone and the underlying granite. Eventually the caves coalesced and captured most of the valley's drainage to become part of the Ōpārara River. The river both dissolved and eroded the caves wider, undercutting the roof supports resulting in progressive roof collapse and daylighting of the river. The three arches are the remnants of the original cave system. Near river level in Ōpārara Arch you can see the contact of the limestone on the old granite surface. Using the strict geomorphologist's definition, the Ōpārara arches should be called natural bridges. Arches are defined as being formed by weathering and wind erosion, whereas structures formed by water erosion and solution with a river passing through them are classified as natural bridges.

Higher up, the Ōpārara Valley is an area with entry restricted to permit-holders. This is to protect New Zealand's best deposits of fossil bird bones in Honeycomb Hill Caves. Here much of the cave network in the limestone is still present, although there are numerous roof-collapse sinkholes. Before the arrival of humans, many specimens of New Zealand's now-extinct flightless birds fell into these sinkholes, where they died and their bones became buried in sediment on the cave floor or coated in secondary calcite deposition. Research on some of these bone deposits has greatly increased our knowledge of our recently lost avifauna, best known of which are the giant moa.

Take a closer look: From the carpark 23 km north of Karamea, separate walking tracks lead to Ōpārara Arch (1-hr return) and Moria Gate Arch (1.5-hr loop track).

71.

Punakaiki / Pancake Rocks

Flaggy-bedded limestone

Pancake Rocks at Punakaiki on the West Coast between Greymouth and Westport is the most visited and easily accessible example of flaggy bedded limestone formation in New Zealand. Visitors are attracted to its blowholes and surge pools when they are active during high tide and a southwest swell.

Like most of the karst limestones in New Zealand (The Big Picture 4, page 83), these rocks started out as shell banks on the floor of a shallow sea that covered much of ancient New Zealand about 30 Myr ago (Oligocene). Over the next 25 Myr the seafloor in the region slowly subsided and up to 2000 m of marine mud and sand was deposited on top of the shell bank (during the Miocene). This deep burial compressed the shell bank and along with percolating groundwaters caused the shells to recrystallise, and insoluble fine-grained minerals such as micas were concentrated into narrow irregular dark seams of mud (called stylolites). This stylobedding is thus the result of pressure solution of the original calcite shells. Today the softer mud stylolites erode more easily than the thick recrystallised limestone layers, thereby accentuating the appearance as a pile of pancakes (71.5).

In the last 7 Myr the oblique compressive forces between the tectonic plates have changed slightly and the sedimentary sequence has been pushed up out of the sea (as in Sites 68–70). At Punakaiki, the huge thickness of overlying softer mudstones has been eroded off exposing the limestone beneath. Further inland, Paparoa National Park is characterised by extensive bluffs of this same uplifted limestone, with karst and cave features and underground streams.

As the limestone has been exhumed, slightly acidic rainwater runoff has seeped along near-vertical cracks (joint planes) that cut through the limestone and dissolved them out

➔ 71.1 Punakaiki / Pancake Rocks from the southwest showing how solution and erosion along the dominant set of vertical joints (running bottom left to top right) have produced numerous narrow rocky fingers (clints) separated by vertical-walled chasms (grikes). Solution and erosion along the second set of joints (parallel to the coast) has cut many of the fingers into short segments and stacks. PHOTO: LLOYD HOMER, GNS SCIENCE, 36438.

71.2 The punch bowl in the centre of Punakaiki / Pancake Rocks is where the roof of a cave has collapsed and the swells surge in along tunnels from several sides. PHOTO: LLOYD HOMER, GNS SCIENCE, 69995.

into elongate caves and gutters (The Big Picture 4, page 83). Two prominent sets of joints are present at Punakaiki, and their orientation (one set perpendicular to the coast and the other parallel to it) has produced the bizarre coastal landforms and rocks (71.1–71.4). Most limestone solution and subterranean passage formation occurred beneath luxuriant forest cover when sea level was much lower than today. When sea level rose after the end of the Last Ice Age, it invaded the caves and guts. Since then, crashing waves and compressed air has been forced along cracks, accelerating erosion. Undercut blocks have crashed into the sea, forming sea-filled crevasses and caves skylighted by roof collapses (71.2).

On calm days and lower tides, the sea flows in and out of the caves and tunnels producing the odd gentle hiss of escaping air. In rougher weather and higher tides, however, a mixture of compressed water and air is forced into the caves producing thunderous booms accompanied by jets of fountaining spray escaping upwards through blowholes in the limestone.

Take a closer look: Take the paved loop track (20–30 mins around) opposite Paparoa National Park Visitor Centre.

71.3 In some places the limestone at Punakaiki has been extensively dissolved out along joint planes, with subsequent roof collapse of the dissolved elongate cave. Solution and erosion along two sets of joints at right-angles to each other has produced a highly dissected rocky surface.

71.4 The West Coast swells thunder in against the limestone rocks at Punakaiki. A combination of solution when sea level was lower and now the erosive force of the waves has produced these limestone pinnacles.

➔ 71.5 The name for these rocks at Punakaiki is derived from the resemblance to a pile of pancakes. Thin irregular mud partings (stylolites) are softer than the rest of the recrystallised limestone and tend to erode out and exaggerate the pancake-like appearance.

➔ 71.6 With big westerly swells on a high tide, waves blast their way along fractures through the pancake rocks and squirt water high into the air through blowholes in depressions in the foreground.

THE BIG PICTURE 12

Alpine Fault and Southern Alps

THE ALPINE FAULT is the largest on-land crustal structure in the New Zealand region, and one of the largest transcurrent faults (with sideways displacement) in the world. The Alpine Fault is a vital part of the obliquely collisional plate boundary that links the west-plunging Hikurangi Subduction Zone beneath the North Island with the east-plunging Puysegur Subduction Zone southwest of the South Island.

Traditionally the Alpine Fault has been said to run through almost the whole length of the South Island from Blenheim on the northeast coast to Milford Sound on the southwest (TBP 12.1). This is the fault line that produced the majority of the sideways displacement of rocks resulting in the same distinctive rock units of east Nelson being displaced 480 km southwards to occur on the opposite side (east) of the fault in West Otago. The straight Wairau Valley in Marlborough has eroded along the route of the northern part of the Alpine Fault, and this section is also known as the Wairau Fault. At the southwest end of the Wairau Fault, west of Nelson Lakes, there is a major bend in the Alpine Fault, probably as a result of the fault displacements jumping from the northeast-striking Wairau Fault, to an existing more north–south-trending fault parallel to the terrane boundaries.

Major sideways displacement on the Alpine Fault began about 25 Myr ago and has become progressively faster in more recent times. Within the last 2.5 Myr (Quaternary) several more northeast-trending faults, parallel to and south of the Wairau Fault, have become active with similar dextral (sideways) displacement. The Hope Fault is now more active than the Wairau Fault, and thus it can be said that the plate boundary through Marlborough now runs along the Hope Fault to where it joins the Alpine Fault inland from Greymouth. Thus at its northern end, the Alpine Fault has two major strands: the Wairau Fault which has been responsible for most of the sideways displacement, and the Hope Fault which is now the more active.

Wairau and Hope Faults are parallel to the direction of oblique compression between the plates, and therefore almost all displacement on them has been sideways. Further south where the Alpine Fault has swung more southward, the direction of compression is more oblique to the fault. Thus along this section there is both sideways and vertical displacement that accounts for the shortening across it. Here there is continental crust on both sides of the collisional plate boundary. Since both are buoyant there is no subduction, and instead the continental crust of the Pacific Plate has both folded down beneath the South Island and been thrust upwards on the east side of the Alpine Fault (TBP 12.2).

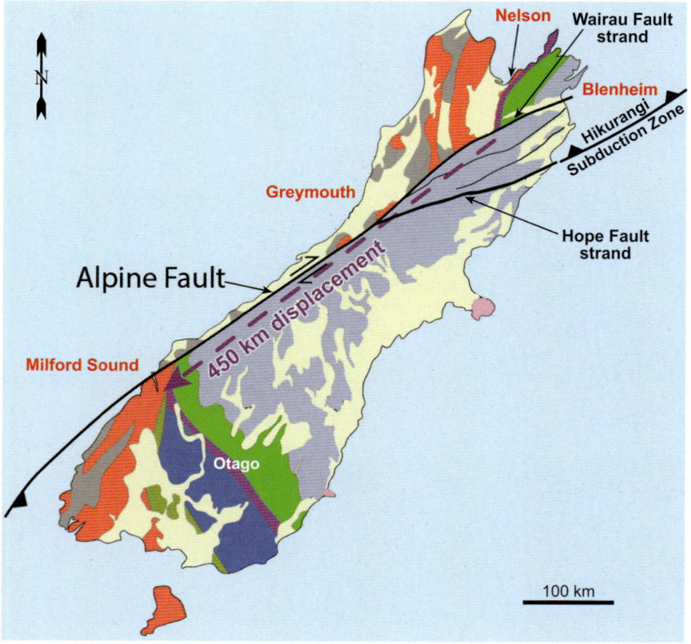

TBP 12.1 Map showing the present distribution of rocks in the South Island. The 450 km sideways displacement of orange, purple and green rocks from Nelson to West Otago on the Alpine Fault can be seen. At its northern end the Alpine Fault splays into several major fault strands that cut through Marlborough. The Wairau Fault strand has had the greatest total displacement, whereas the Hope Fault is now (last 2 Myr) the most active strand and the modern plate boundary linking the southern end of the Hikurangi Subduction Zone with the Alpine Fault in central Westland.

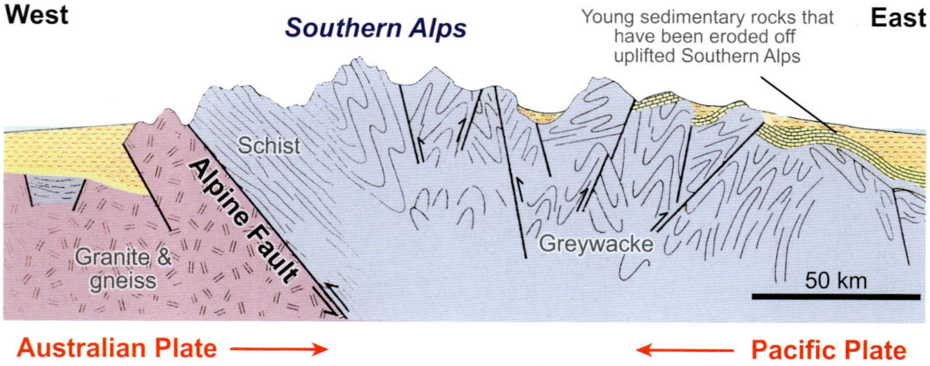

TBP 12.2 West to east cross-section through the South Island in the vicinity of Aoraki / Mt Cook showing how the obliquely compressive forces between the Australian and Pacific Plates have resulted in uplift and erosion of the Southern Alps.

TBP 12.3 Oblique satellite view of the South Island from the southwest showing the straight fault scarp of the Alpine Fault running along the west side of the uplifted, snow-capped Southern Alps and breaking into two major strands as it swings northeast (right) through Marlborough to the east coast as the Hope (red arrow) and Wairau (white arrow) faults. PHOTO: GOOGLE EARTH.

The downward folding has thickened the continental crust to ~35 km beneath the Southern Alps, and being lighter this buoyancy is added to the plate boundary compression to push the mountains upwards. There have been two significant periods of uplift on the east side of the Alpine Fault. The first, 22–20 Myr ago (early Miocene), pushed up an ancestral mountain range composed of softer, younger rocks (Cretaceous–Oligocene age) that were stripped off by erosion as this range was worn down. The second, more major, period of uplift that produced the present-day Southern Alps (TBP 12.3, TBP 12.4), began about 7 Myr ago and the rate has been increasing ever since. Since then there has been 20,000–30,000 m of uplift with more than 80% of the upthrust mountains' height being removed by erosion and carried away as rock and sand by rivers on both sides of the divide. The rising mountain range produced orographic conditions where moist air from the west rises up the sides of the mountain barrier, cools and is precipitated as increased rainfall and snow. The enhanced rainfall accelerates erosion, with increased removal of rock (and lightening of the overburden) from the upper parts of the Southern Alps contributing to an increasing rate of uplift.

Displacement of the rocks across the Alpine Fault has mostly resulted from ongoing build-up of pressure within the rocks that has been relieved at fairly regular intervals by rupture events on the fault. These ruptures produce severe earthquake shaking and displacements across the fault of up to 20 m sideways and 10 m vertically. The current average rate of uplift of the Southern Alps around Mt Cook is 10 mm/yr with 36 mm/yr of sideways movement. The proportion of the Alpine Fault that ruptures in each event is not fully understood, but there is evidence that over 200 km of it ruptured during the last Alpine Fault earthquake in AD1717. Elsewhere around the world, sometimes one section of a major transcurrent fault like this ruptures, relieving pressure in one area and triggering additional ruptures further along the fault within weeks or years of the initial rupture.

Studies near the southern end of the Alpine Fault have recorded 27 major displacements on the Alpine Fault within the last 8000 years. These are fairly evenly spaced apart with an average recurrence interval between each of 290 ± 30 years. Since the last displacement on the Alpine Fault was 300 years ago, in AD1717, seismologists have asserted that the next major rupture and large earthquake (Mw >8) on the straight Westland section of the Alpine Fault is due any time.

TBP 12.4 The steep, straight, western side of the high Southern Alps is the fault scarp of part of the Alpine Fault, seen here in the vicinity of Aoraki / Mt Cook. PHOTO: LLOYD HOMER, GNS SCIENCE, 55278.

72.

Gaunt Creek, Franz Josef

Inside one of the world's largest faults – the Alpine Fault

72.1 The eroding cliff beside Gaunt Creek provides one of the best places in New Zealand to see close-up the actual rupture plane and sheared rocks along the internationally famous Alpine Fault (base of the sloping light-coloured rocks).

72.2 Sketch of the Gaunt Creek cliff section (72.1) showing the precise location of the Alpine Fault overlain by sheared and crushed rocks of the Pacific Plate thrust over stream gravels and scree of the Australian Plate. MODIFIED FROM DIAGRAM BY COOPER AND NORRIS (1994).

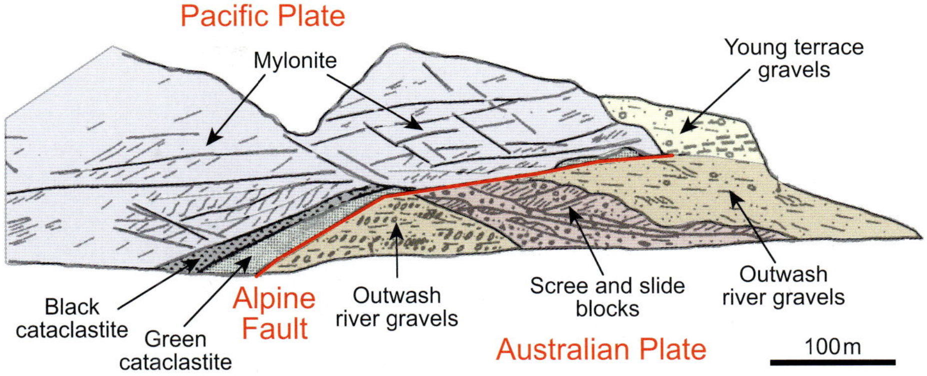

The straight line and immense uplift on its eastern side makes it easy to recognise the incredible Alpine Fault throughout central Westland – one of the great transcurrent (sideways-displaced) faults of the world. In the vicinity of Franz Josef, however, the fault is not precisely straight. Here it has a zigzag map pattern corresponding to fault sections that, of late, have mainly experienced sideways displacements linked together by short fault sections displaying more compressional thrusting of the Pacific Plate (Southern Alps) over the Australian Plate. One of the most easily viewed examples of this overthrusting occurs in cliffs of Gaunt Creek, a tributary of the Waitangitaona River near Whataroa. Because the rocks along the Alpine Fault zone have been subject to so much frictional crushing and alteration they are soft, easily eroded and seldom visible at the surface. Thus Gaunt Creek provides a rare opportunity to 'see inside' one of the world's largest faults (The Big Picture 12, page 248).

In the cliff section, the schist (The Big Picture 17, page 305) of the overthrust Pacific Plate has been recrystallised at depth to finely laminated mylonite rocks (72.5) – a result of the enhanced deformation along the plate boundary. Some of these mylonite rocks nearest the fault have later been intensely crushed to cataclastite rock (often called fault gouge) by fault movements as they got closer to the surface. The wide crush zone along the fault has acted as a pathway for the flow of hot fluids that have additionally altered many of the cataclasite rocks to cream and green clays that are distinctly seen as an east-dipping band directly above the Alpine Fault in the cliff (72.1, 72.2).

The rocks immediately overlying the Alpine Fault rupture plane are the most altered (white then greenish altered cataclasite), containing fragments of green, chlorite-replaced mylonite and angular pieces of thin white quartz veins that have been broken by fault movements. Further away from the fault, up into the cliff, the rocks are less-fragmented dark-grey finely foliated mylonite that retains the texture and features formed by ductile metamorphism at great depths prior to their being progressively uplifted during thousands of earthquake events over the last few million years.

⊙ 72.3 View southwest over the Gaunt Creek cliffs that expose the Alpine Fault with the north-flowing Waitangitaona River beyond. The prominent forest-covered slope above the cliffs (left) is the eroding scarp of the Alpine Fault which has uplifted the Southern Alps to the east. PHOTO: RICHARD JOHNSON, GNS SCIENCE, 157848.

In the cliff, the highly modified schist rocks (cataclasite, mylonite) of the Pacific Plate have been thrust over young outwash river gravels and scree that were deposited over the surface of the Australian Plate during the last 15,000 years. The rounded river gravels, derived by erosion from the nearby uplifted schist and mylonite rocks, were transported and dumped here by glaciers and streams. The more angular rocks in the scree have slid down over the site off the adjacent, episodically growing, Alpine Fault scarp.

In the older, upstream part of the cliff, the Alpine Fault slopes down at ~35° to the east, but halfway up the cliff to the west (downstream) the fault plane flattens off and becomes almost horizontal. This change in orientation of the fault plane is inferred to be a result of the overthrust rocks reaching the surface, whereupon the toe of the Pacific Plate rocks has been pushed westwards across the ground during successive fault movement events (72.6).

Take a closer look: One of the best places to see the Alpine Fault close-up is on a guided tour over private land to Gaunt Creek, 15 km northwest of Franz Josef (Alpine Fault Tours, Whataroa).

⊙ 72.4 The Alpine Fault rupture plane can be viewed close-up at Gaunt Creek. Here a visitor stands astride the plate boundary with their left foot on the Australian Plate and their right foot on the Pacific Plate.

⊙ 72.5 Mylonite from above the Alpine Fault in Gaunt Creek. This rock was metamorphosed under great pressure and temperature induced by the huge frictional stress (ductile metamorphism) between the tectonic plates along the Alpine Fault at depth. Photo 30 cm across.

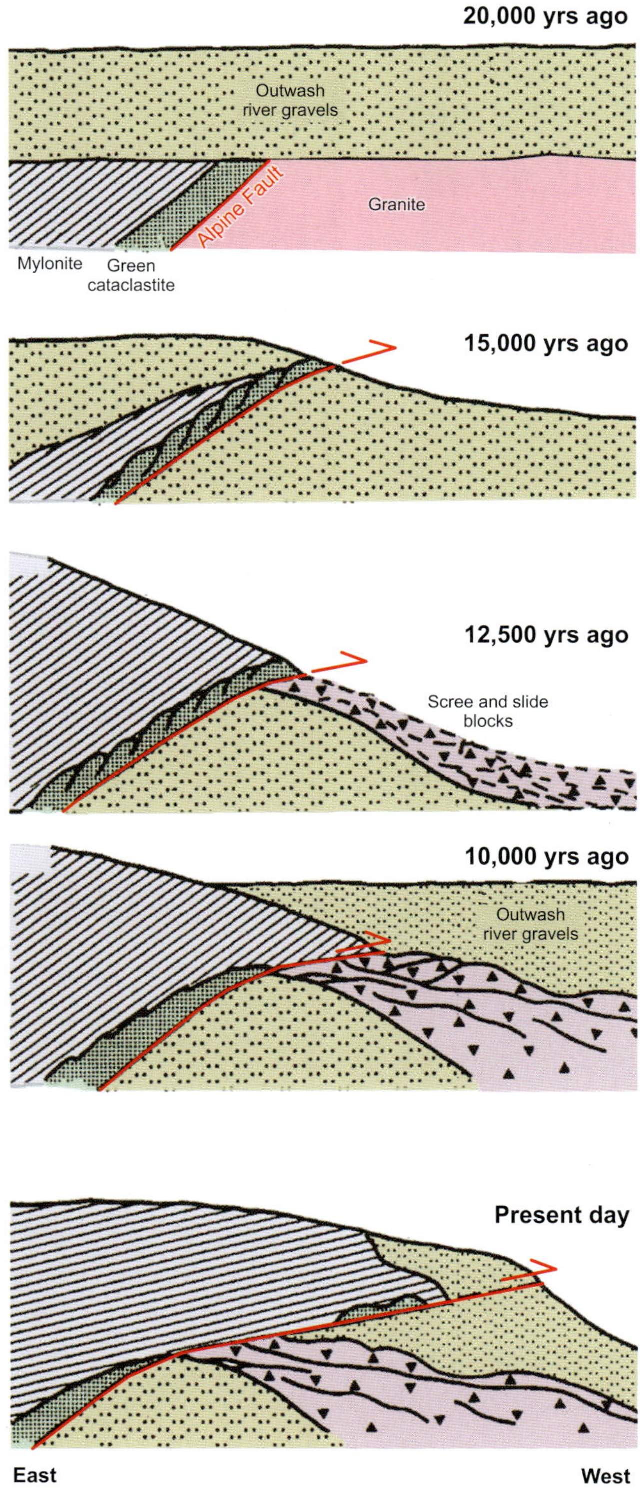

⊙ 72.6 Series of east–west cross-section diagrams through the Gaunt Creek cliff locality of the Alpine Fault showing how it is envisioned that the visible features in the cliff (72.1) developed over the last 14,000 years. MODIFIED FROM DIAGRAM BY COOPER AND NORRIS (1994).

73.

Kā Roimata o Hine Hukatere / Franz Josef Glacier

Abandoned arc of terminal moraine and receding mountain glacier

⊙ 73.1 View from above the névé field down the upper slopes of the Franz Josef Glacier, 2021, with the forested arc of the Waiho Loop terminal moraine on the valley floor below.

73.2 View southeast over the magnificent arc of the Waiho Loop – one of the world's best examples of a classical glacial terminal moraine. It was formed ~12,000 years ago at the toe of the Franz Josef Glacier, which has now receded 11 km back up the Waiho Valley. The snowfield source of the glacier can be seen high on the western side of the crest of the Southern Alps. The straight trace of the Alpine Fault runs along the foot of the forested mountain slopes. PHOTO: LLOYD HOMER, GNS SCIENCE, 3763.

Kā Roimata o Hine Hukatere / Franz Josef Glacier is New Zealand's most visited and most famous glacier. Its source is a vast névé snowfield, 20 km^2 in area, located 2000–2500 m up near the crest of the Southern Alps (73.1). This field of compacted snow is 300 m deep with large quantities feeding into a steep and narrow valley which descends rapidly as a glacier to low altitudes, dropping 2000 m in little more than 10 km.

During the peak of the last few cold glacial periods of the Ice Ages, the glacier was much larger, flowing out across the coastal plain and out beyond the present shoreline. Narrow 200–300-m-high gravel ridges on either side of the lower Waiho River, between Franz Josef and the sea, were lateral moraines on either side of this huge glacier.

The Waiho Loop, 2.5 km downstream from Franz Josef township, is New Zealand's most perfect example of a classical terminal moraine (73.2). The 6-km-long, 60-m-high arcuate and forest-covered ridge extends part-way across the Waiho Valley, 3–4 km north of Franz Josef township and, fittingly, is part of the Te Wāhipounamu – South West New Zealand World Heritage Area. The Loop has been the subject of numerous international scientific studies which have concluded that it was formed about 12,000 years ago around the toe of the receding Franz Josef Glacier. The moraine consists of large angular blocks of rock from high up in the headwaters of the glacier. They were sourced from a huge rock avalanche that

↑ 73.3 The middle reaches of the Franz Josef Glacier splits into two streams around a medial ridge of hard rock, 2021.

← 73.4 Around 2008, at the peak of the most recent advance of Franz Josef Glacier, the ice at its terminus was relatively moraine-free and whitish-blue. Meltwater that flows along beneath the glacier pours out of an ice cave as the source of the Waiho River.

➲ 73.5 Franz Josef Glacier in 2021 showing the vast distance up the valley that the glacier has receded since 1865.

thickly buried the glacier's icy surface. This debris was rafted down to the melting toe of the glacier, where it was deposited as a 140-m-high arcuate ridge of terminal moraine. The glacier has receded up the valley and a lake was formed in behind the moraine. Over time the lake filled with outwash gravels from Tatare Stream, which later eroded a narrow breakout gorge through the middle of the loop.

Over the last 10,000 years the glacier has not advanced out of the mountains beyond Franz Josef township, but there have been many cycles of advance and retreat as the temperature and snowfall high in the source area have varied. The last major advance was during the Little Ice Age, AD1300–1880. Since then the glacier has retreated 3 km back up the valley and its surface level near the present terminus has fallen by 300 m (73.5). When advancing, the glacier often has a fresh white icy surface, but when in recession this surface ice is melting and the glacier becomes almost unrecognisable beneath a thick blanket of grey moraine.

The Franz Josef Glacier flows down from the highest and most uplifted region of the Southern Alps. It has cut through some of the highest-grade metamorphic schists (biotite, garnet and oligoclase zones – The Big Picture 17, page 305; 73.7) in New Zealand as it approaches the Alpine Fault. This major fault runs along the foot of the forested mountains and passes right through the township of Franz Josef (The Big Picture 12, page 248).

Best seen from: Waiho Loop is adjacent to SH 6 between 2 and 5 km north of Franz Josef township. On a clear day the glacier can be seen from Sentinel Rock at the end of Glacier Access Rd (20-min. return walk).

Take a closer look: Walk up the Waiho River gravels to view the terminus of the receding glacier (2 hrs return from carpark) or take a commercial guided walk on the glacier or a flight over it.

➲ 73.6 These beautifully sculpted schist rocks near the terminus of the Franz Josef Glacier (upper left) were carved and smoothed by rocks frozen in the base of the river of ice and moved along by the glacial flow.

➲ 73.7 High-grade metamorphic rocks, like this biotite garnet schist that has been thrust up from depths of 12–15 km, form the walls of the lower Waiho Valley on the walk from the carpark to the glacier.

74.

Kaikōura Peninsula

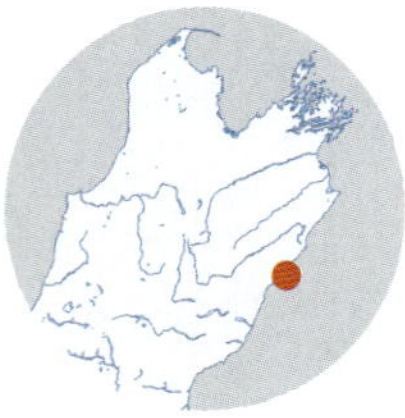

Series of earthquake-uplifted shore platforms

Most of the Kaikōura coast was uplifted during the 2016 earthquake. The Kaikōura Peninsula itself was pushed up 1–1.5 m that evening (74.1). This uplifted the landward parts of the extensive intertidal shore platform around the peninsula's coast to above high tide (74.3). Higher parts of the uplifted platform may become a new coastal terrace unless they are eroded or submerged by the predicted rapid sea-level rise in coming decades. Prior to the earthquake there were already several uplifted levels of shore platform preserved as low terraces between the shingle beaches and the base of the cliffs in a number of bays (74.1, 74.3). The widest is the terrace beneath downtown Kaikōura itself. These low terraces were all uplifted in earlier earthquakes with similar amounts of uplift to 2016 over the last 4000 or so years.

74.1 Kaikōura Peninsula is surrounded by wide supratidal and intertidal shore platforms that were carved by the sea out of the underlying limestone and siltstone rocks in the last 7500 years. At least two major earthquakes (the most recent in 2016) have uplifted these so that the higher parts are now above high water. The farmed crest of the peninsula preserves a flight of uplifted coastal terraces, each of which was once a similar shore platform formed during five intervals of relatively high sea level between 110,000 and 70,000 years ago. PHOTO: ALASTAIR JAMIESON, 2022.

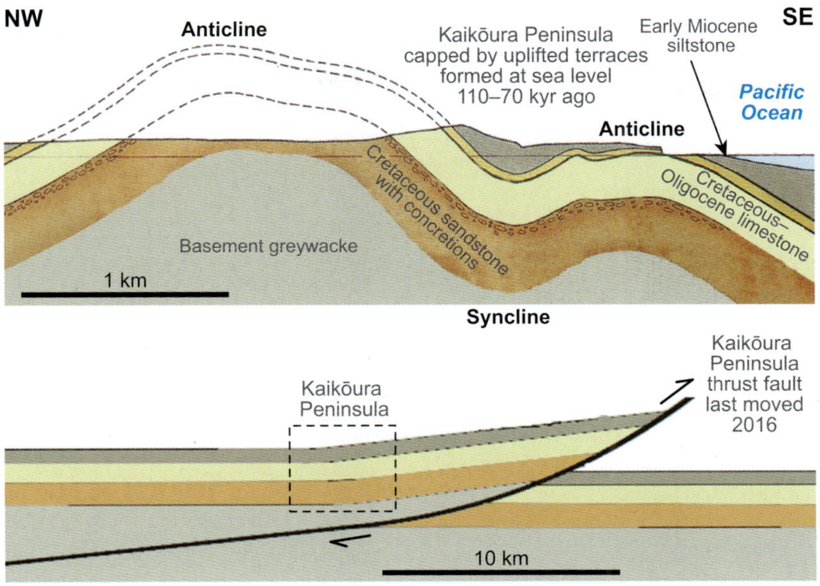

74.2 Above: cross-section through the length of Kaikōura Peninsula showing the folding of the strata that are seen in the cliffs and shore platform. Below: the low-angle thrust fault that is producing the current episodic uplift of the peninsula. MODIFIED FROM COATES ET AL. (1993) AND DUFFY (2020).

74.3 View across Third Bay on the tip of Kaikōura Peninsula after the uplift of 1–1.5 m during the 2016 Kaikōura Earthquake. The grassed terrace between the shingle beach and the base of the cliffs is a former shore platform that had been uplifted during an earlier earthquake a few thousand years ago. The 'rhino horns' (centre) are two former limestone sea stacks now uplifted and permanently joined to the peninsula. The green terrace at the top of the cliff was eroded about 70,000 years ago and has since been uplifted more than 60 m. PHOTO: ALASTAIR JAMIESON.

The grassed top of Kaikōura Peninsula above the cliff tops preserves evidence of the combined total uplift that has occurred since the Last Interglacial, 120,000 years ago (74.1). There are four distinct coastal terraces that were once shore platforms, like those of today, which had been planed down by intertidal erosion during periods when sea level was slightly higher than background levels between 110,000 and 70,000 years ago. During times of high sea level, the sea was between 4 m above and 30 m below present. In the last 120,000 years the Kaikōura Peninsula has been rising (episodically during earthquakes) at a combined rate of about 1 m per 1000 years. This uplift has been slightly faster towards the tip of the peninsula and thus the terraces are tilted gently down towards the west. The reason there are no coastal terraces present that were carved between 70,000 and 10,000 years ago is that they would have been formed when sea level was more than 50 m lower than present and the uplift rate has been insufficient to bring them to the surface since then, if any have survived.

The driving force for the current uplift and tilting of the Kaikōura area is believed to be a northwest-sloping (~30°) thrust fault that reaches the surface offshore from the tip of Kaikōura Peninsula (74.2). During each earthquake-generating rupture of this fault the land above it moves upwards and sideways towards the north.

Prior to ~200,000 years ago the whole of Kaikōura Peninsula was under the sea during higher interglacial sea levels. During the several high sea levels between 110 and 70 kyr ago, the peninsula and adjacent coastal hill behind would have been an island. Further inland the Kaikōura Ranges are rising 10 times faster (10 m per 1000 years) than Kaikōura Peninsula.

The 40–200-m-wide shore platforms around Kaikōura Peninsula were eroded down in the intertidal zone by the breaking and surging waves. The platforms around the seaward end of the peninsula are composed of fine-grained muddy limestone (74.4) and in places softer blue-grey mudstone. The limestone is composed of billions of calcareous shells of minute plankton that accumulated on the floor of the ocean between 70 and 22 Myr ago (late Cretaceous–Oligocene) when little or no sediment was being eroded off low-lying Zealandia and being transported offshore to this area. The overlying siltstone was deposited between 22 and 15 Myr ago (early Miocene) and the change in rock type records the start of uplift of Zealandia with some of the silt eroded off it being carried in suspension out to sea.

Since deposition, these sediments have been buried, compacted and partly cemented. The compressive forces between the plates have folded these strata into two anticlines (upfolds) separated by a syncline (downfold) as seen in the cross-section in 74.2.

Best seen from: Walk the Kaikōura Peninsula Walkway between the end of Fyffe Quay and end of South Bay Parade (2 hrs with return along the foreshore).

74.4 Intensely folded and thinly bedded Oligocene (28–22-Myr-old) limestone forms part of the tip of Kaikōura Peninsula in Spaniards Bay.

74.5 The new intertidal shore platform beside Kaikōura's South Bay marina was hidden beneath the waves until after the 2016 earthquake pushed it up. The large, rounded boulders in the picture are concretions of late Cretaceous age (~100–70 Myr old) that had been eroded out of their parent sandstones and mudstones which form the solid part of the new platform.

74.6 The 2016 uplift of the tip of Kaikōura Peninsula has left wide swathes of former intertidal shore platforms permanently dry, as seen here in South Bay.

THE BIG PICTURE 13

The 2016 Kaikōura Earthquake

JUST AFTER MIDNIGHT on 14 November 2016, central New Zealand was jolted awake by the shaking from New Zealand's largest on-land earthquake (Mw 7.8) since the Hawke's Bay Earthquake of 1931. It made international headlines because it was the most structurally complex earthquake recorded on Earth in modern times. Fortunately it struck the Kaikōura coastal strip which has a low population density, and resulted in just two deaths, far fewer than were lost in the much smaller 2011 Christchurch Earthquake (Mw 6.3). Like all significant earthquakes in New Zealand, the 2016 Kaikōura Earthquake resulted from the sudden release of built-up pressure within the rocks resulting from many centuries of compression between the tectonic plates.

The earthquake rupture began near Waiau in North Canterbury (epicentre) and migrated northwards, taking just 74 seconds to reach Cook Strait, deforming a ~200-km-long coastal strip with displacements on 22 different faults, both onshore and offshore (TBP 13.1). The rupturing travelled through the rocks at 6000 km/h! As the rupturing moved northwards it jumped from one fault to another, leaving areas without displacement.

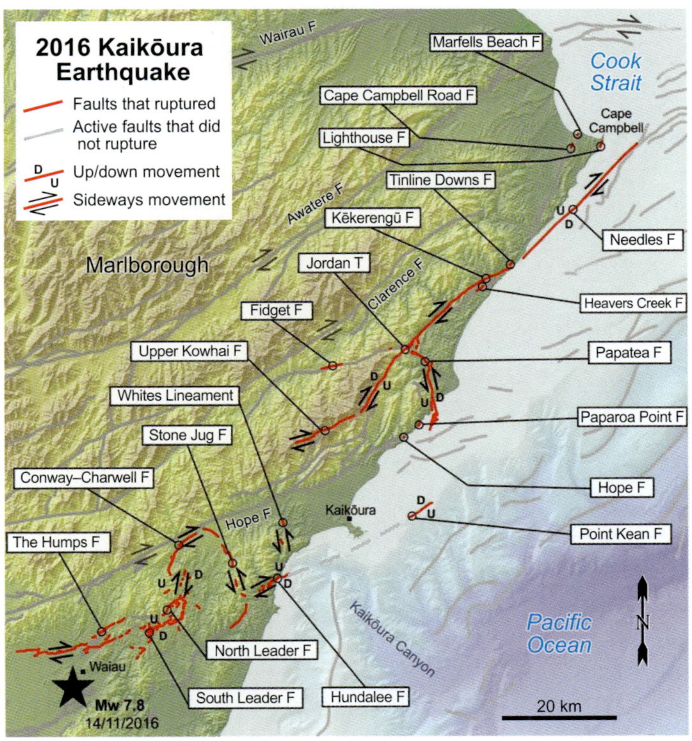

TBP 13.1 The northeast South Island area that was impacted by the 2016 Kaikōura Earthquake, showing the faults that ruptured at the surface (red) as well as those that have been active in the last 10,000 years but did not rupture in 2016 (grey). MODIFIED FROM LITCHFIELD ET AL. (2018).

TBP 13.2 Surface displacement on one of the 22 faults to rupture during the 2016 Kaikōura Earthquake passed right through and effectively destroyed this farmhouse. PHOTO: NICOLA LITCHFIELD, GNS SCIENCE, 210976.

The number of faults that all ruptured in one event over such a length surprised the scientific community, and while all the surface ruptures were on faults that cut through the upper (Australian) plate it is believed that there was also some movement on the subducting interface between the two plates.

The maximum net displacement (vertical and sideways combined) on the surface ruptures was 12 m on the Kēkerengū and Papatea Faults near the Marlborough coast, north of Kaikōura. Displacements were mostly smaller in the south than north of Kaikōura.

It has been calculated that earthquakes like this are 1 in 10,000 year events in this region. All ruptures involved both sideways and vertical displacement components. The ruptures cut right across the countryside no matter what its nature, displacing paddocks, fences, shelter belts, rivers, roads, railway lines – even passing through the middle of a house (TBP 13.2). Several faults offshore left fault scarps across the seafloor.

In addition to the linear fault ruptures, large areas of land were displaced vertically, mostly upwards. Studies show that 80% of the coastline near Kaikōura was uplifted, reaching a maximum of over 6 m of uplift near the Papatea Fault. Kaikōura and its peninsula were uplifted 1–1.5 m. A small tsunami generated offshore by seafloor displacements impacted 200 km of coast with a wave height of 1–5 m.

The intense shaking during the Kaikōura Earthquake, which in places lasted up to 2 minutes, triggered over 10,000 landslides in the coastal belt. The largest and highest-density of landslides occurred close to the main fault ruptures. Many were in more mountainous areas in the Seaward Kaikōura Range, with over 190 creating temporary debris dams across rivers and streams. Landslides along the steep Kaikōura coast caused the most expensive and disruptive damage for the local population and national transportation, requiring rebuilding of the main trunk railway line and main highway.

75.

The Wall of Waiau and Leader River

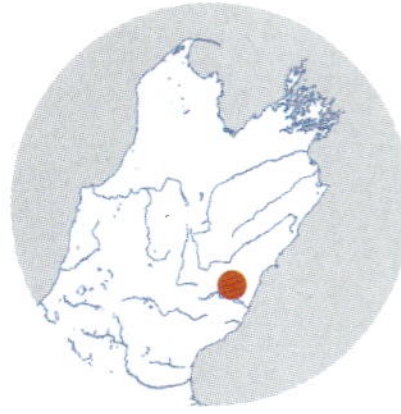

Fault scarp and landslide dam produced by 2016 Kaikōura Earthquake

➔ 75.1 This vertical aerial photo over the 'Wall of Waiau' fault trace shows that displacement was not only vertical but also sideways (sinistral), with sheep tracks, fences and shelter belts moved about 5 m apart. PHOTO: KATE PEDLEY.

➔ 75.2 The 'Wall of Waiau' was the most visually striking fault scarp formed during the 2016 Kaikōura Earthquake, seen here only days after the event. The 3–4-m-high wall of soil and weathered rock extended for many kilometres across the countryside. PHOTO: KATE PEDLEY.

75.3 This huge landslide was triggered by the intense shaking and land displacement during the 2016 Kaikōura Earthquake. It dammed the Leader River, forming a lake that has backed up the valley for several kilometres. The majority of the 10,000 or so landslides triggered by this earthquake occurred in the steep back-country mountains of the Kaikōura coastal strip, but this one occurred in soft sedimentary rocks in farmland.

The intense shaking and immense power of the 2016 Kaikōura Earthquake (The Big Picture 13, page 261) not only shook up the nerves of most of the population of central New Zealand but also resulted in profound changes to the landscape of a large part of the northeast of the South Island. These landform changes comprised widespread coastal uplift, fault line ruptures displacing paddocks, rivers and roads, and landslides scarring every valley and forming temporary dams across many rivers and streams.

One of the most prominent fault rupture scarps (the so-called 'Wall of Waiau'), formed in a matter of seconds, cuts across the undulating hills and valleys of a large farm near Waiau, North Canterbury. A wall of uplifted soil and rock 3–4 m high suddenly appeared along many kilometres of the Leader Fault. Here the fault has cut through soft siltstone rocks which rapidly weather and erode. Thus the scarp, which was fresh and crisp soon after rupture (75.2), is rapidly degrading and being smoothed off by nature and stock (75.4). In most places where fault scarps were formed through farmland they were soon removed by the farmers because they were a hazard to stock and hindered normal farming activities. On Woodchester Station, however, the owner has decided to preserve the fault scarp for its educational, tourist and scientific values.

Also on Waiau's Woodchester Station is one of the largest landslides triggered by the shaking during the Kaikōura Earthquake. An 80-m-thick slide containing approximately 6 million m^3 of soft sedimentary rock slid down into the river as a whole section of valley side gave way (75.3). The slide may have been triggered in part by the rupturing of two faults that meet in this locality. The landslide dammed the river which has backed up several kilometres as a new elongate lake.

Take a closer look: These two features are located on the private Woodchester Station on Leader Rd. The owner permits pre-booked group visits for a small charge.

75.4 The 'Wall of Waiau' fault rupture resulted in the damming of a number of small stream valleys, creating small ponds like this. Fault scarps in soft rock and soil quickly lose their crispness with slumping, frittering and soil creep rapidly rounding them off, as seen happening here just two years after it was created. In years to come the scarp will be grassed over and only recognisable as a rounded rise.

76.

Kura Tāwhiti / Castle Hill

Sculpted and pockmarked limestone tors

The bizarre rounded and pockmarked shapes of these tors (76.1, 76.4) attract numerous visitors to explore and photograph them, but they are also a magnet for South Island boulderers – freedom rock climbers who attempt to conquer low rock faces without the safety of ropes. Castle Hill Basin is an upland depression, 700–800 m above sea level, nestled between two ranges of uplifted basement greywacke rocks (Torlesse Terrane) – the 2000-m-high Craigieburn and Torlesse Ranges. The basin contains an outlier of soft sedimentary rocks all younger than 70 Myr old (Cretaceous and younger) completely surrounded by the older greywacke rocks. The basin contains a transgressive–regressive sequence of strata which started and ended with the accumulation of coal measures and conglomerate in freshwater swamps or on river plains. The peak of the transgression occurred about 30 Myr ago (Oligocene) when fine-grained limestone and greensand accumulated in 100–500 m water depth out in the ocean well away from land.

The limestone that forms the Castle Hill tors accumulated as a shell bank in seawater 10–50 m deep about 25 Myr ago (late Oligocene). It was then buried by up to 1000 m of mud, sand and gravel over the next 10–15 Myr, which compressed it into a moderately hard, but somewhat porous, limestone.

⬇ 76.1 The best-known limestone tors in New Zealand are in Castle Hill Basin in the foothills of the Southern Alps, west of Christchurch. Thousands of tourists and boulderers are attracted to Kura Tāwhiti's magical rocks every year. Their resemblance to battlements led early European visitors to give them their English name of Castle Hill.

➲ 76.2 Rounded limestone tors and boulders in Castle Hill Basin with uplifted greywacke mountains beyond. Note the rounded pockmarks in some of the rocks produced by freeze and thaw and wind erosion that has broken through the case-hardened outer layer of the limestone and is now eating out these small hollows (tafoni). PHOTO: LLOYD HOMER, GNS SCIENCE, 14776.

➲ 76.3 These 25-m-high limestone tors at Castle Hill have their bedding (former sea beds) tilted down to the left as a result of the tight folding they suffered during the squeezing that also uplifted the surrounding mountain ranges.

➲ 76.4 Weirdly sculpted and pockmarked limestone tors, Castle Hill, inland Canterbury.

A change in the angle of collision between the tectonic plates about 7 Myr ago started to push this area up and deform it. The underlying greywacke rocks were shunted both sideways and upwards along a number of faults. This compressive movement at depth squeezed the overlying softer cover rocks, which were crumpled into a series of folds lying in a variety of different directions. As the Craigieburn and Torlesse Ranges were pushed up the younger sedimentary rocks on top were completely stripped off by erosion, but Castle Hill Basin was not pushed up so high and some of the folded younger strata there have not been eroded away.

At Castle Hill itself, the younger strata were squeezed together to form a double fold with a tight anticline (upfold) along the west side of the rocks and a syncline (downfold) on the east. As erosion removed the softer sedimentary rocks from above the limestone layer, percolating near-surface groundwater dissolved away calcareous rock along two sets of vertical joints at right-angles to each other creating narrow crevasses (grikes) through the limestone.

The Castle Hill limestone is granular and relatively porous rather than densely crystalline like many New Zealand limestones which have well-developed solution features, such as fluting. So rather than being dissolved on the surface, the local limestone tends to break up mechanically. Rainwater penetrates into the porous rock, reacting with it and hardening the surface (case hardening). The rocks are slowly being sculpted by the alternating wetting and drying, and freezing and thawing at the surface, which results in slabs of hardened surface rock flaking off. Sometimes this process breaks through the case-hardened outer layer and erosion of the softer rock inside produces the pockmarks (tafoni) on their sides and tops (76.2, 76.4), so beloved as handholds by the boulderers.

Take a closer look: Walk among tors (5 mins from carpark), picnic in Cave Stream Scenic Reserve, or if properly prepared clamber through the 600-m-long cave.

THE BIG PICTURE 14

Banks Peninsula volcanoes

BANKS PENINSULA IS composed of two giant, extinct, composite shield volcanoes and the products from two additional periods of eruption from smaller centres (TBP 14.3). Lyttelton Volcano erupted between 11 and 9.7 Myr ago and Akaroa Volcano followed, erupting between 9 and 8 Myr ago. Between 9.7 and 8 Myr ago extensive basalt lava flows were erupted from the southeast side of Lyttelton Volcano (Mt Herbert, Mt Bradley area) and overlapped in time with eruption of Akaroa Volcano located still further southeast. The final phase of Banks Peninsula eruptions, between 7 and 5.8 Myr ago, was of basalt flows from small vents inside Lyttelton Volcano's crater (e.g. Diamond Harbour) and on its flanks (e.g. Halswell).

The Banks Peninsula volcanoes are classed as intra-plate because they erupted through the crust of the Pacific Plate well away from a plate margin subduction zone or spreading centre. During the period of eruption of these volcanoes the Pacific Plate moved about 120 km southwest over the mantle beneath, yet the volcanoes moved with the plate and hence the magma source of the volcanoes must have been in the Pacific Plate's oceanic lithosphere. Recent research indicates that the basalts and more evolved trachyte magma that erupted from Lyttelton and Akaroa Volcanoes were produced in magma chambers within the lithosphere that were influenced by pulses of magma or related fluids from the deeper mantle beneath. Trachyte is magma with a composition that has changed from basalt over time as a result of fractional crystallisation (and removal from the molten mix) and the addition of wall rock.

Lyttelton and Akaroa Volcanoes are similar in structure and composition, with a variety of small eruptions from numerous vents early in their history followed by growth of a composite shield (TBP 14.4) composed predominantly of extensive basalt lava flows separated by layers of red-oxidised volcanic ash and breccia (broken-up lava flows). These shields had central craters at their tops, 1000–1500 m high. The summit vents moved around and the craters became quite large, partly because some of the erupted rocks subsided back into the emptying magma chambers beneath.

TBP 14.1 The eroded internal part of the trachyte dome of Pulpit Rock that was erupted onto the slopes of Akaroa composite volcano about 8 Myr ago.

These shields also had many other vents dotted around their crest and slopes. Some of these extra vents extruded domes of viscous trachyte lava (TBP 14.1), which, when cool and solid, protruded as 20–50-m-high rounded humps above the shield volcano's slopes (e.g. Castle Rock, Panama Rock, Pulpit Rock, Devil's Gap). The layers of lava flows and ash that formed both shields were also intruded by sheets of lava, mostly trachyte, that radiated outwards from the central craters. These dikes can be seen as erosion-resistant ribs on the surface in some places (e.g. Remarkable dikes) or in cuttings on Lyttelton Volcano's Summit Rd.

Through time the two shield volcanoes developed radiating stream drainage (TBP 14.3). The craters filled with water and became lakes which overflowed down one of these valleys, which was eroded far deeper and wider than the others. These originally crater lake overflow valleys eroded down below sea level when the sea was low during the colder periods of the Ice Ages and were flooded by the rising sea during the warm interglacial periods. The most recent flooding of both occurred about 8000 years ago and created the elongate Lyttelton and Akaroa Harbours, which occupy the eroded-out main craters and the overflow valleys of these two volcanoes. The indented coastline right around Banks Peninsula is a result of high-sea-level flooding up the more-deeply eroded lower parts of many of the smaller radial drainage valleys.

The Banks Peninsula volcanoes erupted well east of the rising land of the Southern Alps region. They created a large massif that was an island when sea level was high and a peninsula when sea level was low. In the last 7 Myr or so the Southern Alps have been growing high. Rapid glacial and river erosion moved gravels and finer sediment eastwards, building up the Canterbury Plains which eventually linked 'Banks Island' to the mainland even during times of high sea level. During glacial periods of the Ice Ages there were extensive unvegetated river plains across and seaward of the present Canterbury Plains. Fine glacially ground sediment (glacial flour) poured over the plains and when dried was blown away as thick dust clouds and deposited over many places, including mantling the eroding hills of Banks Peninsula with thick layers of loess (wind-blown dust).

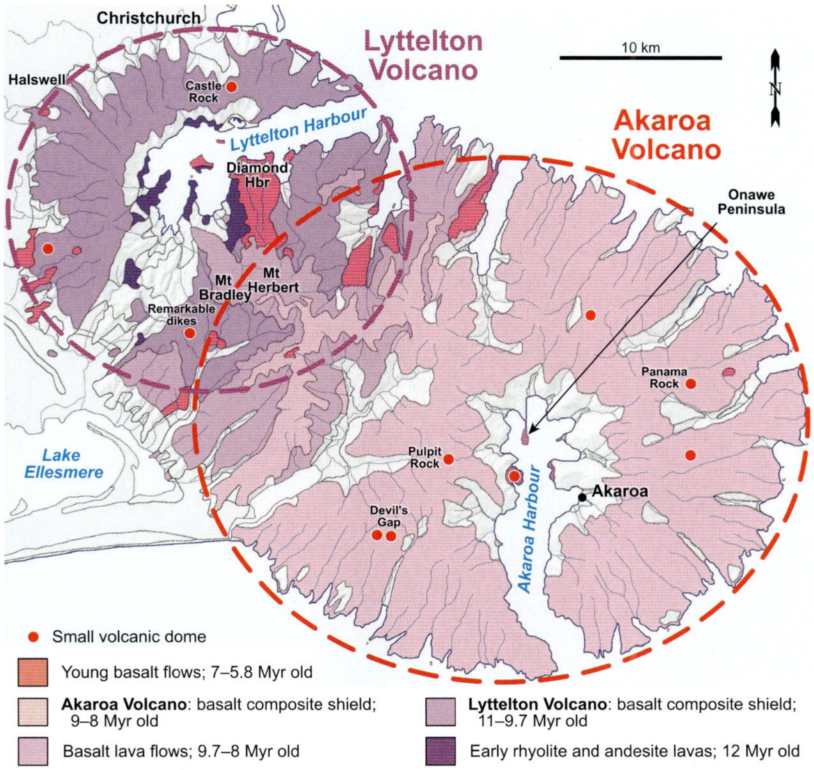

TBP 14.2 The outwardly sloping shield volcano profile of Banks Peninsula as seen from out to sea is still discernible despite extensive erosion since the volcanoes became extinct. The upper third to one-half of the shields has been eroded off.

TBP 14.3 Map of Lyttelton and Akaroa Volcanoes. MODIFIED FROM GNS QMAPS.

TBP 14.4 High-altitude aerial photo of Banks Peninsula and its two composite shield volcanoes: the older Lyttelton Volcano (right) and younger Akaroa Volcano (left). PHOTO: LLOYD HOMER, GNS SCIENCE, 4853.

77.

Ōnawe Peninsula, Akaroa Harbour

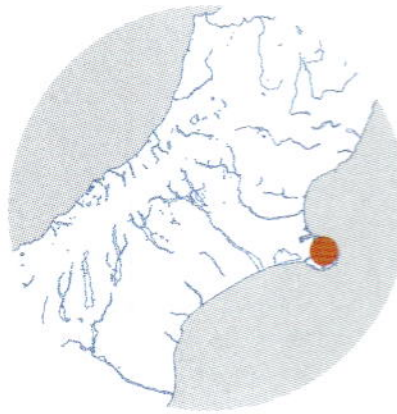

Eroded heart of Akaroa Shield Volcano

Akaroa Harbour is one of the best examples in the world of an eroded crater where you get to see some of the partly eroded remains of the earliest eruptions as well as plumbing beneath the central part of Akaroa Volcano. This is the younger (erupting between 9 and 8 Myr ago) and larger (30 km diameter) of the two composite shield volcanoes that form Banks Peninsula. Akaroa's highest peak is Mt Sinclair (841 m) but the volcano originally may have been nearer 1200–1500 m high.

Remains of some of the earliest eruptions from Akaroa Volcano are now being exhumed by the deep erosion of the valley currently flooded by Akaroa Harbour. Around the harbour's coast and outside its entrance the eroded remains of a number of small basalt scoria cones can be seen in the cliffs. One of these occurs at the narrow landward end of Ōnawe Peninsula where eroding volcanic bombs are visible (77.2). The cliffs and foreshore around Ōnawe Peninsula, at the head of Akaroa Harbour, is also the best place to see some of the complex upper-level plumbing in the heart of this old volcano. The basalt scoria cone and other old lava flows are intruded by a number of near-vertical sheet-like dikes of light-coloured trachyte lava (77.5), some of which may have been conduits for magma rising up from a magma chamber to extrude as a dome or viscous lava flow at the surface 1000 or so m above. Some of these dikes have pushed their way up along fractures with different orientations and so in places you can see one dike cutting across an earlier one. At the seaward end of the peninsula (low tide access; 77.6) you can see part of a magma chamber which sat below the main centre of Akaroa Volcano – when activity ceased it slowly cooled and large crystals

⓿ 77.1 View looking south down beautiful Akaroa Harbour, the erosion of which has cut deeply into the heart of the volcano. The central crater of Akaroa Volcano's composite shield was above the head of the harbour (foreground) with some of its plumbing system exhumed around finger-like Ōnawe Peninsula (centre). PHOTO: LLOYD HOMER, GNS SCIENCE, 17896.

predominantly of alkali feldspar formed to create the plutonic rock called syenite.

The main Akaroa Volcano composite shield appears to have had a large crater situated above the head of the harbour. A crater lake probably overflowed down a stream on its south side. This stream valley eroded much deeper than all the other radial valleys on the shield and now has been partly flooded by the present high sea level, forming Akaroa Harbour.

Best seen from: There are a number of expansive viewpoints over Akaroa Harbour. These include from Hilltop on the Christchurch to Akaroa Rd, and from Summit Rd 3–6 km east of Hilltop.

Take a closer look: Walk the foreshore of Ōnawe Peninsula at the head of the harbour.

➲ 77.2 A 20-cm-long spindle-shaped volcanic bomb in weathered red spatter in the foreshore on the east side of Ōnawe Peninsula. This is part of an eroded basalt scoria cone that erupted early in the life of Akaroa Volcano.

➲ 77.3 This 3-m-wide dike of weathered trachyte is one of numerous intrusive sheets of lava that were forced along joints in the subterranean plumbing at the centre of Akaroa Volcano. The rusty orange patterns in the light-coloured dike are oxidation fronts of iron oxide (limonite) formed during weathering.

➲ 77.4 Ōnawe Peninsula at the head of Akaroa Harbour is composed of the exhumed remains of some of the plumbing from the heart of Akaroa Volcano.

⬆ 77.5 The intertidal platform east of the landward end of Ōnawe Peninsula at the head of Akaroa Harbour. The criss-cross lines are dikes of trachyte that was lava which intruded as vertical sheets through the deeper part of the volcano on its way up from a magma chamber to vents at the surface. The light-coloured rock on either side of the dikes is altered volcanic glass where the lava rapidly cooled and solidified in contact with the cold rocks it was intruding. PHOTO: LLOYD HOMER, GNS SCIENCE, 1690.

⬇ 77.6 The southern end of Ōnawe Peninsula in Akaroa Harbour is made of an eroding magma chamber of syenite that fed trachyte lava to the surface of Akaroa Volcano about 8 Myr ago. PHOTO: LLOYD HOMER, GNS SCIENCE, 30781.

78.

Rakaia River

Globally uncommon braided river eroded into glacial outwash fan

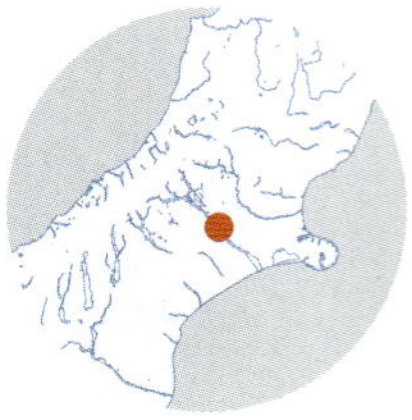

Braided rivers are uncommon globally but New Zealand has quite a few, especially flowing across the Canterbury Plains. One of the largest and best examples is the lower reaches of the Rakaia River (78.1) where it flows across the plains from the Southern Alps to the Pacific Ocean.

A braided river has a dynamic network of small channels separated by islands (78.2). They occur where a river transporting a lot of gravel and sand slows down and spreads out, as on a glacial outwash plain or delta. The Canterbury Plains are essentially a colossal outwash plain made of vast quantities of gravel mostly eroded off the Southern Alps by large glaciers during the Ice Ages. The rapidly rising mountains have been a continuous source of greywacke and schist cobbles and pebbles for more than 5 Myr, and in that time the Canterbury Plains have built out an outwash plain 60–100 km wide. The easternmost 40 km is currently beneath the Pacific Ocean as a result of sea level rise after the end of the Last Glacial period.

Braided rivers erode and deposit gravel depending on the amount of water and its speed. During low flow, gravel and sand may be deposited on the riverbed and the channels may divide into smaller braids with new temporary islands between them. At times of high flow the braids join together into a single main channel that covers or even washes away the gravel islands. With exceptionally high flow rates, during storms and snow melt, a braided river may completely inundate the full width of its braid plain. When water level drops after these events, braided channels and islands develop in new places.

The Rakaia River flows 64 km from where it exits the mountains at Rakaia Gorge to its mouth at the coast. Over this length the present braid plain widens from 1 to 5 km and the number of braids increases significantly. Longshore drift at the coast carries the gravel dumped at the river's mouth northwards forming a 10-km-long gravel spit enclosing an elongate coastal lagoon, or hāpua (78.4). This spit diverts the river through the lagoon to flow into the sea through an ever-

78.1 The Rakaia River is one of the largest braided rivers in New Zealand. Its braids are a beautiful milky blue because of the glacial flour in suspension. Note the dynamic pattern of criss-crossing river channels. PHOTO: GEOFF LEEMING, CREATIVE COMMONS (CC-BY-NC-2.0).

78.2 View down the lower 15 km of the braided Rakaia River before it discharges into the Canterbury Bight. In this area the riverbed is 4–6 km wide from bank to bank, though this is only ever filled with raging water after a huge inland rainfall event. PHOTO: LLOYD HOMER, GNS SCIENCE, 7730.

changing mouth nearer Banks Peninsula. Over thousands of years this longshore drift has moved vast quantities of cobbles and pebbles from the mouth of the Rakaia northwards, building up Kaitorete gravel spit against the south side of Banks Peninsula and enclosing Lake Ellesmere.

In the Southern Alps, the Rakaia River and its many tributaries flow down valleys that were gouged out by the many fingers of the Rakaia Glacier during cold glacial periods of the more recent Ice Ages. During the coldest times, the glacier was up to 80 km long and its toe pushed out about 5 km onto the Canterbury Plains eastward. During the Last Glacial period, that peaked 20,000 to 18,000 years ago, the Rakaia Glacier built out a large outwash glacial fan to the east of the Rakaia Gorge. As the climate warmed, the glacier melted and retreated back into the mountains and the amount of sediment coming down the river decreased markedly. Thus since then, the Rakaia has been eroding down through the outwash gravel fan, and as a result has left behind a magnificent stepped series of erosional river terraces just downstream from Rakaia Gorge (78.5, 78.6).

Best seen from: The braids and terraces are best seen from the air. At ground level view the braids from SH 1 and the river terraces from SH 77, about 3 km east of Rakaia Gorge bridge.

78.3 Thinly layered lake sediment (varves) exposed at river level just upstream of Rakaia Gorge bridge. This was deposited on the floor of a proglacial lake at the toe of the Rakaia Glacier as it was retreating after one of its many advances during the last few hundred thousand years. The angular pieces of gravel are probably drop stones that were transported out onto the lake when it was iced over and fell to the lake floor when it melted in summer.

78.4 Northwards longshore drift has created an elongate gravel barrier across the front of the Rakaia River mouth which diverts river discharge northwards through a bayou (coastal lagoon) to its shifting mouth. The barrier is breached during periods of high river flow but soon rebuilds. PHOTO: LLOYD HOMER, GNS SCIENCE, 17700.

78.5 A sequence of river terraces just upstream of the Rakaia Gorge. These terraces were created as the river cut down into the outwash plain of the Rakaia Glacier after it melted and retreated back up into the Southern Alps after the peak of the Last Ice Age, about 18,000 years ago.

78.6 These massive, 100-m-high river cliffs just downstream from the Rakaia Gorge have been eroded into the outwash gravels brought down by the Rakaia Glacier during glacial periods of the recent Ice Ages.

79.

Aoraki / Mt Cook

New Zealand's highest mountain peak

79.1 Panorama northwards from Mueller Hut, above The Hermitage, over the Hooker Valley with New Zealand's highest peak, Aoraki / Mt Cook (3724 m), towering above it on the ridge crest (left). PHOTO: EGON EBERLE.

Aoraki translates in English to 'cloud piercer', which is most apt for the summit of New Zealand's highest mountain. It is the highest of a number of towering peaks and ridge crests in the central part of the Southern Alps (79.2). This area has the highest average rate of uplift (~10 mm per year) along the whole range. For comparison this uplift rate is about half as fast as the rate of growth of your toenails. With a rate that has been increasing through time, it is estimated that uplift over the last 5 Myr has been in the order of 20,000–30,000 m. As the summit of Aoraki is now 3724 m, it means that an enormous lot of rock (~80%) has been eroded off and carried away by glaciers and rivers during this time.

This period of accelerated erosion is recorded in the obvious construction of the Canterbury Plains on the east side of the Southern Alps, but the amount of gravel and sand forming this plain is small compared with the unseen quantities of eroded sand and silt (glacial flour) that have been deposited offshore on the seabed on both sides of New Zealand. Longshore drift off both coasts of the South Island is from south to north. Sand swept down the west coast rivers into the sea has been transported north by currents and has built out an enormous shallow continental-shelf area west of Cook Strait and Taranaki over the last 5 Myr. This 2-km-thick offshore deposit contains ~50,000 km^3 of sediment, which is about half of the material eroded off the rising Southern Alps during this time period. The 25-km growth of Farewell Spit (Site 65) in the last 8000 years hints at this process but is minuscule in the scale of the whole offshore west coast sand system. Off the east coast, a significantly smaller quantity of sediment has been deposited in the deep sea, carried down submarine canyon systems into the Bounty Trough and Hikurangi Trench.

Today the upper slopes and surrounding ridges of Aoraki are permanently snow- and ice-covered, with several large permanent ice fields feeding a number of major glaciers that grind their way downslope to the west, south and east (79.1, 79.6). Indeed all of New Zealand's largest and best-known glaciers today (e.g. Tasman, Hooker, Franz Josef, Fox Glaciers) are sourced from precipitation on the high mountain plateau around Aoraki. These glaciers are but tiny remnants of the enormous glaciers that must have carved away large portions of this part of the uplifting Southern Alps during each glacial period of the Ice Ages over the last 2.5 Myr.

As the glaciers melted and their surfaces dropped during each warm interglacial period, left behind are steep mountain faces of fractured rock no longer supported by the glacial ice. As a result, one of the main erosive agents in the Southern Alps today is large landslides and avalanches of rock and ice. The best known of these occurred on 14 December 1991 when 14 million m^3 of rock and ice fell off the upper east face of Aoraki, lowering its height by at least 10 m. This rock avalanche thundered down the mountain-side at speeds reaching in excess of 400 km/h. It swept across the Grand Plateau, down the Hochstetter Icefall and deposited dirty-grey rock debris

➲ 79.2 Summer view of the flat-topped summit ridge of Aoraki (left, 3724 m) from the northeast alongside the pointed horn of New Zealand's second-highest peak Horokoau / Mt Tasman (3497 m).

over the full width of the Tasman Glacier, 7 km distant and 2500 m lower than its source (79.5). These large rock-fall events are not unusual, with others recorded from Aoraki's southwestern and northern flanks in 1974 and 1984. Sometimes the unstable rock faces just suddenly fail, whereas at other times falls may be triggered by earthquake shaking.

The rocks of Aoraki massif are composed of fractured greywacke (Torlesse Terrane) that was deposited as sand and mud on the floor of the deep sea off the coast of Gondwana 250–200 Myr ago (Triassic). These layered sediments were buried to a depth of ~10 km, where the increased pressure and temperature greatly hardened the rocks and weakly metamorphosed them with crystallisation of the new minerals, prehnite and pumpellyite (The Big Picture 17, page 305). These are the upper, less-metamorphosed parts of the pile of similar-aged sedimentary rocks that comprise the continental crust of this part of the Pacific Plate. The pile has been transported westwards towards the Alpine Fault plate boundary in the last 25 Myr, with some layers buckled downwards beneath the mountains and others being forced upwards to build the Southern Alps. During this process the greywacke strata of the region have been tilted up at steep angles, bent into folds and cut by numerous faults, some of which are considered to be back faults with the opposite dip to the east-dipping Alpine Fault (TBP 12.2, page 248), located 16 km away to the northwest.

Best seen from: Classic views are from Lake Pūkaki on SH 8, the front lawn of The Hermitage and from the Hooker Valley Track across its icy lake.

➲ 79.3 Aoraki (centre) towers above the other peaks of the central part of the Southern Alps. Photo is taken looking north over the full length of Lake Pūkaki. The milky colour of the water is due to the fine glacial flour from the Tasman Glacier in suspension in the lake.

➲ 79.4 The classic view of Aoraki across the terminal lake of Hooker Glacier. PHOTO: JULIAN THOMSON.

➲ 79.5 Photo taken two days after the large December 1991 rock avalanche on the east side of Aoraki lowered the peak by at least 10 m. Note how the rocky toe of the avalanche has buried the surface of the Tasman Glacier in the foreground. PHOTO: LLOYD HOMER, GNS SCIENCE, 4596.

➲ 79.6 View northwest showing the sharp ridge crest leading up to the peak of Aoraki. PHOTO: ALASTAIR JAMIESON.

80.

Haupapa / Tasman Glacier, Mt Cook

New Zealand's largest glacier

The 26-km-long Haupapa / Tasman Glacier is New Zealand's longest and largest glacier. It is up to 3 km wide and the ice is more than 600 m thick opposite the Hochstetter Icefall. The present-day glacier, however, is a miniature version of the enormous glaciers that carved out the Tasman Valley and the deep straight basin now occupied by 25-km-long Lake Pūkaki during the peak cold periods of the Ice Ages. The last time ice filled Lake Pūkaki Basin down to the terminal moraine, which now dams its southern end, was during the Last Ice Age between 30,000 and 17,000 years ago. By 6000 years ago the Tasman and other New Zealand glaciers were about as small as they are today, and since then there have been three cooler periods that have caused the expansion of glaciers in their upper valleys. The last cool period was the Little Ice Age between AD 1300 and 1900. Warming temperatures

80.1 View south down New Zealand's longest and largest glacier with its proglacial meltwater lake in the distance. The summit of Aoraki / Mt Cook is top right.

80.2 The small Blue Lakes on the west side of Tasman Lake at the terminus of the Tasman Glacier are kettle lakes formed by the melting of large bodies of ice that were left buried in the lateral moraine as the glacier receded several hundred years ago. When the ice bodies melted they left depressions that have subsequently filled with water. The moraine has become covered in matagouri (*Discaria toumatau*) and Spaniard/speargrass (*Aciphylla colensoi*).

in the 20th and early 21st centuries have caused dramatic reduction in the size of the Tasman Glacier. The glacier's terminus has receded 5 km up the valley since the 1980s, with the creation of Tasman meltwater lake (80.6) between the ice front and the arcuate ridge of terminal moraine that was dumped around the toe of the glacier during much of the Little Ice Age. Since the 1890s the surface of the glacial ice has dropped nearly 100 m, well below the crest of the lateral moraine ridges that had built up on either side.

Between the Tasman Glacier's terminus and the head of Lake Pūkaki there is a gently sloping outwash plain composed of rock and sand debris carved out and transported down by the glaciers of the Tasman and tributary valleys (80.4). These gravels have been redistributed out in front of the glaciers by the ever-changing courses of the braided overflow rivers that come out from the meltwater lakes which have now formed downstream from the toes of Tasman, Hooker and Mueller Glaciers.

Best seen from: The best way to appreciate the receding Tasman Glacier is by ski plane or helicopter.

Take a closer look: To see the lateral moraine, kettle lakes and meltwater lake, take the Blue Lakes and Tasman Glacier View Track from the carpark at the end of Tasman Valley Rd (1-hr return).

80.3 The head of the Tasman Glacier is at the Southern Alps' main divide. In its upper reaches the glacier is relatively clean and white on the surface, although at the time of the photograph (March 2021) there had been a recent landslide from the slopes of Mt Darwin that had flowed down the surface of the slowly moving ice.

80.4 Most of the gravel and sand forming the braided outwash plain at the head of Lake Pūkaki comes from the receding terminus of the Tasman Glacier (up valley to right). Lake Pūkaki, filling the straight valley downstream from here, was carved out by much larger versions of the Tasman Glacier during the peaks of the Ice Age glacial periods. PHOTO: LLOYD HOMER, GNS SCIENCE, 12259.

⬆ 80.5 The Tasman Glacier has a number of tributary glaciers like Darwin Glacier, seen here. The lateral moraine on the sides of the two branches joins together at the fork and is carried down to the Tasman Glacier as a line of dark grey medial moraine on top of the Darwin Glacier.

⬇ 80.6 Tasman Glacier has been receding, like most others, at an alarming rate. Since the 1980s the terminal ice face (far left) has receded 5 km, leaving behind a milky blue Tasman Lake dammed by an arcuate ridge of terminal moraine. High ridges of lateral moraine line either side of the lake and indicate how high the glacial ice was 100 or so years ago.

THE BIG PICTURE 15

Mountain glaciers and their landforms

DURING THE COLD glacial periods of the Ice Age cycles of the last 2.5 Myr, most of the Southern Alps and many of the other more mountainous areas of the South Island and southern North Island were permanently ice-covered. Large rivers of ice flowed down from the highest elevations as glaciers and sometimes spilled out of their valleys onto the surrounding plains. During warmer periods, the vast majority of these ice caps and enormous glaciers melted, leaving behind evidence of their former extent in the landforms they had carved and the rubble (moraine) they had deposited.

Mountain glaciers (TBP 15.1) have source snowfields above the permanent snow line, where large volumes of snow fall. The snow is transformed to ice under the pressure of deep burial, and flows downslope under the influence of gravity. Glacier flow is faster down steeper slopes and in the centre of the glacier, and slower on the sides. More deeply buried ice in the lower part of the glacier flows more plastically, whereas nearer the surface it is more brittle and may break up into pinnacles and crevasses, especially where there is a change in slope or speed of flow.

A glacier is a far more powerful erosive agent than a river and as a solid it also transports much larger loads of eroded rock and sand. This debris is carried down on top of and also within the glacier. Rocks trapped deep in the frozen ice may be dragged along the bedrock, carving out deep depressions. Large examples of these deeply eroded valley floors are the flooded fiords of Fiordland and the elongate southern lakes. The immense abrasive power and rigidity of the ice in the moving glaciers straightens the original meandering courses of stream valleys and carves out the classical U-shaped profile of glacial valleys. Larger glaciers can scour their valleys much deeper than those of smaller tributary glaciers, and when the ice has melted a hanging valley may be left high up on the side of the main valley. The U-shaped profiles of glacially carved valleys are better preserved where the rocks are hard and massive like in Fiordland, but less well preserved where the rocks are softer or fractured, like the greywackes of the Southern Alps.

The sand and small rock fragments embedded in the ice at the base of a glacier polishes the bedrock, while larger hard rocks may scratch or gouge out grooves on its surface. When a glacier encounters hard bedrock it may ride up and over it, polishing the upstream, more gently sloping side and plucking out rock on the downstream side creating a steep jagged face. Large and small examples of these asymmetric bedrock hills, called roches moutonnées, are left behind by retreating glaciers and are easily recognised on valley floors today.

At the head of a glacial valley, plucking produces steep, precipitous rock walls forming an amphitheatre known as a glacial cirque. As the glacier creeps downslope away from the cirque walls, meltwater flows down the created crevasse, refreezing and gluing the ice to the rock which is plucked out by further ice movement. Where two parallel glaciers erode back their valley sides until they intersect, they create a sharp intervening ridge called an arête. If several glacial cirques do likewise they will create a steep pyramidal peak or horn on the dividing ridges between them.

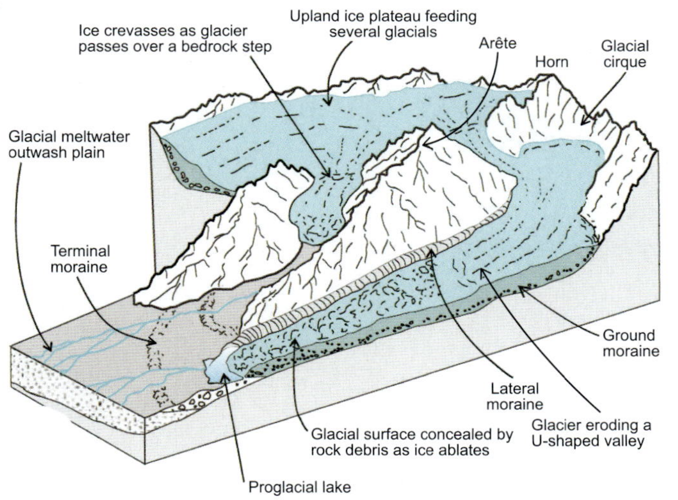

TBP 15.1 Cut-away diagram illustrating the internal and external features of a typical New Zealand mountain glacier. MODIFIED FROM BALLANCE (2017).

The eroded rock and sand debris transported by glaciers is eventually deposited at the glacial terminus forming a terminal moraine ridge, or some may slide off the sides to form elongate lateral moraine ridges. Usually these moraines are unsorted mixes of rock fragments of all different sizes, known as till. When glaciers recede, water may erode the till, removing the fines and leaving behind just the large rocks, called erratics. Often in front of glaciers the moraine is eroded and reworked by meltwater streams or rivers creating braided outwash plains. When a glacier recedes rapidly, a lake is often formed between the uneroded terminal moraine and the melting ice cliff at the glacier front. Commonly the last remnants of a melting mountain glacier leaves behind a ridge of terminal moraine, and the hollow between this and the cirque walls becomes a small lake.

There are considerable differences between glaciers on the wet west side and dryer east side of the Southern Alps. Wet-side glaciers, on the West Coast, descend the steep west face of the Southern Alps and being well supplied with ice they flow fast (metres per day). They advance and retreat on decadal time scales in tune with variations in annual precipitation, and may flow down into the rainforest. Dry-side glaciers are fed by far less ice and flow much more slowly. With recent climate warming they are rapidly thinning and retreating, creating terminal lakes at their rapidly melting toes.

81.

Pareora dolines

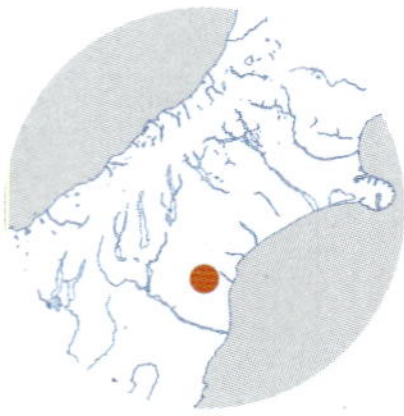

Fields pockmarked with limestone sinkholes

Two flat ridge crests bisected by the South Branch of the Pareora River are pockmarked by New Zealand's most phenomenal fields of dolines (sinkholes). Both the Craigmore and Elworthys ridges are about 10 km long and 1–1.5 km wide and are capped by gently dipping limestone. The limestone accumulated as a mix of sand and shell in a shallow marine shoal area about 24 Myr ago, at the end of the Oligocene. This 30-m-thick shell band was then buried by a further 350 m of silt and sand during the early Miocene (24–15 Myr ago). During this deep burial some of the shell was partially dissolved and recrystallised, creating a moderately dense, sandy limestone. In the last 5 Myr this area has been compressed, uplifted out of the sea and folded into an anticline (upfold) with a north–northeast axis. The crest of the anticline runs along the Craigmore ridge with the layers folded down steeply on its northwest side and much more gently on the southeast side where they extend across the Pareora River South Branch to cap the Elworthys ridge.

As the area was being pushed up, the softer mudstone and sandstone was progressively eroded off, finally exposing the limestone layer underneath. Rainwater that penetrated the forest cover passed down through the soil, becoming slightly acidic, and dissolved out narrow passages along fractures in the limestone down to the base of the limestone. In places water emerges as springs where these underground passages reach the surface along the foot of some of the limestone bluffs that run along the edge of the ridge crests (81.5).

81.1 View northeast over Craigmore ridge, host to one of two awesome fields of dolines on either side of the South Branch of the Pareora River (upper right). These are solution dolines (sinkholes) in the gently sloping surface of a 30-m-thick layer of limestone. They are dissolved along fractures in the limestone. PHOTO: LLOYD HOMER, GNS SCIENCE, 42743.

⊙ 81.2 View northeast across Craigmore ridge showing the lines of dolines following two trends that correspond to two sets of fractures that were preferentially dissolved along to form subterranean passageways through the limestone. The dolines range in size between 3 and 72 m across. PHOTO: LLOYD HOMER, GNS SCIENCE, 4871.

⊙ 81.3 View south over the Valley of the Moa, which has been dissolved and eroded through the limestone layer that caps Craigmore ridge. The limestone layer is exposed in the bluffs. Dry limestone faces beneath overhangs still contain rock drawings made by pre-European-period Māori. PHOTO: LLOYD HOMER, GNS SCIENCE, 93096.

Where the water entered a fracture beneath the forest soil, solution gradually widened the entrance and the passing water carried the dissolved carbonate and insoluble grains away down through the underground passages. Over tens of thousands of years these entrances grew wider and deeper and numerous solution depressions developed on the limestone crests of the ridges. The funnel-shaped depressions (dolines) have inner slopes of 30–40° and depths of 1–13 m. The dolines vary in diameter between 3 and 72 m, and average 20 m across. The dolines are aligned along two trends that presumably correspond to sets of fractures through the rock. One trend is parallel to the axis of the upfold and the second set is northwest at about 70° to the main set. Both sets of fractures were probably formed at the same time as the fold was developing.

Today the doline fields are easily seen from the air because the forest has been cleared and replaced with grass (81.1, 81.2). We can see that the ridge crests and doline slopes are smooth, but it was not like this when the solution dolines were forming. The original rocky limestone surfaces and blocky limestone ridge crests have been buried by 2–5 m thickness of wind-blown loess. This is glacial silt that was eroded by the scraping of the glaciers and ice caps during the peak of the glacial climate periods of the Ice Ages. The eroded silt was picked up by strong winds from the outwash coastal plains and distributed across the countryside. These ridges would have been above the bush line and covered by winter snow and ice during these cold periods.

Best seen from: The dolines on the flat ridge crests are on private farmland and cannot be seen from most roads because they are in the valleys. A view of a few of the sinkholes is possible from Frenchmans Gully Rd between Craigmore Valley and Gordons Valley Rds.

⊙ 81.4 The Pareora dolines were formed by solution of the limestone bedrock beneath a forest cover. During the peak cold periods of the last few Ice Ages these ridge crests would have been devoid of forest and covered by ice and snow each winter. Voluminous dust (loess) derived from the increased erosion of a deforested landscape was blown up onto the rocky dimpled ridge crests, burying them in a 2–5-m-thick blanket of loess which gives the landforms their current smooth appearance. PHOTO: LLOYD HOMER, GNS SCIENCE, 4871.

⊙ 81.5 Limestone containing the dolines at Pareora is exposed in bluffs around the sides of the Craigmore and Elworthys ridges. The dry rock faces beneath overhangs sometimes still contain rock drawings made by pre-European-period Māori. These are protected from vandalism by wire-netting cages. PHOTO: JILL KENNY.

82.

Clay Cliffs, Ōmārama

Pastel-coloured pinnacles and fluted cliffs

⊙ 82.1 Badlands erosion, more typical of Arizona or Utah, can be seen at Clay Cliffs near Ōmārama in North Otago. Rain runoff down the steep uplifted scarp of the active Ostler Fault has produced gullies and pinnacles in pastel-coloured, partly weathered, soft sedimentary rocks.

⊙ 82.2 Parts of the uplifted Ostler Fault scarp at Clay Cliffs have been eroded into a dense mass of tapering pinnacles. The soft sedimentary strata have been tilted down to the west (right) during the succession of earthquake uplift ruptures over the last 100,000 or so years.

⊙ 82.3 These west-tilted layers of gravel and sand at Clay Cliffs have been eroded into pinnacles and flutes. The top of the largest pinnacle is composed of layers of alluvial fan gravel that were deposited over the top of the lower layers after they had been tilted up and were eroding down.

Clay Cliffs, west of Ōmārama in the Waitaki District, is one of the best examples of badlands erosion in New Zealand. Badlands develop by rain and wind erosion of relatively soft sedimentary rocks in relatively dry regions with limited vegetation cover. In this instance the sedimentary rocks consist of soft sandstone and siltstone overlain by conglomerate (82.5). These were deposited as silt, sand and gravel eroded off the rising Central Otago mountains about 4–5 Myr ago. The lower, finer-grained sediments may have been deposited in a small lake, whereas the gravels appear to have been deposited in an alluvial fan where a stream came down from the hills and dropped its load around its mouth on the flat floor of the ancient Maniototo Valley.

These sediments were not deeply buried by overlying sediment and so they are only weakly consolidated and therefore prone to rapid erosion on steep slopes. At this locality the steep slope has been created by a 140-m-high fault scarp along the active Ostler Fault. This fault runs along the foot of the Clay Cliffs with the braided flats of the Ahuriri River to the east and the fault scarp on the west side (82.6). The Ostler Fault has ruptured in major earthquakes about every 3000 years over the last 100,000 or so years. These uplift events have tilted the strata from their original horizontal orientation to 30–40° to the east.

Rapid water runoff down the fault scarp during heavy rain events has eroded out gullies and narrow ravines, leaving behind a pinnacled landscape of 10–30-m-high spires and turrets (82.3, 82.4). Weathering of the sedimentary rocks has produced a variety of pastel orange-brown, pink and cream colours that are accentuated in certain light conditions (82.1, 82.3).

Take a closer look: Accessed off Henburn Rd on private land (small access fee). Walk from carpark (30-min. return).

82.5 Visitors to Clay Cliffs can walk up one of the narrow ravines between the towering pinnacles. The lower portion is composed of sandstone and siltstone that results in the smoother fluted surfaces on the pinnacles. This is overlain by conglomerate that produces a much rougher surface (top right).

82.6 These hills have been uplifted in the middle of the Ahuriri Valley along the Ostler Fault, which runs along their east (right) side. They are tilted down to the west (left) and erosion is biting deeply into the fault scarp, especially at Clay Cliffs in the nearest hill. PHOTO: LLOYD HOMER, GNS SCIENCE, 15864.

82.4 The northern area of badlands erosion at Clay Cliffs.

83.

Cape Wanbrow, Ōamaru

New Zealand's most photogenic pillow lavas

Ōamaru pillow lava is famous as the most photogenic lava pillows in New Zealand. The dark grey of the basalt pillow lava lobes enveloped by thin skins of black glass contrasts with the bright yellowy-orange of the limy sediment that fills the irregular gaps between the individual pillows (83.1). The exquisite internal structure of the pillow lava is best seen in the sea cliffs of Cape Wanbrow, Ōamaru, where erosion has cut a vertical cross-section right through the flow.

Pillow lava forms when basaltic or andesitic lava is erupted into water, usually on the seafloor. The resulting flows consist of numerous elongate, interconnecting fingers (83.6), each elliptical to circular in cross-section and 0.2–2 m in diameter, which may cover the seafloor as an extensive sheet or grow into a steep-sided heap of overlapping lobes (83.3). Pillow lava lobes have a distinctive structure when seen in cross-section. Each has a thin black glassy outer skin (selvedge) formed by rapid chilling of the surface of the molten lava when it encountered cold seawater. This skin confined the lava to finger-like lobes as it flowed along, sometimes breaking through the skin and 'budding off' another lobe. Inside the glassy skin the lava slowly cooled and contracted as it solidified, forming a characteristic radial pattern of joints or fractures (83.4). The upper surface of each lobe (pillow) is usually convex, and the lower part is concave or irregular, fitting snugly over the underlying pillows (83.1).

The Ōamaru pillow lavas are part of the eroded Ōamaru Volcano (83.7), which erupted on the seafloor in relatively shallow water of about 100 m depth. Much of the volcano's products consist of layers of volcanic material (tuff and breccia) ejected by violent explosive eruptions generated by the interaction of molten magma with the cold seawater. The cliffs either side of the pillow lava are made of volcanic breccia (with numerous angular pieces of basalt), which was produced when lava entering the sea was instantaneously chilled and shattered into small glassy pieces. In several places

⊙ 83.1 Erosion of the Cape Wanbrow sea cliffs has exposed a cross-section through an undersea lava flow – known as Ōamaru pillow lava. Photo 5 m across.

83.2 Ōamaru pillow lavas are exposed in the shore platform and lower cliffs below the track that leads down to Second Beach at Cape Wanbrow. Above the track the cliff is composed of thick frittering deposits of loess that was blown in from outwash plains of the glaciers during the Ice Ages.

83.3 A cut-away diagram showing how pillow lava was formed by fingers of lava gently squeezing out onto the seafloor and building up a heap of overlapping lobes. When eroded through and seen in a vertical cliff section, the lobes are said to resemble a pile of pillows.

83.4 A cliff cross-section through several pillow lava lobes of different diameters at Cape Wanbrow, Ōamaru. Note the black glassy selvedges around each pillow (lobe), the radiating cooling joints inside each pillow, and the yellow limy sandstone filling the gaps between the lobes. Photo 1.2 m across.

there are layers of light-coloured limy sediment (sandstone and mudstone) that accumulated around and over the volcano during quiescent periods in the volcano's life (83.5). This sediment often contains fossils of organisms that lived on the volcanic rocks in the shallow sea.

Ōamaru Volcano was one of a number of small basalt volcanoes that erupted around what is now coastal North Otago during the late Eocene–early Oligocene (about 40–30 Myr ago). These volcanoes form the intra-plate (away from a plate boundary) Waipiata Volcanic Field dominated by tuff mounds and cones. It erupted above an anomalous area of partially melted mantle rather like the younger monogenetic volcanic fields in northern New Zealand.

Take a closer look: Do the short walk through the blue penguin colony around the base of the rocks at low tide, or down the track to Second Beach from Ōamaru Lookout (45-min. return).

83.5 The Ōamaru pillow lava flow (bottom left) is overlain by layers of orange and yellow fossil-bearing sandstone and pink volcanic breccia.

83.6 In the low-tide shore platform seaward of the cliff exposure, the finger-like lobe structure of the interior of the Ōamaru pillow lava can be clearly seen. Note the black glassy skin that formed around each lobe. Photo 2 m across.

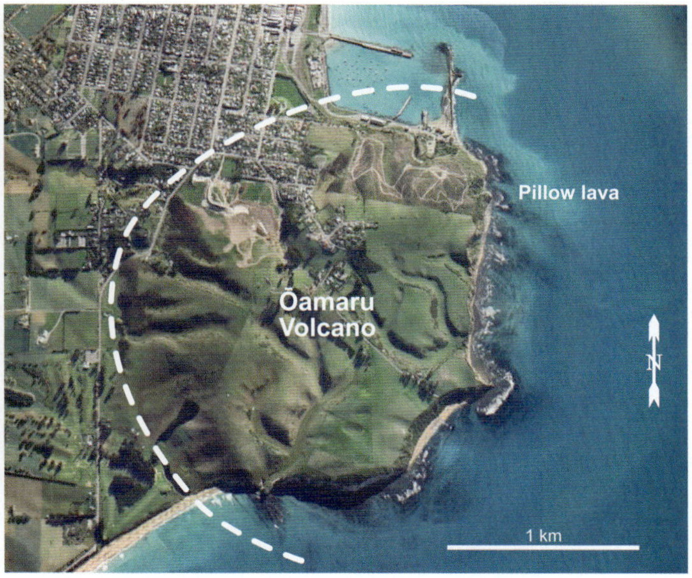

83.7 Map showing the approximate on-land extent of the eroded 35–30-Myr-old Ōamaru Volcano, which forms all of Cape Wanbrow headland, and the dissected ridges and small valleys on the landward side.

84.

Moeraki Boulders

Large spherical concretions on the beach

84.1 Moeraki Boulders consist of spherical concretions, each about 1.5 m across. They sit on the beach and are surrounded by water at high tide, but are high and dry at low tide.

⊙ 84.2 Moeraki Boulders are eroded out of the soft enclosing grey (when fresh) mudstone that forms the low cliff at the back of the beach.

⊙ 84.3 A complete, 1-m-diameter, septarian concretion or 'turtleback' on Moeraki Beach.

Several clusters of perfectly rounded boulders on the beach at Moeraki have attracted attention and evoked stories of their origins ever since the first humans set their eyes on them. The local iwi Ngāi Tahu explained the boulders as the washed-up food baskets from the wreck of a great canoe. Now with modern science we know the actual, if less colourful, history of how the Moeraki Boulders were formed. Their perfectly spherical shapes are not because they have been turned over and over on the beach but because of the way they grew outwardly around a small piece of shell, pebble or wood within the enclosing rock.

The enclosing rock was deposited as mud on the floor of the sea about 60 Myr ago (Paleocene). As more and more sediment accumulated on top of this mud, it was compacted and somewhat hardened. Over tens to hundreds of thousands of years the hard Moeraki Boulders grew within the mudstone. Groundwater started circulating through the rocks and dissolving minerals such as calcite. When the conditions were right, calcite crystals started to form around a small pebble, plant fragment or shell, and progressively grew outwards to form the hardened concretion. The crystals grew in the small voids between the grains, cementing them together. Because the minerals crystallised at the same rate in all directions, these concretions were formed as large spheres.

About 15 Myr ago these rocks started to be uplifted out of the sea and the layers of rock above them were progressively stripped off by erosion. Today that erosion continues and the boulders that we see on the beach today are those that have eroded out of their mudstone host rock that forms the low cliffs at the back of the beach (84.2), in just the last few hundred years.

Take a closer look: In Moeraki Boulders Scientific Reserve, freely accessible by walking 400 m along the beach from the end of Moeraki Boulders Rd. Closer access via private café.

⊙ 84.4 The mosaic-like pattern on the outside of most Moeraki Boulders tells us that they have a complex interior structure that gives them the name turtleback, or septarian. The concretions first formed by calcite crystallisation out of percolating groundwater. Sometime later the hard spheres developed radiating and concentric fractures through them, possibly as a result of the rock drying. Later still these fractures were invaded by more calcite-saturated groundwater, and veins of larger calcite crystals precipitated along them. This Moeraki Boulder has split open along some of these fractures, exposing the orange veins of coarse calcite crystals within.

85.

Matanaka sea caves, Waikouaiti

Longest sea cave in the world

85.1 Beach in large cavern at the end of the main 100-m-long passage of Pink Cathedral Cave, Matanaka. Note that the speleothems do not descend below high-tide surge level. PHOTO: DAVE BUNNELL.

Inside the calcareous sandstone cliffs of Matanaka / Cornish Head on the north side of Waikouaiti Beach, a little north of Dunedin, are numerous astonishing networks of sea caves. The most extensive network, Matainaka Cave, was surveyed by geologist Nicolas Barth and colleagues in 2011, and is currently recorded as the longest sea cave in the world with a total length of connected passageways of 1540 m (85.7). Speleologists (cavers) measure the size of caves by adding up the lengths of all the connected passageways. Matainaka Cave has 18 separate entrances from the sea (85.5), and the furthest point from any entrance is 210 m into the cliff. An adjacent sea cave, named Skull Head Cave, is claimed to be the second longest in the world with a network of over a kilometre of passageways, and there are many other shorter ones.

These caves have been eroded by large Southern Ocean swells and breaking waves, particularly during storms. The 30–40-m-high cliffs are made of fairly massive sandstone that was deposited as sand on the seabed at depths of 100–200 m about 19–16 Myr ago (Miocene). Burial by later sediments and some volcanic rocks compacted these sands into the moderately hard sandstone we see today. Stresses within the buried rocks have created three sets of fractures (joint planes) through the sandstone. It is along these steep fractures that the sea has eroded out the passages. The largest passages have been eroded along joints oriented east–west and northeast–southwest, whereas passages eroded along tighter joints oriented north–south are usually much narrower and smaller (85.7). Most of the cave passages are partially filled with seawater, which rises and falls with the tides and especially as large swells roll in.

↑ 85.2 The coarse golden sand beach inside the low but wide Arch View Cave, Matanaka. In storms this cave would be filled with surging seawater and thus no speleothems could grow in it. The green-algae-covered stalactites that can be seen were presumably formed some time ago when sea level was lower. PHOTO: NICOLAS BARTH.

↓ 85.3 Speleologists explore a wide flooded passageway at low tide within the world's longest sea cave, Matainaka Cave. Note the fracture planes in the ceiling where blocks have been undercut then fallen out. PHOTO: JONATHAN CARR.

➲ 85.4 A kayaker explores a narrow passage with extensive flowstone down the sides in Skull Head Cave, Matanaka, currently the second-longest sea cave in the world. PHOTO: DAVE BUNNELL.

➲ 85.5 The low sea cliffs of weakly bedded sandstone north of Matanaka. Note the erosion scallops and sea cave entrances along steep joint planes through the rock. PHOTO: NICOLAS BARTH.

➲ 85.6 The Matanaka sea caves have only been explored and surveyed by intrepid speleologists since 2010. The caves are only safe to enter during low tide on rare calm days with little swell. PHOTO: NICOLAS BARTH.

There seem to be several reasons why the waves preferentially erode along the joints. The first is that it is an existing fracture that the water can be injected into under the pressure of surging waves. The second is that near-surface groundwater has flowed along these joints over hundreds of thousands of years, when sea level was lower, and penetrated and softened (weathered) the sandstone, as can be seen in orange-stained zones of iron oxide on either side of them. Erosion along close-spaced joints may eat out the rock sideways within the tidal zone, thereby undercutting the roof which may then drop out as a block and be broken up by the surging waters. In this way some large caverns have been created near the end of passages inside the cliff, and some of these caverns have sandy beaches that are dry at low tide (85.1, 85.2).

One of the most unusual and attractive features of the Matanaka caves is the development of speleothems within some of them. Fresh groundwater passing down through the sandstone, particularly along some joints, has dissolved out lime (calcium carbonate) contained within the sandstone and then reprecipitated it on the ceiling and down the walls of some caves. Speleothems cannot grow under seawater and so the lowest flowstone and stalactites are usually above high-tide level within the caves (85.1, 85.2). In several places, however, there are stalagmites on cave floors surrounded by seawater at high tide. These must have grown when sea level was lower, prior to 8000 years ago, and provide evidence that some parts of the Matanaka cave networks are much older. The low height of ceilings (less than 7 m) and horizontal nature of these caves attests to them having been formed intertidally by the sea, and therefore some parts were eroded during earlier warm interglacial periods when the sea was up at a level similar to today.

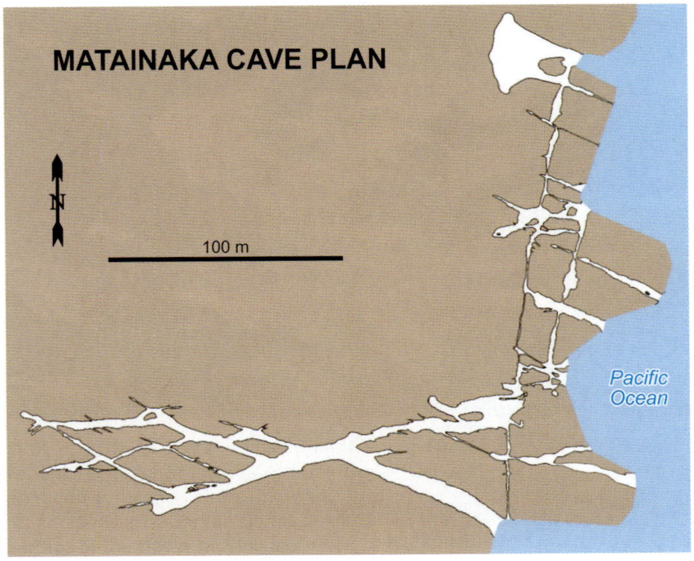

➲ 85.7 Simplified plan of the world's longest sea cave, Matainaka Cave, with a total 1540 m of passageway. Note that the passageways have three orientations, which reflect the three sets of joints they have been eroded along. PLAN: NICOLAS BARTH.

Take a closer look: The sea cave entrances can only be viewed from the ocean side, and can be explored by kayak only on calm days. Guided kayak tours can be arranged locally.

THE BIG PICTURE 16

Dunedin Volcano

DUNEDIN VOLCANO IS a large intra-plate shield volcano (40 km diameter) that erupted between 16 and 11 Myr ago (middle Miocene). An intra-plate volcano erupts through a tectonic plate away from a plate boundary. The magma that erupted was highly variable in composition. Geochemical studies suggest it was initially derived from intermittent melting of the lithosphere at 50–60 km below the surface, influenced by pulses of magma rising from the aesthenosphere at greater depth. This initially basaltic magma stewed and was altered in composition in deep magma chambers by crystallisation of some minerals and mixing with melted surrounding rock. This led to the production of more silica-rich and alkalic (unusually sodium- and potassium-rich) andesite, trachyte and phonolite magma. All of these compositions were erupted from a variety of vents at different times.

Eruptions began in the sea in shallow water where the molten rising magma hit cold water, resulting in explosive (phreatomagmatic) eruptions of glassy volcanic ash, breccia, and minor pillow lava. Early in the volcano's life there were many small eruptions from a number of volcanic vents, but later the volcano built up a sizeable shield volcano of numerous basaltic lava flows centred above what is today Port Chalmers. The youngest activity was the eruption of thick flows in the northeast. The highest part of the shield was probably about 1000 m higher than present in the vicinity of Mt Cargill. In many places small, steep-sided domes of viscous phonolite lava were extruded out on the surface of the shield. Today they form many of the high points around Dunedin, such as Mt Cargill, Saddle Hill, Scroggs Hill and Highcliff Hill (TBP 16.1). During the eruptions there were quieter periods when the volcano was forested, as remains of some of these trees are found fossilised in swamp and lake deposits within the volcanic sequence.

Since eruptions ceased, the shield volcano has been eroded by streams radiating outwards down the slopes from the top and by the sea around its lower slopes. During the more recent Ice Ages, some of the stream valleys eroded down below sea level during cold periods when sea level was a lot lower. The two deepest valleys are oriented northeast–southwest, possibly aligned along the northern end of the Akatore Fault. These two deep valleys eroded into the heart of the volcano, exposing some of the plumbing and a large diatreme (volcanic neck) infilled with breccia at Port Chalmers. When sea level rose to its present level, most recently by 7500 years ago, it flooded up both of these valleys and over the low saddle in between, creating a strait separating the elongate Otago Peninsula island from the mainland. Since then a wide flat sand tombolo has been thrown up by the sea at St Clair, turning the island into a peninsula and the strait into elongate Otago Harbour (TBP 16.3).

TBP 16.1 Saddle Hill (right) and Jaffray Hill (left) in South Dunedin are steep-sided domes of viscous phonolite lava extruded out on the surface of the Dunedin Volcano. PHOTO: LLOYD HOMER, GNS SCIENCE, 10324.

TBP 16.2 The Organ Pipes near the top of Mt Cargill were produced by cooling and contraction of a basalt lava flow as it solidified.

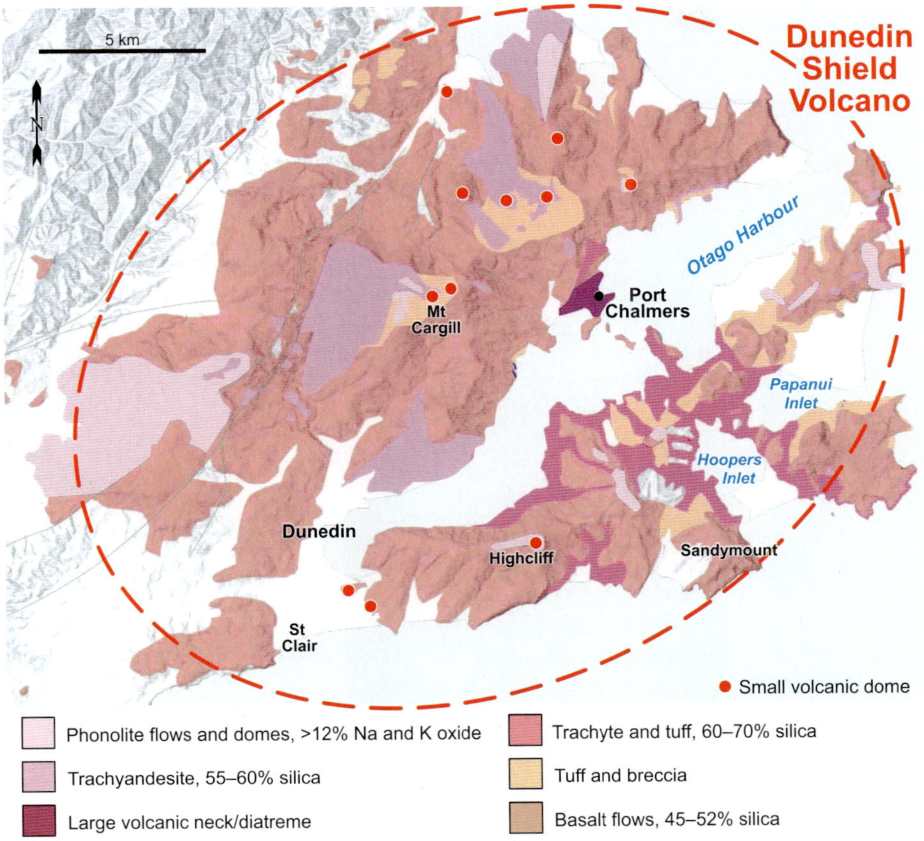

TBP 16.3 Map showing the extent of the eroded footprint of the Dunedin Shield Volcano. BASE MAP FROM JAMES SCOTT.

TBP 16.4 View east along the length of Otago Harbour and Peninsula (right). All the visible land is the partly eroded remains of Dunedin Shield Volcano. The centre of the volcano was in the vicinity of the islands halfway down the harbour. The harbour is the drowned floors of two radial valleys. PHOTO: LLOYD HOMER, GNS SCIENCE, 10395.

86.

Pikiwhara / Sandymount, Otago Peninsula

Natural bridge eroded along phonolite dike

86.1 Lovers Leap is the narrow chasm at the inland end of a remarkable 50-m-high sea tunnel on Otago Peninsula's south coast. The tunnel has eroded along a dike – a sheet of lava that intruded up along a near-vertical fracture through a sequence of thick lava flows on the south side of Dunedin Volcano, about 12 Myr ago.

The impressive coastal cliffs around Pikiwhara / Sandymount and Harakeke Point are eroding remnants of the southeast part of the Dunedin Volcano. The lower cliffs contain volcanic ash and breccia that was explosively erupted in the sea during the early life of the volcano. These are overlain part-way up the cliffs by two thick basalt flows that were erupted about 13 Myr ago out of a vent that now forms Sandymount cone itself. These flows are cut by numerous vertical columnar joints that formed as the lava contracted as it cooled from the molten state to solid rock (86.2).

Since the volcano became extinct about 11 Myr ago, it has been subject to erosion by streams draining radially from the high centre and by wave erosion around the eastern side. In the last 7500 years after sea level had returned to near its present level after the low levels of the Last Ice Age, the waves have pounded the cliffs and eroded out a sea cave along a vertical dike (a sheet of lava that intruded along a vertical fracture through the sequence of tuffs and flows). The dike rock has been altered by weathering and is now softer and perhaps slightly more jointed than the rocks it cuts through, and therefore has been eroded more readily by the waves. The roof inside the back of the cave has collapsed, forming Lovers Leap, and the debris has been flushed away by the waves so that now there is a natural bridge spanning the seaward part of the cave that has not collapsed (86.1). The dike is a lighter colour than the basalt flows it intrudes because it contains more silica (55–60%) as well as a lot more sodium and potassium (12–15%). This dike is one of eight known phonolite dikes that radiate 3–4 km outwards from an inferred volcanic centre in the middle of Hoopers Inlet.

Five hundred metres east of Lovers Leap is The Chasm, which was also a sea cave eroded by waves along a highly jointed phonolite dike. This deep, vertical-sided chasm was formed when the roof of the cave collapsed but the debris has not yet been flushed out by the sea (86.3).

Take a closer look: Walk Sandymount Track from the end of Sandymount Rd for impressive views of the rugged south coast of Otago Peninsula (1-hr loop). Unfortunately the viewing platforms and tracks to Lovers Leap and The Chasm have recently been removed.

86.2 A 40-m-high cliff cut through two columnar-jointed lava flows beside the recently removed Lovers Leap lookout platform (top left). Rubbly breccia occurs between the flows and is distinguished by its lack of vertical jointing, halfway up the cliff.

86.3 The Chasm, near Lovers Leap, was formed by roof collapse of another sea cave that had eroded along a phonolite dike, but unlike Lovers Leap the collapsed debris has not yet been washed out by the sea.

86.4 View east along the southern coast of Otago Peninsula with the peak of Sandymount above and to the right of the area of active sand dunes. Lovers Leap and The Chasm are located in the shade seaward (right) of Sandymount. PHOTO: ALASTAIR JAMIESON.

87.

Tunnel Beach, Dunedin

New Zealand's widest sandstone sea arch

87.1 View north up the coast from the Tunnel Beach access track. The magnificent sea arch is cut through part of the small finger-like peninsula (lower right). The three rock stacks are also made of sandstone and probably were formed when similar sea arches collapsed, separating them from the adjacent cliffs.

87.2 **This dramatic 20-m-wide sea arch has been carved by the waves along vertical joints through 20-Myr-old sandstone at Tunnel Beach, south of Dunedin.**

87.3 **When the tide is full the sea reaches all the way to the base of the sandstone cliffs at Tunnel Beach (to the right of the small sea-arch peninsula). The public access track wends its way down the slope and disappears into a man-made tunnel. The cliffs and most of the slopes are composed of sandstone, capped near the top of the slope by weathered basalt lava flows of the Dunedin Volcano. PHOTO: LLOYD HOMER, GNS SCIENCE, 29605.**

The widest sandstone sea arch in New Zealand can be seen in the cliffs at Tunnel Beach on the southern outskirts of Dunedin city (87.1). It is quite unusual for a New Zealand sea arch, because it has been eroded through massive calcareous sandstone rather than the more common arches in limestone, ignimbrite or other volcanic rock types.

This sandstone (Caversham Sandstone) was deposited on the seafloor in shallow water, about 50 m deep, during the early Miocene about 20–16 Myr ago. The sandstone is grey-white when fresh, but weathers to a yellowy-orange colour where erosion is not rapid. It is a mix of quartz grains and 40–70% lime, which has helped cement the sand grains together into a calcareous sandstone or sandy limestone. Rare fossil lampshells, sea urchins and even rarer whale bones have been found in the sandstone.

Caversham Sandstone underlies all of the Dunedin Volcano, but is mostly seen at the surface and in the sea cliffs on the northern and southwestern outskirts of Dunedin city. Just south of Dunedin the sea cliffs are composed of an uplifted, eroding block of Caversham Sandstone. On either side of Tunnel Beach, these sandstone cliffs and steep slopes are overlain with slight erosional unconformity by weathered basalt lava flows from the Dunedin Volcano (87.3). Some of the volcano's fresh, columnar-jointed basalt lava flows are down-faulted to sea level and form the cliffs at either end of the Tunnel Beach sandstone block at St Clair and Black Head.

The sandstone exhibits weak layering (bedding) and is cut by a number of near-vertical, widely spaced joint planes. These joints are zones of weakness in the rock, and the breaking waves exploit them by blasting out the sandstone, grain by grain, as they erode their way along the fractures. This erosion has created numerous guts, sea caves, pinnacles and the dramatic sea arch at Tunnel Beach (87.2, 87.4).

Best seen from: View from the Tunnel Beach Walking Track (10 mins) off Blackhead Rd, or explore the beach and sea caves at low tide (40-min. return).

87.5 This narrow man-made tunnel, dug in the 1870s, is the only access down the vertical sandstone cliffs to secluded Tunnel Beach.

87.4 This narrow, 10-m-high blade of sandstone towers above Tunnel Beach when seen from sea level. Erosion along two close-spaced parallel joints through the sandstone has produced this unusual feature.

87.6 The man-made tunnel passes down through the sandstone cliffs and emerges at the back of the boulder beach.

THE BIG PICTURE 17

South Island metamorphic rocks

IN NEW ZEALAND, higher-grade metamorphic rocks, like schist, slate, gneiss and marble, primarily occur in the South Island (TBP 17.1). Metamorphic rocks are sedimentary or igneous types that have been greatly changed by enormous pressures, heat and/or hot solutions, usually as a result of deep burial, contact with high-temperature magmas, or under extreme pressures such as along the Alpine Fault.

The widespread Otago Schist in the South Island was formed by regional metamorphism resulting from deep burial in the crust. These rocks were deeply buried, in part by the accumulation of a thick pile of sediment on top of them, but mostly when plate-tectonic forces caused them to collide with and buckle up against the edge of the continental crust of Gondwana about 150–110 Myr ago (Cretaceous). This was during the suturing process when these terranes were added to the coast of Gondwana before they later split away to form Zealandia.

The deeper the burial, the higher the temperature and pressure and the greater the grade of metamorphism. With progressively increasing metamorphism a succession of new minerals crystallise within the rock or grow in veins that cut through it. The mineral assemblage and increasing degree of foliation produced allows us to determine the pressures and temperatures that the various rocks have been subjected to, and thereby the approximate depth they were buried and have subsequently been pushed up from. With burial depths of ~5–10 km, sandstones and mudstones are compressed and hardened into greywacke and argillite rocks with the crystallisation of additional, rather inconspicuous, low-temperature minerals such as zeolites, prehnite and pumpellyite. With further burial to depths in excess of 12 km, weakly layered grey schist develops with the formation of quartz segregations or veinlets, or in volcanic sandstones the development of green chlorite. With progressively deeper burial and recrystallisation, zones of more foliated biotite schist, garnet schist and oligoclase schist form down to depths in excess of 18 km.

Grey schist is the most widespread metamorphic rock in the South Island, forming a vast swathe of country right across Otago and displaced across the Alpine Fault to the Marlborough Sounds. The higher grades of biotite schist, from the deeper levels, have come to the surface only next to the Alpine Fault, and then only in the central reaches where uplift is fastest (TBP 17.2). Although the central parts of the Southern Alps have been pushed up by 20,000–30,000 m, with erosion removing the upper 80–90% of this, the rocks themselves have not been buried this deep. They have instead been conveyed westwards within the upper 10–20 km of the Pacific Plate's continental crust, before curving upwards to form the Southern Alps along the east side of the plate boundary.

The second major kind of metamorphic rock in the South Island is gneiss (pronounced 'nice'). It is widespread in Fiordland and parts of Westland where it, too, has been pushed up by the collisional forces between the plates. It forms by recrystallisation of quartz, feldspar and mica under higher temperatures and pressures than the schist is formed at, and is coarser-grained and less foliated. The South Island gneisses are often closely associated with granites of the Median Batholith (The Big Picture 11, page 207) which form from the same rocks if temperatures exceed melting thresholds.

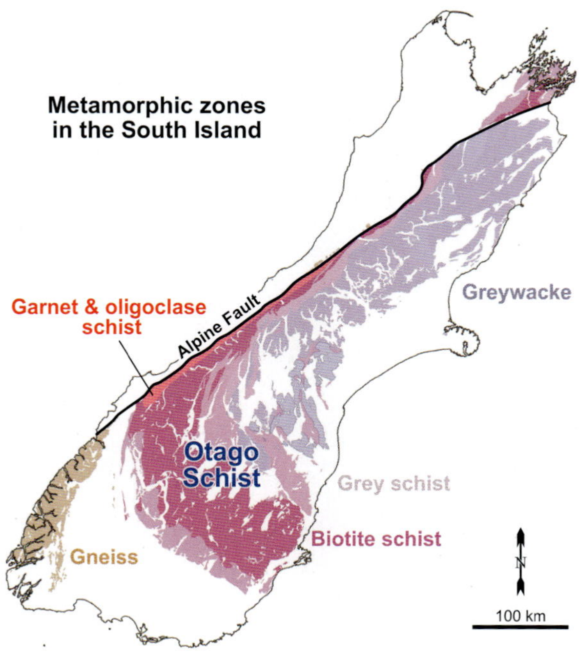

TBP 17.1 Distribution of the various metamorphic rocks at the surface in the South Island. MODIFIED FROM GNS 1:1,000,000 GEOLOGICAL MAP.

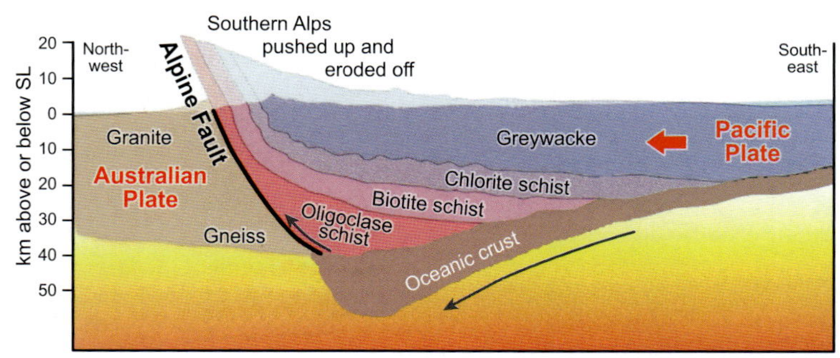

TBP 17.2 Cross-section through the central South Island showing the distribution of the various zones of metamorphic rock and their relationship to uplift on the Alpine Fault. MODIFIED FROM COATES (2002).

88.

Upper Taieri River valley

New Zealand's only large scroll plains

The Upper Taieri River valley contains the two best and largest examples of scroll plains in New Zealand. They are also among the most stunning examples anywhere in the world. To truly appreciate the features' dramatic grandeur they need to be viewed from the air (88.1–88.3).

A scroll plain is an alluvial plain characterised by the presence of meandering river scrolls with many cut-off ox-bow ponds, old braids and wetlands (88.1). Meandering rivers on alluvial plains like these usually have deep pools on the outside of bends, where the current is faster and scours out and erodes the riverbed and bank. This eroded sediment is transported downstream and deposited on the edge of the slow-flowing water on the inside of bends. This process backfills the meanders as they migrate downstream and builds the distinctive scrolls you can see from the air (88.2). Slowly but surely these meander belts migrate downstream because of this cut and backfill process.

Two stretches of the Upper Taieri Valley have well-developed scroll plain features. The downstream (northeastern) belt is 1–2 km wide and extends 25 km between Patearoa and Waipiata, where the Taieri River flows through the Maniototo Basin at an elevation of ~370 m above sea level. The upstream (southwest) belt is also 1–2 km wide and extends 16 km from Canadian Hut to Paerau in the Styx Basin (88.1), 200 m elevation above the downstream belt. The two belts are separated by a 12-km-long gorge where the river has incised 100–150 m into an uplifted block of schist. A delightful 2-km-long, 200–300-m-wide miniature version of the Taieri Scroll Plains occurs 12 km southwest of the Styx Basin, where it has been partly drowned by the artificial Lake Onslow.

These scroll plains also have high values for indigenous wetland biota. The key to the long-term survival of the wetlands is the dynamic natural forces that occur during seasonal flooding which create new wetlands and change existing ones, but these forces are under threat by the numerous water reservoir dams that have been built in the river's upper catchment and reduce the natural flooding events.

88.1 Part of the Upper Taieri Scroll Plains in the Styx Basin showing the low-gradient alluvial valley floor and the upfolded Rock and Pillar Range beyond. The flat top of this schist mountain range is the Waipounamu Erosion Surface, which is found throughout New Zealand and was eroded at about sea level when New Zealand was at its most submerged and low-lying. PHOTO: ALASTAIR JAMIESON.

88.2 The Upper Taieri Scroll Plains have been constructed by progressive migration of the meanders – being eroded on the outside of the bends and sediment deposited in a succession of arcuate ridges on the inside of the bends. PHOTO: ALASTAIR JAMIESON.

88.3 The amazing Upper Taieri Scroll Plains are among the best examples in the world of this unusual kind of dynamic alluvial landform. PHOTO: ALASTAIR JAMIESON.

The Taieri River is the fourth-longest river in New Zealand, almost as long as the nearby Clutha. It follows an inverted U-shaped course (88.4) from the Central Otago block mountains to the sea, just south of Dunedin. The upper catchment is a mosaic of tussock grasslands, farmland, wetlands and bogs, with the scroll plains an integral part of the river's natural drainage system. The low gradient of the valley floor that resulted in the development of the scroll plains, and also the strange route of the river, is a result of major folding and faulting of the mountain blocks in the area over the last 3–2 Myr.

Prior to half a million years ago the Taieri River was about half its length, consisting of the southwards-flowing part, below Hyde. At that time the Upper Taieri was the lower section of a much larger Kye Burn River that flowed southwards and joined the Clutha River near Millers Flat (88.4). Over the last 3–2 Myr or so, oblique compression across the plate boundary through New Zealand produced upfolding (anticlines) of mountain blocks on either side of the Upper Taieri valley. These forces also upfolded the Lammerlaw Range across the downstream end of the south-flowing Kye Burn, causing it first to pond as a lake and subsequently form a new course around the northern end of the Rock and Pillar Range near Hyde and flow into the ancestral Taieri River. Thus only in fairly recent geological time (last 500,000 years) has the Upper Taieri River reversed its flow direction, and in places the gradient of the valley floor is still very low as a result, allowing the development of the scroll plains. Today the Upper Taieri River flows along the axis of a downfold (syncline) between the upfolded northeast-trending Rough Ridge on the northwest side and the Rock and Pillar Range on the southeast.

Best seen from: To fully appreciate these extraordinary features you really need to fly over them.

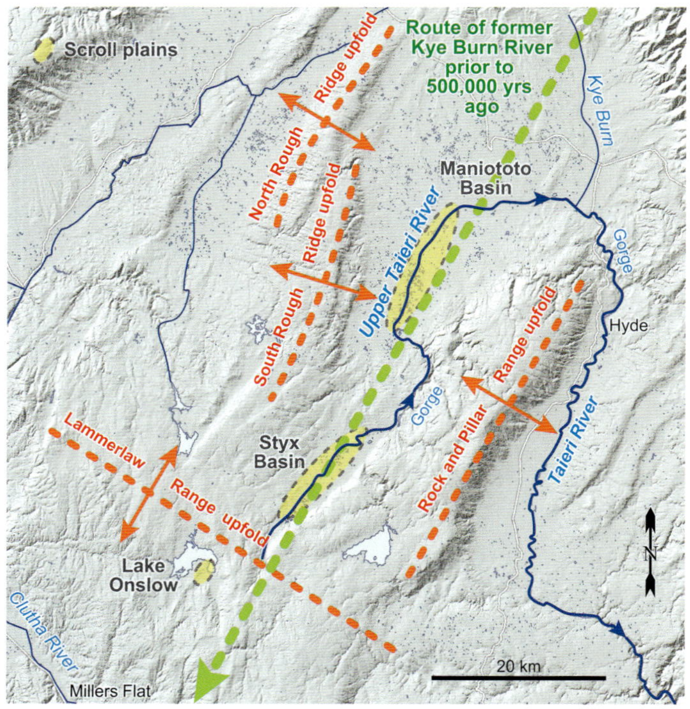

88.4 Map showing the structure of the Taieri Valley with the U-shaped route of the river (blue) swinging around the end of the Rock and Pillar Range. Prior to about half a Myr ago, the Kye Burn (green arrows) flowed southwards through what is now the Upper Taieri as a tributary of the Clutha River. In the last 3–2 Myr the surrounding mountain ranges have been formed by upfolding (anticlines) of the schist, which blocked the Kye Burn and caused it to reverse its flow direction and erode a new course (gorge) through volcanic rocks near Hyde to join the ancestral Taieri River. MODIFIED FROM DIAGRAM BY DAVE CRAW.

THE BIG PICTURE 18

The Waipounamu Erosion Surface and Otago's Basin and Range topography

THE WAIPOUNAMU EROSION SURFACE is sometimes known as the Great New Zealand Unconformity as it can be recognised in many parts of the country. The erosion surface is named from the southern South Island where it is best preserved, especially in Central and East Otago where it forms the smooth tops of most of the uplifted ranges (TBP 18.2).

The surface was eroded across Zealandia between 80 and 25 Myr ago, during and after Zealandia's separation from Gondwana. This was a tectonically quiet period as Zealandia's lithosphere slowly cooled and subsided after the high heat flow associated with the opening of the Tasman Sea. As the land subsided, the sea advanced across it and the irregular land surface was planed down by erosion in the coastal zone.

In Otago, the sea spread across the land from east to west. The marine erosion began near the east coast in the Cretaceous (90–66 Myr ago), and by 25 Myr ago it had eroded down virtually all of southern Zealandia. By this time (late Oligocene) most of Zealandia was submerged and there were only small areas of land left above the waves on which New Zealand's ancient terrestrial biota was able to survive. Beneath the sea, shelly sediment (that became limestone) was deposited over most of the Waipounamu Erosion Surface followed by sand and mud.

In the last 10–15 Myr, and particularly the last 5 Myr, the compression between the tectonic plates has deformed and uplifted the Waipounamu Erosion Surface and its covering of younger sedimentary rocks. As they were uplifted, much of the softer sedimentary rock was stripped off and the erosion surface exhumed over large areas of Central and East Otago. Uplift was much greater in West Otago, west of Cromwell, where not only the marine cover rocks have been removed but the erosion surface itself has been largely destroyed by deeper terrestrial erosion. In a strip along the east coast of Otago and in some of the elongate basins between the uplifted ranges, the land has not been pushed up very much at all and the young marine sedimentary rocks still bury the erosion surface.

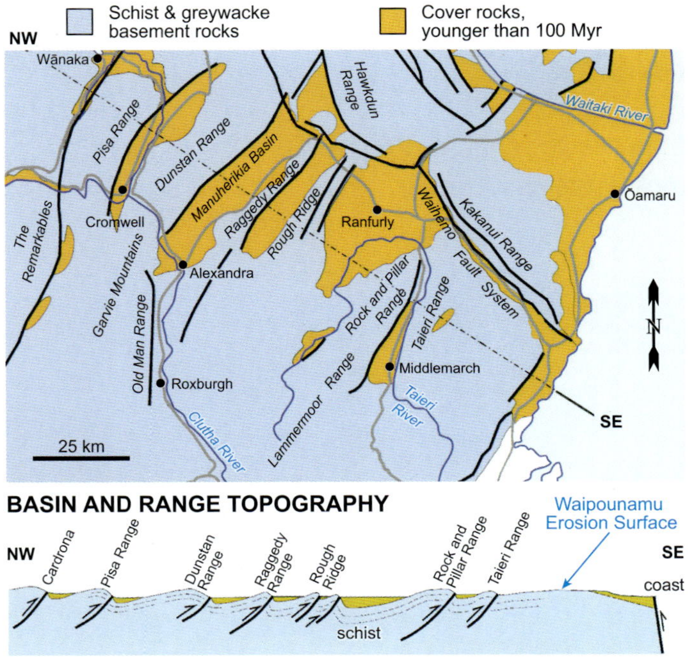

TBP 18.1 Map and cross-section of the basin and range area of Central and East Otago showing how folding and faulting has uplifted a series of northeast-trending ranges and in so doing deformed the Waipounamu Erosion Surface that runs along on top of the basement schist. MODIFIED FROM BALLANCE (2017).

The compression has gently folded the land and erosion surface with many northeast-trending anticlines (upfolds) forming elongate ranges, and synclines (downfolds) forming basins in between. This basin and range topography has been accentuated by displacements along reverse faults running along the southeastern sides of the ranges (TBP 18.1). The elevation, folding and fault displacement that can be seen on the Waipounamu Erosion Surface that forms the land surface through much of Central and East Otago provides clear visual evidence of the deformation that has occurred through this region in the last 5 Myr or so.

TBP 18.2 The Waipounamu Erosion Surface can be seen forming the smooth updomed tops of many of the schist ranges in Central and East Otago.

89.

Kopuwai / Old Man Range, Alexandra

Impressive schist tors formed by Ice Age weathering

↑ 89.1 Tors like these are scattered around the tops of all the Central Otago ranges. They are the harder core stones left standing after the surrounding weathered schist has been blown and washed away. These 5-m-high tors are on the crest of the Old Man Range. Note the horizontal schistosity and vertical joints which have strongly influenced their erosional shapes. PHOTO: NICOLAS BARTH.

↓ 89.2 The Obelisk, 1682 m above sea level, on the crest of the Old Man Range, is one of New Zealand's largest and most impressive schist tors. The flat sides of the 20-m-high monolith are joint planes where surrounding blocks of softer schist have split off and been weathered away by the harsh conditions during the most recent Ice Age periods. The indentations in the sides have been produced by freeze and thaw erosion (cavernous weathering) of softer schist layers. PHOTO: JULIAN THOMSON.

Tors are isolated rock pillars or piles that stand above the surrounding terrain. The smooth unforested ranges of Central and East Otago (east of Queenstown) that are capped by the Waipounamu Erosion Surface (The Big Picture 18, page 309) have pimply surfaces studded with thousands of these schist tors. Some of the tallest and most impressive tors occur near the crest of the Old Man Range, west of Alexandra (89.1, 89.2) – a popular wilderness mountain biking area in summer.

People are often puzzled as to how the tors formed. They were all once part of a continuous block of schist, with their tops not far beneath the Waipounamu Erosion Surface. This was prior to the start of the deep-seated movements that progressively pushed up these mountain ranges in the last 5 Myr or so. At that time the low-lying Waipounamu Erosion Surface was slowly developing a deep (~10 m) weathering profile. Oxygen-saturated groundwater penetrated down into the flat-lying schist along slightly open, steep joint planes, and began weathering inwards from these conduits. Progressively the schist minerals were altered to kaolinite clay and only the more inert quartz grains were unaltered.

When the ranges were uplifted these zones of soft weathered schist and clay were easily stripped off by rain and wind. In some places the whole rock zone beneath the erosion surface was weathered and it was all eroded down. In other places, where the original schist was harder or silicified or where the joints were further apart or tighter, weathering was not complete and fresh hard schist remained as core stones inside the rectangular blocks defined by the joints. When these areas were uplifted, erosion removed the soft weathered rock and clay, and the hard, resistant core stones were left standing high on the surface as tors (89.3, 89.4).

These tors have been formed and moulded during the Ice Age climate cycles. They were not scraped off by glaciers because there was no ice sheet on top of these mountains, even in the coldest glacial periods. This was because of the low precipitation of Central Otago, lying as it does in the rain shadow of the Southern Alps. The mountains were affected by the very cold temperatures, with freeze and thaw of the waterlogged ground breaking up the near-surface rock which gradually moved off downslope in a process called solifluction. The freeze and thaw of rainwater (when present) on the sides and top of the tors also broke their crystalline structure apart, and the wind blew away the frittered rock fragments. In this way cavernous weathering has slowly eaten into the tors, creating notches, hollows and overhangs (89.2, 89.5).

Best seen from: Schist tors can be seen on most of the ranges in Central and East Otago.

Take a closer look: In summer use a 4WD or mountain bike or walk to the Obelisk and other tall tors on the Old Man Range from Symes Rd at Fruitlands (12 km each way).

⊙ 89.3 The flat Waipounamu Erosion Surface tilts down towards the Clutha Valley on the eastern side of the Old Man Range, but is still capped with numerous schist tors.

⊙ 89.4 On this section of the Waipounamu Erosion Surface on the Rock and Pillar Range, 60 km east of the Old Man Range, the schist tors are elongate and aligned in rows. They are believed to be fresh core stones from inside blocks of rock that weathered inwards from two sets of parallel joints. One set of joints is parallel to the tor alignment with each joint located between the lines, and the less prominent joint set is oriented perpendicular to the first. PHOTO: LLOYD HOMER, GNS SCIENCE, 156784.

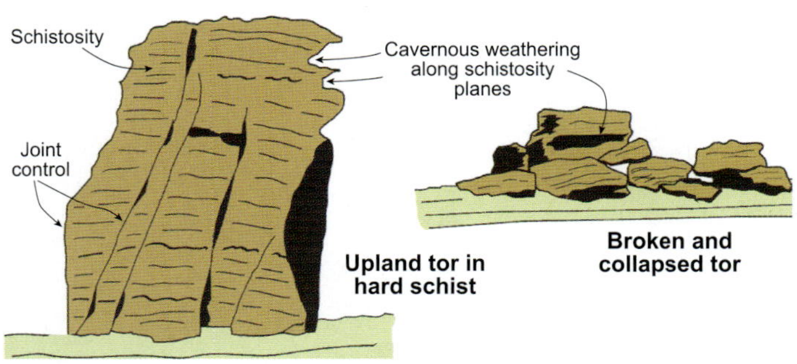

⊙ 89.5 Two kinds of schist tor typically seen on the Old Man Range and other Otago mountain tops. FROM BALLANCE (2017).

90.

Tititea / Mt Aspiring

Pyramidal peak left by glacial erosion

⊙ 90.1 Tititea / Mt Aspiring is often referred to as the 'Matterhorn of the South'. Both have similar rocky summit profiles because they were formed by the same style of erosion at the heads of several glacial cirques. PHOTO: LLOYD HOMER, GNS SCIENCE, 5602.

⊙ 90.2 The distinctive horn shape of Mt Aspiring, seen here in winter, rises 300–500 m higher than all the surrounding ridges and peaks. It is the highest mountain (3033 m) in New Zealand outside of the Aoraki / Mt Cook area. In the foreground is the 7-km-long, gently west-flowing (to the left) Bonar Glacier, which sits at an elevation of 2100–1700 m. PHOTO: LLOYD HOMER, GNS SCIENCE, 3508.

The Māori name Tititea translates to 'glistening peak', most appropriate for the sun reflected off the summit's ice in winter and crystalline schist rocks in summer. Like the Matterhorn in Italy's Alps, the summit of Mt Aspiring is a steep pyramidal peak, or horn, produced by the headwards retreat of three or more glacial cirques (90.1). In this instance there are four glacial cirques that have come together to create Aspiring's conical peak. Mt Aspiring lies just 2 km west of the South Island's main divide. Meltwaters from the glaciers that carved out Mt Aspiring drain northwest to the Tasman Sea.

Within 10 km of Mt Aspiring there are more than a dozen pyramidal peaks produced by similar glacial erosive processes (90.4, 90.5). They are mostly linked together by knife-edge ridges, called arêtes, which were formed between two adjacent mountain glacial valleys. Where two glaciers have eroded near-parallel U-shaped valleys or two glacial cirques meet, an arête is often all that is left separating them. The narrow ridge crest or pyramidal peak is often sharpened by freeze and thaw weathering processes. The steep mountain slopes may also be steepened by rock avalanches or erosion of unstable rock, especially after glacial ice that was supporting a rock face has melted away.

The Mt Aspiring area of the Southern Alps has several kinds of glacier. The majority flow down steep mountain valleys and have a glacial cirque at their head. A few are hanging glaciers, seemingly plastered across the upper slopes of a mountain or ridge, like Bonar, Kitchener, Avalanche and Maud Francis Glaciers (90.2, 90.5). These slowly creep downslope and have broad glacial fronts, maybe several kilometeres across, with steep or even overhanging ice faces. Periodically, especially with summer thaw, large ice and rock avalanches break off from these glaciers and sweep down into the valleys below.

During the peaks of the major cold glacial periods of the Ice Ages the whole Mt Aspiring area would have been a large ice plateau with only a few of the highest rocky peaks exposed above its surface. Enormous glaciers fed from this ice plateau flowed out from the mountains and eroded the deep valleys now occupied by the Arawhata, Waiatoto, Matukituki and Dart Rivers and by Lakes Wānaka and Wakatipu.

Mt Aspiring has been carved out of grey schist (Torlesse Terrane; 90.3). The boundary between two of the major terranes (Caples and Torlesse) that were added onto the side of Gondwana before Zealandia split away runs through the Mt Aspiring massif. Aspiring's schist was deposited as sand and mud on the floor of the deep sea between 280 and 250 Myr ago (Permian). These sediments were buried to a depth of about 12–14 km, compressed, hardened, and metamorphosed into schist rock, 150–110 Myr ago. Within the last 7 Myr these schists have been pushed upwards by plate-tectonic forces and exposed by removal of the overlying rocks by rapid erosion.

Best seen from: Mt Aspiring can be seen on the horizon on a clear day from a number of viewpoints around Wānaka and the Wānaka Mt Aspiring Rd. Closer views require a long tramp and mountaineering skills.

⊙ 90.3 In summer much of the winter snow on the summit pyramid of Mt Aspiring melts, exposing the hard foliated grey schist it has been carved out of. PHOTO: LLOYD HOMER, GNS SCIENCE, 3812.

⊙ 90.4 Just 7 km northeast of Mt Aspiring are two more pyramidal peaks, Fastness and Sisyphus, also formed by the headwards erosion of three to four glaciers.

⊙ 90.5 Maud Francis Glacier is a classic hanging mountain glacier, and clings to the west side of the arête (knife-edge ridge) between the pyramidal peaks of Mt Avalanche (left) and Rob Roy Peak (right), 3–7 km south of Mt Aspiring. The junction of the east and west branches of the Matukituki River can be seen in the distance.

THE BIG PICTURE 19

The former Wakatipu Super Glacier

QUEENSTOWN'S LAKE WAKATIPU is New Zealand's largest proglacial lake (TBP 19.1), formed by the melting and recession of the Wakatipu Super Glacier after the peak of the Last Ice Age about 18,000 years ago. The 80-km-long lake fills a valley that has been over-deepened by glacial scour and dammed by a thick pile of terminal moraine at the southern end, at Kingston, its former outlet. The lake level is 309 m above sea level and at its deepest is 380 m, or about 70 m below sea level. The lakebed is nearly perfectly flat with original glacier-bed undulations buried by sediment that has flowed into the lake since it formed.

Lake Wakatipu is an unusual 'S' shape with two north–south-oriented arms joined by a central east–west arm (TBP 19.3). Each of these arms is nearly perfectly straight and U-shaped in profile, having been carved out by the relatively rigid river of ice. The lake is 5 km wide at its northern head, tapering to 3 km across at its southern end. At its peak, during the coldest glacial periods of the last 600,000 years, Wakatipu Glacier's surface was about 1000 m above present lake level and nearly 150 km long from the head of the Dart Valley to its terminus south of Kingston. The 5-km-long Dart Glacier at the head of the Dart Valley, 60 km up from the lake, is the only remnant of the Giant Wakatipu Glacier of the past. Today the main water flowing into the lake comes down the Dart and Rees Rivers near Glenorchy. They transport vast quantities of eroded sediment that has built out a large delta capped by a braided river plain near Glenorchy.

The sides of Lake Wakatipu's valley change from lower slopes smoothed off by glacial ice to more rugged topography that has always been above the level of the glacier (TBP 19.2). After the glacier receded and the lake formed, lake level was ~50 m higher than present up until about 10,000 years ago. In many places around the lake, a bench can be recognised that was eroded at this higher lake level. In some places there are up to another nine benches at progressively lower elevations (TBP 19.2) that were formed as the newly created outlet through the Arrow Basin and down the Kawarau River was eroding down.

TBP 19.1 Lake Wakatipu, at 80 km long, is New Zealand's largest proglacial lake, dammed behind a terminal moraine barrier at Kingston and filling the elongate depression left behind by the melting Wakatipu Super Glacier. View north up the northern arm of the lake.

TBP 19.2 View of the Von River delta on the south side of Lake Wakatipu showing lake benches, formed when the lake was up to 50 m higher, 15,000–10,000 years ago. Most of the valley sides have been smoothed by the passing Wakatipu Super Glacier. Only the more rugged rocky tops, such as those seen here in Mt Turnbull, were above the upper surface of the glacier during the coldest parts of the recent Ice Ages.

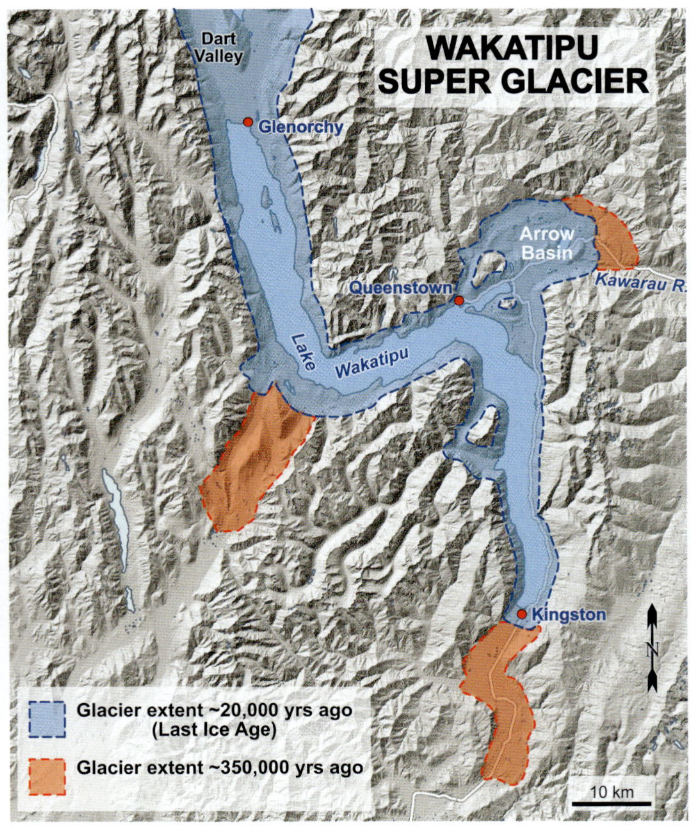

TBP 19.3 The former extent of large glaciers during recent cold glacial periods of the Ice Ages. MODIFIED FROM LEE AND FORSYTH (2008).

91.

Arrow Basin, Queenstown

Best examples of large roches moutonnées in New Zealand

At its peak size, the Wakatipu Super Glacier had a lobe extending eastwards into the Arrow Basin from Queenstown (TBP 19.3 in The Big Picture 19, page 315). The basin floor was not carved out as deep as the main valley and so little of it is now flooded as an arm of Lake Wakatipu. The glacial ice reached 1000 m thick over the Arrow Basin (91.4). The five largest hills in the Arrow Basin (Queenstown, Peninsula, Ferry, Slope, Morven) are the most outstanding cluster of large roches moutonnées in the country, best appreciated from the road up to The Remarkables Skifield (91.1). The floor of Arrow Basin is irregular with heaps of moraine, some containing small kettle lakes, left behind by the receding glacier and meltwater channels eroded through heaps of moraine (91.5). Lake Hayes is not a kettle lake, but fills a glacial depression dammed by sediment from the Shotover River delta about 10,000 years ago.

Best seen from: The best panoramic views are from The Remarkables Skifield road.

➲ 91.1 Slope Hill (left) and Morven Hill (right), seen here on either side of Lake Hayes in the Arrow Basin, are two classic large roches moutonnées of hard schist rock that were overridden by the Wakatipu Super Glacier. The upstream slopes were smoothed off by the moving rock frozen in the base of the glacier, whereas the steep downstream slopes (right) are more jagged, where glacial ice has stuck to the bedrock then plucked it out as it moved along.

➊ 91.2 View south across the west end of the Arrow Basin with the large roches moutonnées of Ferry Hill (left) and Peninsula Hill (right) rising from the basin floor with Lake Johnson in between. The Remarkables range is in the background.

➋ 91.3 View east over Queenstown and the Arrow Basin. During most of the recent glacial periods of the Ice Ages, a lobe of the Wakatipu Super Glacier filled the basin and overrode and smoothed off the hills of harder schist within it – like Queenstown Hill (left) and Peninsula Hill (right) on either side of the entrance. The distinct change from smooth slopes to rugged on The Remarkables range (far right) shows the upper limit of the icy glacier during its maximum size about 600,000 years ago. PHOTO: LLOYD HOMER, GNS SCIENCE, 12120.

91.4 Fog fills the Arrow Basin up to the level that it would have been filled with the Wakatipu Super Glacier at the time of its greatest thickness, about 600,000 years ago. View from Coronet Peak carpark.

91.5 The northwest portion of the Arrow Basin showing a deep meltwater channel (containing Downeys Dam) that was eroded through the valley-floor moraine as the Arrow Arm of the large Wakatipu Super Glacier was retreating after the end of the Last Ice Age. Malaghans Rd runs diagonally across the photo near the Skippers Rd turnoff.

91.6 This erratic boulder at Frankton marina, Queenstown, was transported here from the Routeburn Valley area, 70 km away, by the Wakatipu Super Glacier during the Last Ice Age, about 20,000 years ago.

THE BIG PICTURE 20

Murihiku Terrane strata

Murihiku Terrane is one of the basement terranes of older rocks that built up the continental crust along the edge of Gondwana, which later split off as Zealandia when the Tasman Sea opened up (83–53 Myr ago). Unlike most of the other terranes, this group of related sedimentary strata may have not been moved far from where they originally accumulated on Gondwana's coastal plain and in the shallows just offshore during the late Permian to earliest Cretaceous periods (270–140 Myr ago; TBP 20.1). A total thickness of 6000–7000 m of sedimentary strata accumulated as the coastal fringe was slowly subsiding.

Murihiku Terrane is a 1200-km-long, elongate belt of rocks stretching from Southland to Northland with a gap of 450 km in the middle, where they have been displaced sideways by movement on the Alpine Fault (The Big Picture 12, page 248). In the North Island, Murihiku Terrane rocks outcrop at the

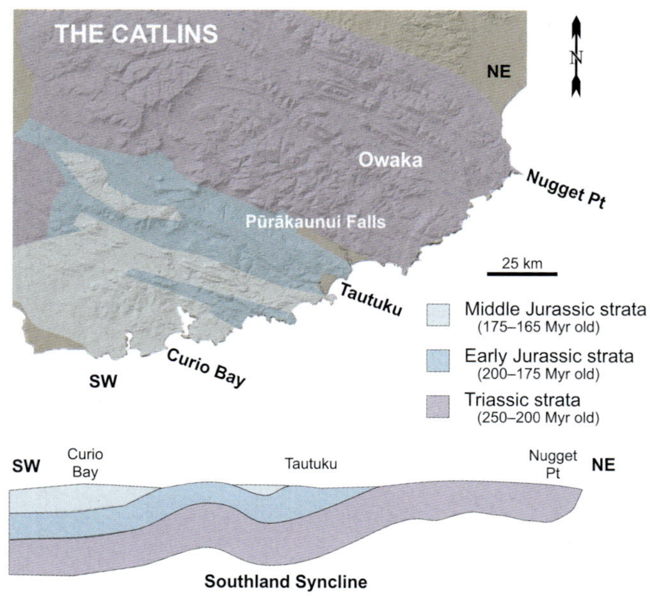

TBP 20.1 Map and northeast to southwest cross-section showing the Southland Syncline (downfold) of the Murihiku Terrane strata that form the Catlins coast.

surface in a wide belt along the west coast from Awakino to Port Waikato. In the South Island there is a thin sliver in east Nelson, and then a wide belt from Te Anau swinging around to form the hills and coast of the Catlins, South Otago.

The belt of sediments accumulated in the fore-arc basin between an active volcanic arc (above the subduction zone) and the deep trench on the edge of the Pacific Ocean. These rocks are mostly mudstone and sandstone, with some conglomerate, which have been hardened into argillite and greywacke as a result of burial to depths of 2–5 km. This level of burial has weakly metamorphosed them with the growth of new zeolite minerals. This low level of deformation and metamorphism means that the marine rocks contain rich and diverse fossil assemblages, mostly of molluscs (bivalves, belemnites, ammonites). The non-marine coastal plain sediments contain fossil wood, sometimes leaves, and three well-known Jurassic fossil forests (at Curio Bay, Kāwhia and Port Waikato).

Late in the history of accumulation of the Murihiku rocks, several additional terranes were accreted onto its seaward edge (between 160 and 105 Myr ago). The compressive forces that accompanied these additions along the plate margin squeezed and folded the Murihiku rocks into several giant, open folds (synclines and anticlines) perpendicular to the coast. Today, one of these large synclines (the Southland Syncline; TBP 20.1) passes through the Catlins region and extends northwest (TBP 20.3) towards the southern end of the Alpine Fault near the west coast. The Nuggets (Site 92) are on the steeply tilted north limb with the strata younging southwards, and Curio Bay (Site 93) is on the opposite limb with north-dipping strata younging northwards.

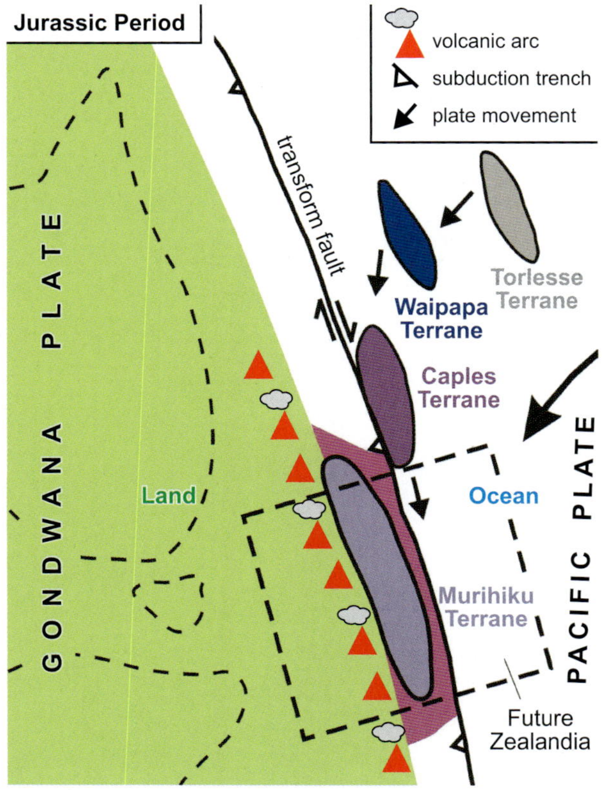

TBP 20.2 Schematic map of the eastern margin of Gondwana, 180 Myr ago, showing the elongate Murihiku Terrane fore-arc basin accumulating sediment, much of which was eroded from the adjacent volcanic arc. Caples, Waipapa and Torlesse Terranes are shown in the process of being moved southwards along the coast by transform faults before they were accreted onto the edge of Gondwana seaward of the Murihiku Terrane.

TBP 20.3 The Hokonui Hills are an inland extension of the Catlins, composed of steeply tilted Triassic strata (seen here) on the north limb of the Southland Syncline. View southeast from south of Lumsden. PHOTO: LLOYD HOMER, GNS SCIENCE, 43841.

92.

Tokata / Nugget Point

Chunky sea stacks eroded from vertically tilted Triassic strata

⬇ 92.1 The photogenic Nuggets rock stacks off the end of Nugget Point, South Otago. The lighthouse was erected in 1870 and stands 75 m above the waves. It was cared for by a lighthouse keeper until being automated in 1989. PHOTO: NICOLAS BARTH.

Tokata / Nugget Point gets its European name from The Nuggets, a group of rocky sea stacks located off the end of this narrow peninsula (92.1). The point and stacks extend 2 km out into the stormy Southern Ocean (92.4). Waves batter and bash the stacks, sending up huge plumes of spray during big storms. These coastal landforms have been eroded out of hard sedimentary rocks that were deposited as layers of sand and mud on the floor of the sea about 240 Myr ago (middle Triassic). These sediments were compressed and hardened by the weight of another 5000–6000 m of layered sediment that accumulated on top of them during the rest of the Triassic and Jurassic periods (240–150 Myr ago). This whole rock sequence is known as the Murihiku Terrane (The Big Picture 20, page 318).

Nugget Point and The Nuggets exist, in part, because the rocks they are made of are slightly harder than the surrounding strata. The orientation of Nugget Point and The Nuggets is northwest–southeast, parallel to the strike (orientation) of the layering. Here the strata have been tilted up to vertical (92.2, 92.3) and are on the northeast limb of a large syncline. The waves have eroded along weaker vertical layers within the sedimentary sequence, and also along a number of cross-cutting joint and fault planes of broken-up rock. This has resulted in the shapes of The Nuggets.

You can see some of these rocks close up in Roaring Bay immediately south of The Nuggets. Moving south the strata become slightly younger and are of late Triassic age (240–200 Myr old). If you walk to the south end of the Roaring Bay beach and clamber around the base of the cliffs at mid–low tide you should see rocks packed with fossil lampshells (92.6), bivalves (92.5) and trace fossils that accumulated just offshore from the coast of Gondwana.

Best seen from: Lighthouse walk (15-min. return) from end of The Nuggets Rd.

Take a closer look: Take the track down to Roaring Bay from its carpark 800 m before the end of The Nuggets Rd to see the rocks and fossils at the south end of the bay (low tide, 30-min. return).

◐ 92.2 The vertical strata are clearly seen in the nearshore Nuggets. These strata were deposited as horizontal layers on the floor of the sea and have subsequently been folded up to vertical at this locality.

◐ 92.3 Erosion has preferentially eroded along some of the softer vertical layers creating separate nuggets and guts. Erosion on cross-cutting joint planes has also helped separate The Nuggets.

92.4 Elongate Nugget Point and The Nuggets offshore owe their orientation to the strike of the vertical sedimentary layers that they are composed of. PHOTO: LLOYD HOMER, GNS SCIENCE, 3957.

92.5 This boulder on the beach at Roaring Bay, just south of Nugget Point, is packed with fossil shells of a 220-Myr-old (late Triassic) mussel-like clam called *Manticula*. With deep burial the shells have been squashed together, producing the swirling patterns in the rock.

92.6 This boulder at the south end of Roaring Bay contains beautiful fossil examples of an ancient, winged brachiopod (lampshell), called *Clavigera*, that lived in shallow water off the coast of Gondwana about 220 Myr ago (late Triassic). Photo 20 cm across.

93.

Tumu Toka / Curio Bay, The Catlins

Best example of a Jurassic fossil forest in New Zealand

At Curio Bay, in the south Catlins, is New Zealand's most famous fossil forest. This short stretch of coastline has been set aside as a Scientific Reserve in an effort to protect the fossil forest. The silicified logs and tree stumps are best seen in the shore platform that is awash at high tide (93.1). The forest grew here on a river plain near the coast of Gondwana during the Age of Dinosaurs in the middle Jurassic (about 170 Myr ago). Indeed, dinosaurs most likely wandered through this forest although no fossil bones nor footprints have yet been found here.

The shoreline cliffs and shore platform from Curio Bay and around Waikawa Harbour South Head (to the east) are composed of a 40-m-thick sequence of nearly flat-lying sandstone beds with thin interbedded mudstone layers and occasional gravel-filled river channel deposits (93.3). They are some of the younger beds of the Murihiku Terrane that forms the Catlins area (The Big Picture 20, page 318). In this sequence at least 10 separate fossil forests have been recognised that grew on the flood plain at different times (93.3).

The famous and best exposed fossil forest at Curio Bay is the oldest of these. It was buried and killed by a thick layer of sand that swept through the forest during a flash-flood coming off the adjacent volcanic hills. After some time, another forest became established in the top of the sand before it, too, was

93.1 Curio Bay fossil forest has been exhumed by marine erosion on the southern coast of the Catlins. Here on the intertidal platform is the forest floor, with numerous long, slender fallen tree trunks and small conical mounds that surround petrified tree stumps.

⊙ 93.2 Numerous in-situ fossilised tree stumps protrude above the level eroded shore platform, which is the exhumed forest floor.

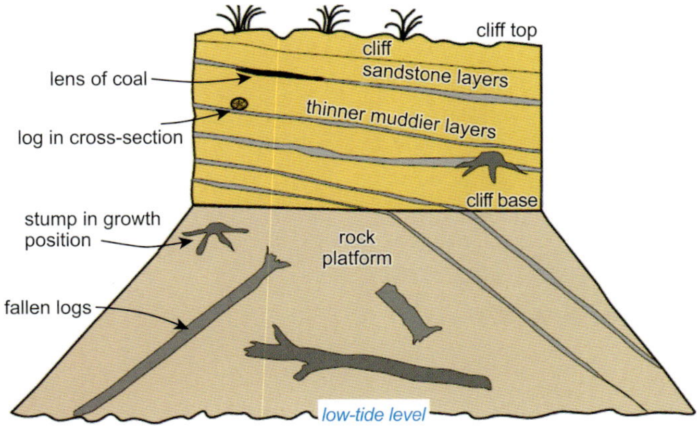

⊙ 93.3 Diagram illustrating how the Curio Bay fossil forest remains are seen on the bedding plane (forest floor) on the intertidal platform and in cross-section through time in the cliffs to the east. MODIFIED FROM BALLANCE (2017).

buried and killed by the next blanket of thick sand brought down by another massive flood event. This cycle was repeated many times over a matter of just a few millennia. The preserved tree stumps (93.2) and slender trunks and branches (93.1) are mostly of relatively young trees, seldom more than 20–30 cm in diameter, and thick soils had insufficient time to develop. Thus we infer that these flash-floods possibly occurred every few hundred years.

Because the layers are flat-lying here, when you walk on the shore platform among the logs and stumps you are actually walking on the exhumed Jurassic forest floor. After deposition, the sand layers must have remained waterlogged to preserve the wood they had buried. Once the fossil forests were even more deeply buried, percolating groundwater passed through the rocks dissolving minerals and becoming supersaturated with silica. As this water flowed into the wood, silica was precipitated and turned it to stone (petrification). The silica replaced the wood cell by cell, retaining the internal structure (93.5), which allows paleobotanists to identify what kinds of trees they were. The majority are conifers (similar to modern kahikatea and tōtara) that would have formed a 15–30-m-high forest canopy along with some tall tree ferns. The understorey appears to have been dominated by ferns and lycopods. Flowering plants (angiosperms) did not appear on Earth for another 140 Myr after this Jurassic forest was growing. This region was located close to the South Pole at the time, and yet forest grew down there because there was no ice cap and the global climate was much warmer than today.

Best seen from: The carpark on Waikawa Harbour South Head and from the top of the stairs down to Curio Bay.

Take a closer look: Visit the intertidal platform at mid to low tide to walk among the in-situ fossil tree stumps and fallen logs. Walk 100–400 m east to see the several levels of fossil forest and tree stumps in cross-section.

⊙ 93.4 View west along the coast of the southern Catlins with Curio Bay and its extensive intertidal platform with the fossil forest in the centre. Slope Point, the southernmost part of the South Island, is at the top of the picture. PHOTO: LLOYD HOMER, GNS SCIENCE, 1283.

⊙ 93.5 This cone of rock contains the petrified stump of a gymnosperm tree that grew here about 170 Myr ago (middle Jurassic). The annual growth rings of many of the trees are preserved in the silicified wood. Photo 80 cm across.

94.

Key Summit, Milford Road

High glaciated ridge crest between glacier-carved valleys

If you are not up to tramping or mountaineering, then a half-day hike up Key Summit on a clear day is the next best thing for getting the feel of a formerly glaciated mountain landscape. Key Summit is a 1000-m-high ridge crest sitting 600 m above the glacier-carved Hollyford Valley and lying between the heads of the Eglinton and Greenstone Valleys, which also were glacial ice pathways during the cold peaks of the Ice Ages in the last million or so years.

At the peak of the Last Glacial period (18,000 years ago) and several before that, the upper parts of the Hollyford Glacier covered Key Summit with as much as 500 m of moving ice. Rocks embedded in this moving ice ground down and rounded off the top of Key Summit. Shallow irregular hollows eroded in the Key Summit surface are now small water-filled tarns, and erratic boulders lying around are of gabbro rock transported across here by the glacier from the Darran Mountains at the head of Hollyford Valley to the west (94.1, 94.2).

Looking around at the spectacular mountain-top views the evidence of glacial erosion is everywhere. Jagged, pointed mountain peaks and ridges are horns and arêtes produced by headwards glacial erosion on two or more sides (94.1). Hanging valleys with wide U-shaped profiles are visible on both sides of the Hollyford (94.3). They were carved by small tributary glaciers that did not erode as fast or as deep as the major Hollyford Glacier between them. Marian Valley, which is easily seen from Key Summit and a lookout on the Milford Rd, has often been referred to as a hanging valley, but closer examination shows it is not. It has steep valley sides all the way down to its junction with Hollyford Valley (94.5) and the lower portion is filled with a large landslide that has dammed picturesque Lake Marian (94.6).

Geologically, Key Summit is in a unique faulted squeeze zone between several ancient terranes produced in part by the enormous sideways displacement along the Alpine Fault. These terrane belts (Caples, Dun Mountain–Maitai, Murihiku, Median Batholith) contain rocks of Permian–early Cretaceous age (formed 270–100 Myr ago) that reappear on the opposite (west) side of the Alpine Fault, 400 km away in Nelson. Rivers and glaciers erode more rapidly along the crushed rock in these fault zones, producing the relatively straight courses of the Hollyford, upper Eglinton, and Pyke Valleys. Key Summit itself is composed of hard grey and green sandstone and siltstone of the Dun Mountain–Maitai Terrane that was deposited off the coast of Gondwana about 250 Myr ago.

Best seen from: To experience the magic and panoramic views from Key Summit you have to be reasonably fit to ascend the track from the Milford Rd (3–4 hrs return).

➔ 94.1 In clear, fine weather Key Summit is one of New Zealand's most magical places. The summit (918 m) has been rounded off by the glaciers that passed over it during the peak of several Ice Ages. The ice eroded small hollows on the summit that are now attractive tarns. In the distance to the west are the Darran Mountains with the horn of Mt Christina (2474 m) in the centre formed by erosion of glacier valley cirques on all sides.

94.2 These erratic boulders on the top of Key Summit are made of gabbro and were left behind here by a glacier that picked them up in the Darran Mountains, in the vicinity of the Homer Tunnel, during the cold peak of an Ice Age period.

94.3 To the east of Key Summit are the Ailsa Mountains that form the main divide between the west-draining Hollyford Valley (left) and east-draining Caples Valley. Note the small glacier-carved hanging valleys separated by razor-sharp ridges (arêtes). The mountains are made of hard sedimentary rocks that were deposited on the seafloor about 250 Myr ago (Permian). PHOTO: MARIAN PREDESCU, WIKIMEDIA COMMONS (CC0).

94.4 View south on the east side of Key Summit ridge looking down the glacier-carved Greenstone Valley containing Lake McKellar dammed by landslide debris. PHOTO: MARIAN PREDESCU, WIKIMEDIA COMMONS (CC0).

➔ 94.5 View west from Key Summit across the Hollyford Valley to the steep-sided glacial Marian Valley. It is not a hanging valley, but the lower portion has been partly filled by a large landslide after its glacial ice had melted.

➔ 94.6 Lake Marian in the steep-sided Marian Valley, across the Hollyford from Key Summit, is impounded by a large bouldery landslide. PHOTO: MICHAL KLAJBAN, WIKIMEDIA COMMONS (CC-BY-SA-4.0).

95.

Piopiotahi / Milford Sound

Deep glacier-carved valley drowned by sea-level rise after the Last Ice Age

⭕ 95.1 Rāhotu / Mitre Peak (1683 m, centre) is the most iconic landform at Milford Sound. Its diorite rock has been carved by glaciers on all sides into its glacial horn.

Milford Sound is the most accessible of New Zealand's spectacular drowned glacier-carved valleys, or fiords (95.1, 95.2). Hundreds of thousands of tourists are drawn to it every year to cruise its length and marvel at some of the highest sea cliffs in the world (up to nearly 2 km high), especially on a gloriously fine day. Thunderous waterfalls tumble over the lips of smaller, tributary hanging valleys (95.3). During and after heavy rainfall, which this place is famous for (annual rainfall is ~7 m), the entire sides of the Sound become a curtain of hundreds of white waterfalls, giving it the well-earned name of 'land of waterfalls'.

Fiordland is a region that has undergone significant uplift and erosion over the past 7 Myr. Before the Ice Ages got into their 100,000-year cycle of cold and warm, rivers drained in all directions off the updomed Fiordland region. Erosion progressively stripped off the upper rocks and uncovered coarsely crystalline plutonic rocks of the Median Batholith, and rocks that had solidified or been metamorphosed to foliated gneiss at depths of 10–20 km, and sometimes even 30 km, beneath the surface. As the Ice Ages set in, mountain glaciers formed during each cold glacial period, flowing down the routes of the original river valleys and eroding them wider and deeper into the U-shaped profile so characteristic of glacier-carved valleys.

The more recent glacial periods of the Ice Ages have been more severe than those prior to 700,000 years ago, and since then many of the Fiordland glaciers exited their mountain valleys and spread out on the plains beyond. Sea level was much lower than now during the coldest times, and the west-flowing glacier toes spread out across the newly exposed, narrow continental shelf and dumped copious moraine (glacier-transported debris) beyond the valley mouths.

The Milford Glacier, and many of the others to the south, built up to several kilometres thick during the cold glacials, and the extra weight of ice forced rocks frozen in the base of the glaciers to gouge out valley floors lower than at their mouths. The Milford Glacier was not just a single valley glacier but a network of glaciers in its headwaters that came together to form a super-glacier around the head of the present Sound. It may be because of the extra thickness and vigour of this

95.2 The Harrison Valley is one of a number that were deeply carved into characteristic U-shaped profiles by their glaciers. These glaciers combined together to form one large super-glacier.

super-glacier that it eroded out the Stirling Basin (the middle section of Milford Sound) deeper than the valley floor down near its mouth to the open sea (95.6).

The most recent Milford Glacier was at its peak size and extent by about 25,000 years ago, with its terminus reaching the coast 4 km beyond the present shore and about 100 m lower than the sea is today. From then through to 15,000 years ago the Milford Glacier began melting and receding back up the valley. As it receded, the glacier dumped a huge heap of moraine inside the valley mouth, and this now forms a 70-m-deep sill beneath the water (95.6). The over-deepened middle section of Milford Sound initially became a freshwater lake behind this moraine barrier. By 12,000 years ago rising sea level, as a result of the climate warming, had overtopped the moraine barrier and saltwater invaded the Milford Valley and turned the lake into a fiord. The floor of Stirling Basin is now almost 400 m below sea level. About 40% of the floor of Milford Sound is now covered in landslide debris that has come down off the steep sides during major earthquake shaking events in the last 15,000 years. The Alpine Fault crosses the entrance to Milford Sound and it appears to have ruptured on a number of occasions since the glacier melted, with over 400 m of sideways movement since then.

Best seen from: Driving down the Milford Rd from the seaward portal of Homer Tunnel to the foreshore of Milford Sound.

Take a closer look: Hire a kayak or take a tourist boat cruise along the full length of Milford Sound to the open sea and back.

95.3 The hotel buildings at Milford Sound (on shoreline to the right) are dwarfed by the enormity of the cliffs on either side of the head of the Sound. Bowen Falls (162 m high) drops over the lip of a hanging valley on the left, and peaks of the Darran Mountains can be seen at the head of one of the glacial valley tributaries.

⊖ 95.4 The 150-m-high Stirling Falls plunge over the front of a small hanging valley into the Milford Sound beneath. The Milford super-glacier eroded its valley bottom 450 m deeper than this small tributary glacier.

⊖ 95.5 Many of the vertical rock cliffs of Milford Sound bear the scars of rocks being scraped across them as they were moved along by the glacier and gouged out prominent grooves. Photo about 200 m across.

⊖ 95.6 Cross-section down the length of Milford Sound showing the Stirling Basin carved deeper than the entrance (left), and the entrance sill composed of a huge heap of terminal moraine and landslide debris left behind by the most recent glacier when it receded. MODIFIED FROM DYKSTRA (2012).

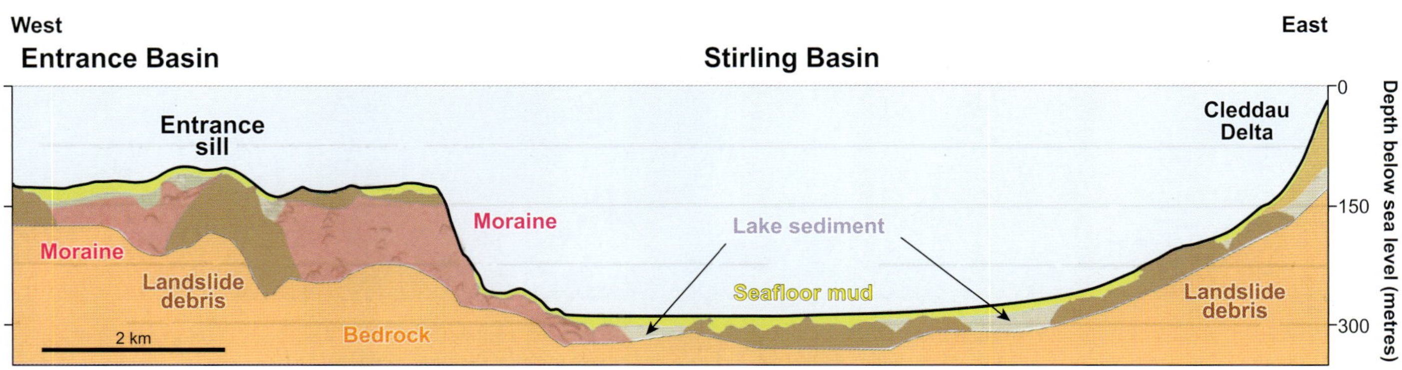

96.

The Chasm, Milford Road

Unworldly river-carved potholes in diorite rock

96.2 When the river is not in flood, most of the potholes are exposed. In their bottoms, many still retain the pebbles that are whirled around by the flowing water to erode the smooth basins.

96.1 These weird, irregular yet smooth basins in the bed of the Cleddau River are among the most extraordinary fluvial potholes in New Zealand. They have been eroded by pebbles being swirled around by eddies in the river as the water passes over the bedrock when in flood.

96.3 View upstream from the walkway bridge over the head of The Chasm on the Cleddau River during typical low flow levels. The unusual potholes have been eroded in the bedrock as the river starts its plunge down into The Chasm. The watercourse itself has eroded along a fracture through the diorite rock.

One of the most impressive examples of fluvial potholes in New Zealand occurs near the Milford Rd in the bed of the Cleddau River as you descend from the Homer Tunnel (96.5) to Milford Sound. Here, at The Chasm, the glacially carved valley floor is crossed by a relatively hard ridge of diorite rock. The glaciers that carved out the two branches of the Cleddau Valley during each glacial period came together just upstream from The Chasm ridge. Where they joined together, their combined more vigorous erosive forces carved a large depression in the valley floor and then they rose up over the diorite ridge as they moved downslope. When the most recent glacier melted and receded up the valleys, this depression became a lake about 15,000 years ago. With the enormous rainfall of this region, the Cleddau River overflowing from this lake has eroded a narrow gorge through the ridge and the lake has mostly drained away.

At The Chasm itself, the river has partly eroded along fractures in the rock creating the slot-like chasm (96.4), which it gushes through with enormous force, especially after heavy rainfall. At the upstream entrance to The Chasm the smoothed grey crystalline rock has been carved into the most unworldly potholes (96.1). A pothole is a smooth, bowl-shaped hollow carved in the rocky bed of a watercourse. These are created by the grinding action of a stone, many stones or coarse sand, swirled around by eddies in the water as it flows over the irregularities in its rocky riverbed (96.2). Exactly how these potholes start to erode seems rather mysterious, but presumably a pebble gets caught in a small crack or depression and its circular movement begins the process. The potholes we see are out of the stream flow for most of the time, and thus their erosion by the swirling stones only occurs during the times of high flood and maximum current speed.

The Chasm and potholes have eroded into diorite, a coarsely crystalline plutonic rock that was formed between 170 and 130 Myr ago (Jurassic–Cretaceous) by slow cooling and crystal growth within a huge intrusive pluton many kilometres below the surface. It is part of the Median Batholith – a series of large plutons that were intruded into the edge of the continental crust of Gondwana before New Zealand (Zealandia) split away (The Big Picture 11, page 207).

Take a closer look: Easy walking track from carpark on Milford Rd (20-min. return).

96.4 Tourists peer down into The Chasm and the potholes at its upstream end.

96.5 The Chasm is 5 km downstream of the Milford end of the Homer Tunnel (seen here). Here you are surrounded by vertical rock walls of the armchair cirque that was formed by ice erosion at the head of a succession of Cleddau Glaciers that carved out the present-day valley during each of the more severe glacial periods of the Ice Ages.

97.

Green Lake, Fiordland

The world's largest terrestrial rockslide

The largest terrestrial rockslide of its type on Earth (by volume) is in the Hunter Mountains in eastern Fiordland, south of Lake Manapōuri. About 12,500 years ago, 27 km^3 of rock slid up to 2.5 km down into the Grebe Valley. The debris and enormous blocks now cover 45 km^2 of mostly hummocky land with numerous ponds and lakes filling hollows on the surface (97.1). The largest of these is the 5 km^2 Green Lake that gives its name to this giant landslide (97.4).

When it came down, the debris bisected the ancestral Lake Monowai, which at that time extended a further 15 km up the valley (now mostly the Grebe Valley) above its present head. At 12,500 years ago the upper northern end of Grebe Valley was occupied by the Monowai Glacier and its meltwaters flowed south into the head of Lake Monowai. The landslide created a second, 11-km-long lake which rose 60 m above the level of Lake Monowai. The new 'Lake Grebe' spilled over the debris dam and eroded out a winding overflow gorge down to Lake Monowai. This new lake rapidly filled with glacial flour (silt) from the Monowai Glacier, but as the glacier receded it opened up the lower Grebe Valley to the north and 'Lake Grebe' drained out into Lake Manapōuri, leaving behind a flat, silt-filled section of swampy valley floor (97.1).

The landslide is composed of Jurassic granodiorite and gneiss (97.6). The 9-km-long southwest-sloping head scarp and failure surface of the rockslide runs along the Mt Cuthbert Fault Zone of weak hydrothermally altered gneiss (97.5). It was down this surface that the landslide moved (97.3). The southern end of the landslide is the straight southern shore of Green Lake and appears to be a steep fault, oriented at right-angles to the head scarp, which fractured as the giant blocks of rock slid down into the lake below (97.2, 97.4). These zones of weak rock explain why the rockslide broke away where it did, but it is believed to have been the Monowai Glacier that had undercut the slope by its deep grinding erosion during the Ice Ages that made the whole mountain-side unstable. As the world warmed after the Last Glacial peak, the glacier melted and its support of the mountain-side disappeared.

⬇ 97.1 The Green Lake Giant Landslide (forested area in the upper middle of the photo) is New Zealand's largest landslide and the largest rockslide in the world. All the hummocky forested land in the centre of the photo comprises the 45 km^2 landslide that slid into the Grebe Valley from the east (left). Green Lake sits in a hollow on top of the landslide debris in the centre distance with the head of Lake Monowai to its right. The flat-floored Grebe Valley with ponds in the foreground is underlain by lake sediment that accumulated in a temporary lake created by the landslide. PHOTO: LLOYD HOMER, GNS SCIENCE, 5577.

Thus conditions were right for a large landslide to occur by about 12,500 years ago. All that was needed was a trigger mechanism, and this is believed to have been a very large earthquake, possibly on the southern end of the Alpine Fault, which shook Fiordland with a magnitude of Mw 7.5–8 or more. All large earthquakes that have shaken the mountainous areas of New Zealand in historical times have brought down numerous landslides, large and small, and this seems the most likely scenario in the case of New Zealand's largest landslide at Green Lake. The Green Lake landslide was so enormous that it was not recognised to be a slide by geoscientists until the 1980s.

Best seen from: The Borland Saddle on the Borland Rd or as you descend the road down into Grebe River valley.

Take a closer look: From the tramping tracks up Mt Burns or to Green Lake hut from the Borland Rd.

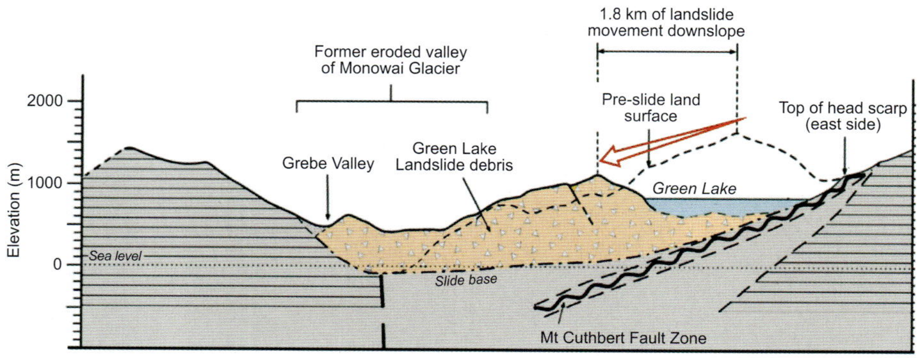

97.2 Cross-section through the Green Lake Landslide showing how it was formed. MODIFIED FROM HANCOX AND PERRIN (2009).

97.3 View south from Mt Burns across part of the 9-km-long southwest-sloping head scarp and failure surface of the Green Lake Landslide that occurred about 12,500 years ago. The forested hummocky area is the top of the landslide debris, with Island Lake (middle) and part of Green Lake (distant left) formed in hollows.

⊙ 97.4 View north over the Green Lake Giant Landslide which occupies all the hummocky lower-elevation land in the middle of the picture. The sloping head scarp is in the shade on the right, and the southern boundary is a sloping fault in the bottom left that forms the straight section of shore of a blue Green Lake (2 km diameter). PHOTO: LLOYD HOMER, GNS SCIENCE, 5516.

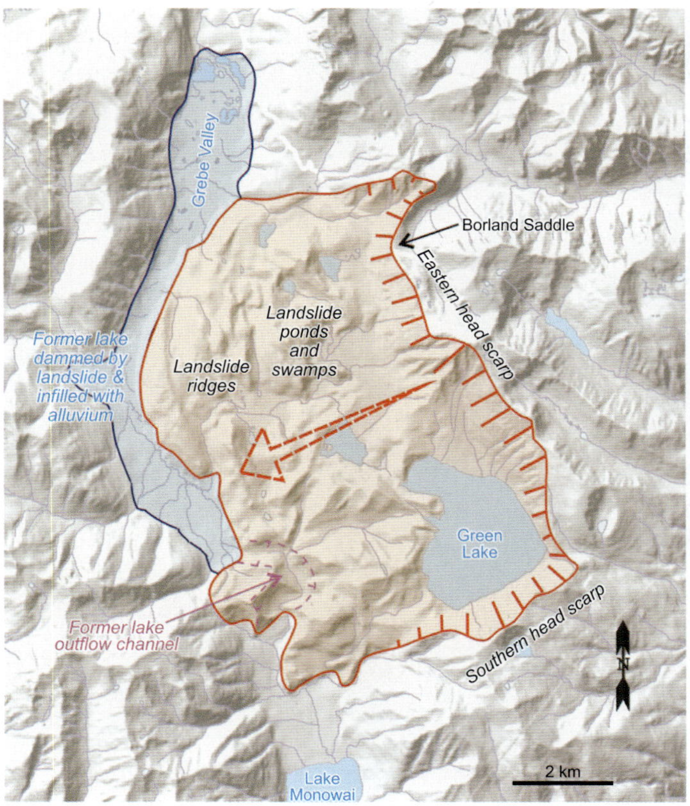

⊙ 97.5 Map showing the extent of the Green Lake Giant Landslide and the former extent of Grebe Lake that it dammed. MODIFIED FROM HANCOX AND PERRIN (2009).

⊙ 97.6 The Green Lake Giant Landslide occurred within gneiss (speckled black and white) and intrusive granodiorite and granite (pinkish white) as seen here beside the Borland Road.

98.

Awarua / Bluff coastal ventifacts

New Zealand's largest ventifacts, abraded by wind-driven sand

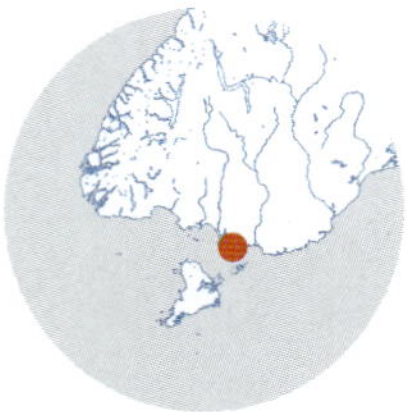

The Foveaux Strait coastline of Southland, between Colac Bay and Awarua / Bluff, is the largest area of ventifacts in New Zealand, and also contains some of the largest einkanter (one sharp edge) ventifacts in the world (98.2). A ventifact (meaning 'made by wind') is defined as a wind-faceted stone that has been abraded, pitted, etched, grooved or polished by wind-driven sand or ice crystals (98.1). The Southland ventifacts are mostly composed of locally eroded rock types, but particularly the finer-grained sedimentary and igneous rocks that occur within the volcanic-dominated Brook Street Terrane. These rocks were erupted and deposited on the deep ocean floor off the coast of Gondwana about 270 Myr ago (Permian). They have subsequently been deeply buried and metamorphosed by the increased pressure and temperature at depth and also partly cooked by the hot temperatures associated with emplacement of the nearby Median Batholith (The Big Picture 11, page 207). More recently these rocks have been uplifted and exposed by erosion.

98.1 Wind-abraded and grooved ventifacts of pebble and cobble size that have been eroded out of glacial sand deposits and now lie exposed on the intertidal flats of Bluff Harbour.

⊙ 98.2 New Zealand's largest ventifact (2.0 × 1.3 × 0.7 m), from Colac Bay coast, now on display outside Riverton Museum.

⊙ 98.3 A large 80-cm-long pyramidal dreikanter ventifact superbly created by the forces of wind-blown sand and ice on the shore of Bluff Harbour. PHOTO: LLOYD ESLER.

The Foveaux Strait ventifacts are no longer forming, and are relicts from earlier colder, glacial times when there were active sand-dune areas along the coast and very little vegetation.

During the long, cold glacial periods of the Ice Ages, and particularly those of the last 500,000 years, sea level was 50–130 m lower than present and Stewart Island was joined to the South Island. At these times there was little if any forest cover over the Foveaux Strait valley and surrounding lowlands of Southland. The climate here was colder and drier and gales were more frequent, and large areas of semi-bare sand dunes developed. It was in these conditions that numerous blocks and pieces of rock that broke off local outcrops were sculpted by the bombardment of sand grains propelled against their sides by the howling winds.

In-situ outcrops and the sides of huge boulders became scoured and scarred by the flying sand, whereas smaller boulders, cobbles and pebbles of the harder and finer-grained rock types were gradually sculpted into characteristic ventifact shapes. The most recognisable shape for a ventifact is a dreikanter (three-edged), having a pyramidal form with three wind-abraded facets and a base (98.3). Some ventifacts are elongate Brazil-nut-shaped einkanters with just two facets above ground and one underneath (98.2). In areas with a prevailing wind direction, the blown sand will often abrade the windward side into a polished flat surface. To develop other abraded faces the ventifact must move. This may happen if the sandy ground on which the ventifact is sitting is scoured out on the lee side by wind eddies. Now unsupported, the ventifact may simply roll over, exposing its underbelly to the blast, and another flat surface develops.

Today, many of the ventifacts are buried by soil that accumulated under coastal forest after the peak of the Last Ice Age, about 15,000 to 18,000 years ago. These ventifacts are encountered by farmers and road-makers disturbing the ground. Places where they can be seen at the surface are mostly in intertidal areas around the edges of sheltered harbours and inlets where recent sea erosion has exhumed the buried ventifacts that now lie loose on the shore (98.4). The early Māori visitors to this area used some of the finest and hardest argillite ventifacts as preformed blanks which they shaped into adzes for a variety of uses. They are legally protected and should not be collected.

Take a closer look: View New Zealand's largest ventifact on display outside Riverton Museum.

⊙ 98.4 In places, the intertidal flats of the estuaries and harbours around Bluff are littered with cobbles and boulders that were in various stages of ventifact formation when the process was stopped by a warming climate. They have recently been unearthed by erosion of sand and soil. Photo 1.5 m across. PHOTO: LLOYD ESLER.

99.

Kaka-kura / Gog and Tu-pouri / Magog, Stewart Island

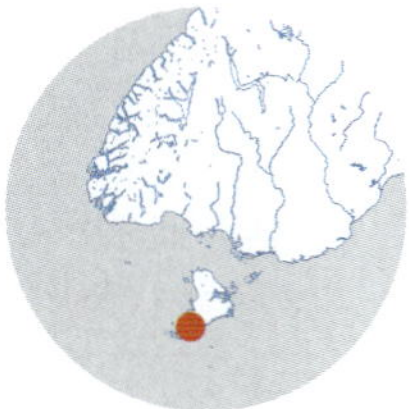

Classic exfoliation domes made of granite

99.1 Kaka-kura / Gog (left) and Tu-pouri / Magog (right) are the two most prominent granite domes on Stewart Island. PHOTO: LLOYD HOMER, GNS SCIENCE, 95318.

99.2 Unusual solution basins in granite and a balancing rock, looking north from the ridge between Gog and Magog with the conical peak of Magog beyond. PHOTO: NICOLAS BARTH.

Kaka-kura and Tu-pouri / Gog and Magog are two rounded, rocky dome-like peaks that dominate the otherwise rather subdued topography of southern Rakiura / Steward Island (99.1, 99.3). They are composed of quartz-rich granite with scattered black plates of biotite. They are eroded out of the Gog Pluton which is part of New Zealand's Median Batholith (The Big Picture 11, page 207). Most of Stewart Island is made of granitic rocks that were intruded as a number of separate plutons between 350 and 105 Myr ago (during the Carboniferous, Jurassic and Cretaceous periods).

Gog Pluton is one of the youngest of these granite intrusions, dated as cooling and solidifying between 110 and 105 Myr ago. Since intrusion of the plutons ceased, the Stewart Island area has been slowly uplifted and a 5–10-km thickness of overlying rocks has been stripped off by erosion. In the last few Myr there has been greater uplift in the west, and the young erosion surface and stream drainage has been tilted down to the east. In southern Stewart Island most streams flow to the east coast, and their eroded valleys have been drowned by rising sea level after the Last Ice Age to form the intricately embayed Pegasus Inlet.

Gog and Magog are rounded peaks known as exfoliation domes. The solidified plutons were formed under considerable pressure at great depths within the crust. When this rock is uplifted and exposed at the surface the decrease in pressure allows their crystal structure to expand ever so slowly. As a result, convex, relatively thin, sheets of rock crack and slough off the exposed surface, aided by the freeze and thaw of ice crystals in the cracks in winter. This onion-skin-like mass wasting produces the rounded dome shape of many granite peaks such as these two.

Best seen from: Located in one of the more remote parts of the country with no formal tramping track access. Easily seen from the South Arm of Port Pegasus, southern Stewart Island, if you can get there.

99.3 Two trampers approach Gog (left, 407 m) and Magog (right, 282 m) granite peaks from the east. PHOTO: KEN SMITH.

99.4 Various forms of weathering of the exposed granite can be seen on the ridges of southern Stewart Island. Here a granite core stone has been hollowed out from the inside by some poorly understood process that includes case hardening of the outside shell of the boulder. This kind of weathering is called tafone and produces 'tortoiseshell' boulders such as this. PHOTO: LEE SAWYER.

99.5 Looking north over granite tors to Magog's conical dome. PHOTO: NICOLAS BARTH.

100.

Ohira Bay, Chatham Island

Columnar-jointed basalt containing lumps of mantle rock

Chatham Island has one of the best examples of columnar-jointed basalt in New Zealand. It is a tourist attraction and photographer's mecca on the coast close to the Port Hutt Road (100.4). The rock formation is an eroding lava flow that was erupted about 80 Myr ago (Cretaceous). The lava that erupted was sourced from about 60 km beneath the ground in the upper part of the mantle. It rose quite rapidly to the surface and erupted from a small fissure. A thick lava flow accumulated in a small valley and slowly cooled and solidified to basalt rock.

As liquid lava solidifies, it shrinks and cooling cracks develop within the solid rock. These cracks often create five- to six-sided columns that are oriented perpendicular to the cooling surfaces (100.1, 100.5). In lava flows the cooling surfaces are the top and bottom of the flow, and so the columns are generally vertical (100.2). Columnar-jointed basalt like this is quite common around the world

⊙ 100.1 Chatham Island's version of the Giant's Causeway at Ohira Bay consists of columnar-jointed basalt that was erupted as a lava flow about 80 Myr ago. Here we see the five- to six-sided columns end-on in the foreground and side-on beyond. PHOTO: JULIAN THOMSON.

➲ 100.2 A mound of lichen-festooned basalt columns above high tide at Ohira Bay, Chatham Island. PHOTO: HAMISH CAMPBELL.

➲ 100.3 Mantle xenoliths of green peridotite, dominantly composed of olivine and orthopyroxene with lesser quantities of chromium spinel crystals, within the Ohira Bay basalt.

and elsewhere in New Zealand, but at this locality the columns are particularly sharply defined and well exposed by the erosive forces of the Southern Ocean.

Within the grey basalt rock there are scattered pebble-sized lumps of yellow-green harzburgite – a variety of peridotite composed of the minerals olivine and orthopyroxene (100.3). These are lumps (xenoliths) of mantle that were incorporated into the basalt magma at depth and carried to the surface and erupted with it.

Take a closer look: Walk 400 m to the foreshore from the Port Hutt Rd, 7 km east of Port Hutt. Access is through a private paddock with a small charge.

➲ 100.4 Ohira Bay, Chatham Island. PHOTO: HAMISH CAMPBELL.

➲ 100.5 Mostly pentagonal basalt cooling columns end-on. PHOTO: HAMISH CAMPBELL.

Other contenders for the top 100 natural wonders

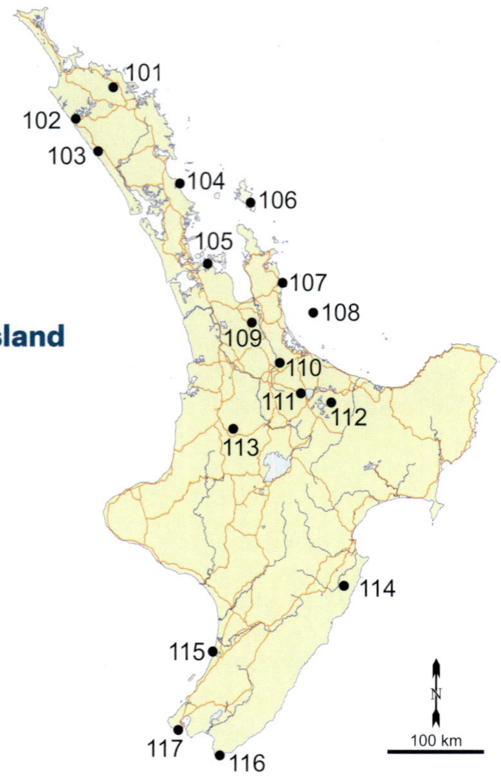

102. Hokianga sand dunes, Ōmāpere

The golden sands of the 150-m-high sand dunes on the northern side of the entrance to the Hokianga Harbour provide an unforgettable backdrop to the quiet tourist hamlet of Ōmāpere across the harbour. The sand has been thrown up on the beach and blown inland to form the dunes over the last 7000 years. PHOTO: ALASTAIR JAMIESON.

103. Kai Iwi Lakes

Sand dunes are usually thrown up on land during times of falling sea level, and as a result there can be numerous belts of coastal sand dune that record these times. Between the belts, there are often elongate depressions that become dune lakes. The Kai Iwi Lakes area on Northland's west coast, just north of Dargaville, contains the largest dune-dammed lakes in New Zealand. Their shorelines are surrounded by golden sand beaches.

101. Lake Manuwai basalt karst, Kerikeri

Lake Manuwai has slightly drowned the best examples of fluted basalt in New Zealand, but in so doing has created a magical setting that can be explored by kayak. Basalt karst like this is rare globally and was formed by slow solution of the basalt beneath a forest cover. The lake is a man-made water reservoir for supplying horticultural establishments around Kerikeri.

104. Dacite columns, Mangawhai Heads

The best example in New Zealand of large cooling columns in a rhyolitic volcanic dome occurs on the Mangawhai Coastal Walkway. The dome was extruded onto the surface of ancient Northland about 19 Myr ago.

105. Motukorea Volcano, Auckland

The small Waitematā Harbour island of Motukorea is the best example of a small basalt volcano in New Zealand. It has examples of the three main landform types formed by different styles of eruption from monogenetic basalt volcanoes: a tuff ring arc (back right) produced by the initial wet explosive style of eruption; a central scoria cone with crater formed by later dry-fountaining eruptions; and a large flat area on top of lava flows (foreground) that poured out from the base of the central cone. It erupted 24,500 years ago. **PHOTO: ALASTAIR JAMIESON.**

106. Te Ahumata Plateau, Great Barrier Island

This flat-topped tableland, at 320–400 m altitude, surrounded by vertical bluffs, is the most distinctive landform on Great Barrier Island. The flat top is the land surface formed by some of the last eruptions on the island. The surface is composed of erosion-resistant ignimbrite that was erupted about 10 Myr ago and subsequently silicified by hydrothermal fluids that also deposited sinter over the top.

107. Paku Island, Tairua

This double-peaked cone on the east coast of Coromandel at Tairua is the eroded central part of an 8–7-Myr-old rhyolite dome. Excellent examples of flow-banded and spherulitic rhyolite can be seen in the intertidal rocks around its base. It is joined to the mainland by a sand tombolo thrown up in the last 7000 years since sea level returned to this elevation. **PHOTO: ALASTAIR JAMIESON.**

108. Tuhua / Mayor Island volcano

Tuhua / Mayor Island is the upper part of a dormant volcano situated out near the edge of the continental shelf in the Bay of Plenty. It has erupted at least 50 times in the last 130,000 years with the last eruption between 1000 and 500 years ago. It is the only volcano in New Zealand made of peralkaline rhyolite or pantellerite (deficient in aluminium), which explains why the black volcanic glass obsidian is better developed here than anywhere else in the country. The obsidian was highly prized as a cutting implement in pre-European Māori times and was extensively traded around New Zealand. The volcano has a gentle shield profile with a large central collapse crater containing two lakes and a small, young volcanic dome. **PHOTO: LLOYD HOMER, GNS SCIENCE, 10664.**

109. Kopuatai peat dome, Hauraki Plains

The largest (102 km^2) and most impressive peat dome in New Zealand is at Kopuatai, 10–20 km north of Te Aroha and Morrinsville on the Hauraki Plains. The dome rises 6 m above the level of the alluvial plain. It has grown to this height over just the last few thousand years as a result of the accumulation of wetland peat (partly decomposed vegetation). A sediment core on the edge of the dome shows that this area was on the intertidal fringe of the Firth of Thames just 6000 years ago. **PHOTO: ALASTAIR JAMIESON.**

110. Hauraki Fault scarp, Matamata

The 400-m-high, 130-km-long scarp of the Hauraki Fault separates the uplifted Kaimai and Coromandel Ranges from the alluvium-filled Hauraki Plains and Firth of Thames. The fault line is the eastern border of the young Hauraki Rift that foundered in just the last 2–3 Myr. The relatively flat, uplifted and east-tilted top of the Kaimai Ranges (seen here) is underlain by ignimbrite that was erupted 2 Myr ago. **PHOTO: ALASTAIR JAMIESON.**

111. Hamurana Springs, Rotorua

These are one of the largest and clearest freshwater springs in the North Island. The water flows through the less-welded ignimbrite and pumice sands of the Mamaku Plateau, and emerges here near the base of the collapsed Rotorua Caldera wall.

112. Tarawera Falls, Kawerau

Tarawera Falls, on the Tarawera River that drains the lake of the same name, is the best example in New Zealand of a waterfall that pours out from a hole in a cliff (right). When the flow is high, water also cascades from a slot higher up on the left.
PHOTO: C. RODLIFFE, WIKIMEDIA COMMONS (CC-BY-3.0).

113. Mangaokewa ignimbrite castle, Te Kuiti

This unusually shaped hill in the Mangaokewa Valley, east of Te Kuiti, resembles a European castle atop a high point. It consists of cliffs of a 1.2-Myr-old sheet of welded, columnar-jointed ignimbrite sitting on top of softer, more erodible rocks. This landform has been created by stream erosion in the short time since the ignimbrite was erupted as a fast-moving cloud of incandescent ash (pyroclastic flow).

114. Silver Range Sandstone, Southern Hawke's Bay

The Silver Range is a similar strike ridge, or cuesta, to Te Mata Peak, 15 km to the north. The ridge has been pushed up by the collision forces between the tectonic plates and tilted steeply (50°) to the west. The range is held up by a thick bed of erosion-resistant sandstone that was deposited on the seafloor about 8 Myr ago (late Miocene). Thick layers of softer mudstone above (to right) and below (to left) the sandstone have been eroded down, leaving the upstanding ridge with smooth bedding plane surfaces tilted to the west (right).
PHOTO: EGON EBERLE.

115. Horowhenua coastal parabolic dune field

The Horowhenua coastline has New Zealand's longest strip of coastal parabolic sand dunes. Along most of its length there is a wide strip of these recently active hairpin-shaped dunes. Orientation of their axes indicates a dominant wind direction from the west-northwest blowing sand inland from the beach. The area also contains numerous lakes that have been dammed by moving sand-dune belts over the past 8000 years.
PHOTO: LLOYD HOMER, GNS SCIENCE, 11589.

116. Kupe's Sail, Cape Palliser

Kupe's Sail, near Wairarapa's Cape Palliser, is a huge tilted triangular bedding plane of hard limestone that has resisted erosion as softer rocks have been eroded away from on top of it. The straight reef in the foreground is part of the same bedding plane along the strike where the section upslope (above) has been removed. **PHOTO: JULIAN THOMSON.**

South Island

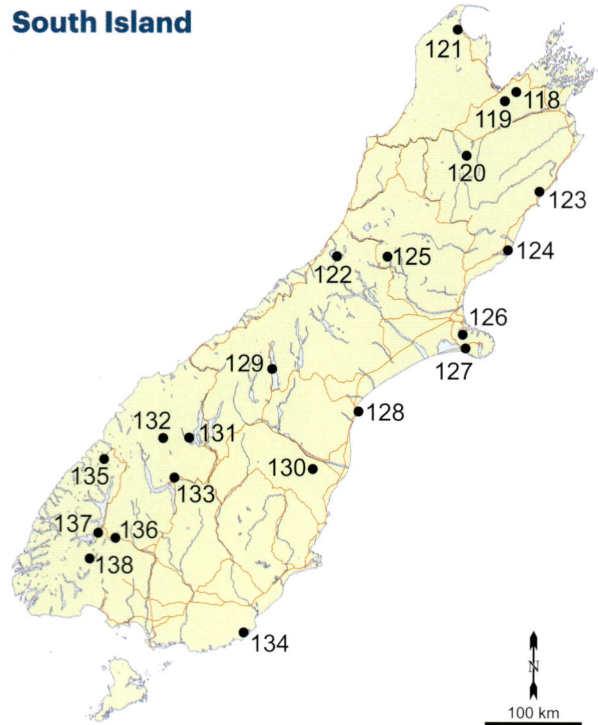

117. Red Rocks, Wellington

Red Rocks Scientific Reserve, on the south coast of Wellington, is the best-known example of red chert and argillite in New Zealand. Here the red rocks overlie pillow lava. The lava was erupted from a spreading centre and the sediment deposited on the floor of the ocean well away from land about 250 Myr ago (Permian). The chert is composed of the recrystallised silica skeletons of a microscopic group of plankton called radiolaria.

118. Dun Mountain, Nelson

Dun Mountain, east of Nelson city, is part of a belt, or terrane, of ultramafic igneous rocks that runs north–south through east Nelson, and is then displaced 450 km south along the Alpine Fault to occur again in northwest Otago. Recognition that these were the same rock belt that had been displaced was a crucial clue that helped geologists Harold Wellman and George Grindley first recognise the Alpine Fault in the 1940s. Dun Mountain is where Ferdinand von Hochstetter in 1859 collected an unusual rock made entirely of olivine and gave it the name dunite, based on the colour of its weathered surface.

119. Whispering Falls travertine, Nelson

These falls, in east Nelson's Aniseed Valley, cascade over the largest deposit of travertine in New Zealand. The travertine (calcium carbonate) has been deposited on the forested hillside from water flowing out of a spring that passes underground through limestone.

121. Devil's Boots, Collingwood

The Devil's Boots are two limestone pillars on either side of Devil's Boots Road, Collingwood, that resemble the gumboots of a giant buried in the ground with only his feet protruding. This photo shows the size of one of the 'boots' with its long overhanging 'toe'. They are the remnants of a former cave system that has been eroded out on the banks of the Aorere River, northwest Nelson.

120. Lake Angelus Basin, Nelson Lakes

Lake Angelus, at the north end of the Travers Range, between the heads of Lakes Rotoroa and Rotoiti in Nelson Lakes National Park, is among the best examples in the northern South Island of a mountain lake filling a glacier-deepened cirque basin. The lake is surrounded by the steep greywacke slopes of the cirque which enclosed the head of a tributary glacier to the large Travers Glacier during the Last Glacial period. PHOTO: ALASTAIR JAMIESON.

122. Hokitika Gorge blue pools

The gorge cuts through granite rock immediately west of the Alpine Fault. The slightly milky blue colour of the water is due to scattering of light by suspended glacial flour – rock fragments that have been finely ground up by glaciers in the Southern Alps upstream.

123. Puhipuhi Syncline, Kaikōura

Looking north along the eroded-out axis of New Zealand's most photogenic syncline, located in the Seaward Kaikōura Ranges just north of Kaikōura. The eroded core of the 10-km-long, tightly folded Puhipuhi Syncline was composed of soft Miocene mudstone deposited between 20 and 15 Myr ago, with the inwards-dipping ridges (limbs of the syncline) made of resistant limestone. PHOTO: ALASTAIR JAMIESON.

124. The Cathedrals, Gore Bay

This is one of the three best examples of badlands erosion in New Zealand. These cathedral-spire-like pinnacles are eroded into sandstone and conglomerate that were laid down on the shallow seafloor about 3 Myr ago. They can be seen from a lookout beside the road just inland and south of Gore Bay, North Canterbury.

125. Falling Mountain, Arthur's Pass

Shaking during the 1929 Arthur's Pass Earthquake (Mw 7.1) caused this 900-m-high hillside to collapse onto Taruahuna Pass, 10 km east of Arthur's Pass. Because the landslide deposit is on the divide between two river catchments, there has been little erosion of it since then. PHOTO: LLOYD HOMER, GNS SCIENCE, 4969.

126. Remarkable Dikes, Banks Peninsula

These two 5-m-wide dikes are large parallel blades of solidified magma on the eroded rim of Lyttelton Volcano's large summit crater. They were intruded into the volcano about 10 Myr ago. PHOTO: LLOYD HOMER, GNS SCIENCE, 1495.

127. Gravel-dammed Lake Forsyth, Banks Peninsula

Lake Forsyth (upper left) is a partly drowned radial valley on the side of Akaroa Volcano that was eroded below sea level during the long periods of lower sea level during the Ice Ages. It was flooded by rising sea level as it reached its present level about 7500 years ago, and subsequently cut off from the ocean by the accumulation of a series of gravel storm beaches that built up the northern end of Kaitorete Spit across its mouth. Most of this gravel was carried down the Rakaia River to the sea and transported north by longshore drift. **PHOTO: LLOYD HOMER, GNS SCIENCE, 1575.**

129. Lake Pūkaki moraine

The lateral moraines on either side of Lake Pūkaki are among the highest and most extensive in New Zealand. Here on the west side of the lake we see a terrace on top of the lateral moraine dating back to the peak of the Last Glacial period (18,000–20,000 years ago). The steep slopes down to the lake show the thickness of the glacier ice at that time. Since the glacier retreat from this locality, by about 15,000 years ago, Whale Stream has eroded through the loose moraine and built out a small alluvial fan on the valley floor. When the level of Lake Pūkaki was raised for hydroelectric water storage, the fan became a delta extending into the lake. **PHOTO: LLOYD HOMER, GNS SCIENCE, 3633.**

128. Dashing Rocks loess, Timaru

Dashing Rocks at the north end of Timaru's Caroline Bay is made of a dark, columnar-jointed basalt flow mantled by a thick covering of orange loess. The flow comes from the South Island's youngest volcano at Mt Horrible, which erupted about 2.5 Myr ago. This locality is of considerable historical importance as it is where trail-blazer John Hardcastle was the first person in the world to document loess stratigraphy and to interpret its origin as layers of wind-blown glacial flour from a series of cold Ice Age periods.

130. Elephant Rocks limestone tors, Duntroon

This much-visited site is a field full of 'elephants', or more correctly rounded limestone tors. The tors are made out of limestone that was deposited as a shell bed on the seafloor about 25 Myr ago (Oligocene). It was buried by further sediment and hardened into limestone. Following uplift and removal by erosion of the overlying rocks, weathering has worked its way along several sets of widely spaced joints and left behind these core stones of fresh limestone that are now being moulded by rain, ice and wind.

131. Matukituki roche moutonnée hillside

This hill on the side of the lower Matukituki Valley has been shaped by the passage of a giant glacier during the Last Ice Age. The lumpy hillside was produced by the glacier ice freezing to the schist rocks, progressively plucking out rocks and carrying them away as the glacier slowly moved down the valley.

132. Landslide-dammed Lochnagar

The many small lakes in the Southern Alps and adjacent mountains are mostly of one of two origins. They either fill cirque basins ponded behind moraine barriers, or are dammed behind large landslides brought down from the steep mountain-sides by earthquake shaking. Lochnagar is a superb example of the latter origin, having been dammed by a huge landslide that came down from the northern slopes (far left). The lake (loch) is 3 km long and 1 km wide and the landslide deposit is 400 m thick and 1.5 km across.

133. Skippers Canyon, Queenstown

This is one of New Zealand's best examples of a narrow meandering gorge, deeply incised into schist rock. It is the upper part of the Shotover River, north of Queenstown, and is navigable by tourist jet boats. Gravel deposits forming river terraces high above the gorge were extensively sluiced in the search for gold.

134. Pūrākaunui Falls, The Catlins

The picturesque Pūrākaunui Falls, in the Catlins, cascades over sedimentary strata that were deposited on the seafloor off the coast of Gondwana about 170 Myr ago (Jurassic). These rocks have been folded and uplifted, but at this locality the sandstone beds are flat-lying, creating the horizontal ledges that the water flows over.

135. Sutherland Falls, Milford Track

Sutherland Falls (three drops totalling 580 m) are often claimed to be the highest waterfall in New Zealand, but there are several other falls in Fiordland with higher single drops. Sutherland Falls is the overflow from Lake Quill, which fills the carved-out depression in the cirque at the head of the Arthur Valley. The Arthur is one of the major tributaries of the super-glacier that carved out Milford Sound. **PHOTO: LLOYD HOMER, GNS SCIENCE, 3830.**

136. Te Anau erratic boulder

This boulder beside the road from Queenstown to Te Anau is the focus of the smallest Scientific Reserve in New Zealand. The boulder is made of a distinctive granite that can be traced back to the mountains 50 km away north of Lake Te Anau. It was transported here by the Te Anau Super Glacier during the peak of a cold glacial period of the recent Ice Ages.

137. Balloon Loop, Waiau River, Te Anau

The upper reaches of the Waiau River link Lakes Te Anau and Manapōuri. Because of the only slight elevational difference between the two lakes, the river has developed a meandering course typical of a mature river in a wide alluvial plain. The tightest of these meanders is the 500-m-diameter Balloon Loop. Its neck has been breached by the river (foreground), creating an island encircled by the river channel. This loop will eventually become abandoned as an arcuate lake cut off from the straightened river channel by deposition of sediment across both ends of the loop. **PHOTO: LLOYD HOMER, GNS SCIENCE, 8022.**

138. Mt Titiroa granite tor, Manapōuri

A number of weirdly sculpted granite tors sit on the top of Mt Titiroa (1715 m), in the Hunter Mountains south of Lake Manapōuri. This tor is distinctive because its top consists of numerous cylindrical panholes, the origin of which is not fully understood. The ground is covered in white quartz sand derived from physical weathering of the granite. **PHOTO: ROSE TURNBULL.**

Explanation of some geological terms and concepts

Types of rock

Most rocks are made of minerals, although a few are composed entirely of organic matter (**coal**). **Minerals** are composed of elements in varying combinations and arrangements and each has its own characteristic properties (e.g. crystal shape, colour, hardness). The most common rock-forming minerals are compounds of silicon and oxygen called silicates, such as feldspars (potassium, sodium and calcium silicates), micas (sheet silicates), augite, hypersthene and hornblende (iron and magnesium silicates). Quartz (silica) and calcite (calcium carbonate) are major mineral constituents of sedimentary rocks.

The rocks that make up the Earth's crust are classified into the following three groups based on their origins.

Igneous rocks (volcanic and plutonic)

Igneous rocks are formed by the cooling and solidifying of molten magma which originates deep within the Earth. This molten magma moves upwards through the crust, often collecting in large underground chambers that solidify as plutons. **Plutonic rocks** are formed when the magma cools and solidifies slowly, resulting in the growth of large crystals of the various mineral constituents. Magma that rises to the surface erupts lava, which cools and solidifies quickly to form fine-grained to glassy **volcanic rocks**. Magma is a mix of liquid rock, crystals (that have grown within the melt as it is cooling) and dissolved gas. Magma with low gas content erupts as lava flows or domes. Magma full of gas might erupt in frothy fountains at the surface. This lava froth cools to form rocks full of holes such as **scoria** and **pumice**. If the rising magma encounters cold water near the surface, the resulting explosive eruptions might throw out vast quantities of ash (fragmented lava and country rock) that accumulates in layers that harden into **tuff**.

Igneous rocks are classified on the basis of their chemical composition, mineral components and texture. Those with a high silica content (more than 63% by weight) are called **acidic** or **felsic**. They are usually light in colour, with feldspar and quartz as the dominant minerals. Those with low silica content (45–52%) are called **basic** or **mafic**. They are dark grey and composed dominantly of augite and feldspar (plagioclase). Igneous rocks with **intermediate** silica content (52–63%) are usually grey or grey-green with dominant minerals feldspar, augite, hypersthene or hornblende.

Texture	PLUTONIC Coarsely crystalline	VOLCANIC Finely crystalline	Typical colour of fresh rock
Acidic or felsic (>68% silica)	GRANITE	RHYOLITE or IGNIMBRITE	White, pink or orange
Acidic or felsic (63–68% silica)	GRANODIORITE	DACITE or TRACHYTE	Light grey
Intermediate (52–63% silica)	DIORITE	ANDESITE	Medium grey
Basic or mafic (45–52% silica)	GABBRO	BASALT	Dark grey
Ultrabasic or ultramafic (<45% silica)	PERIDOTITE	NEPHELINITE	Black or green

X.1 Classification of igneous rocks.

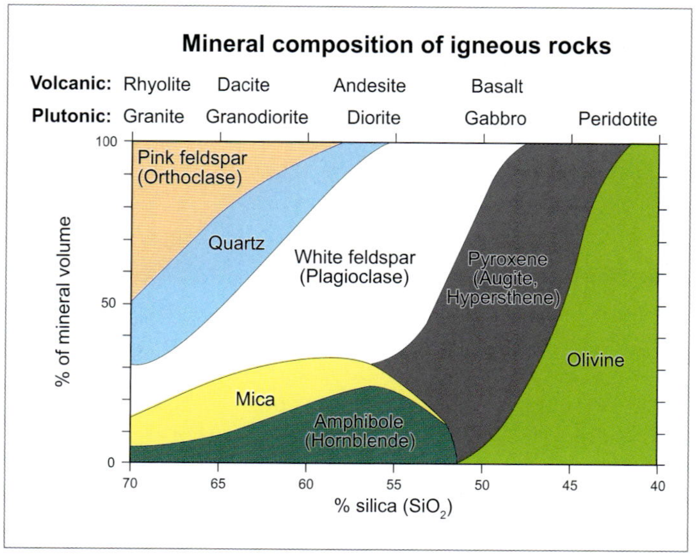

X.2 Diagram illustrating main mineral assemblages present in plutonic and volcanic (igneous) rocks with decreasing amount of silica in their composition.

Grain size	Sediment	Rock name	Name of hardened rock
>25 cm	Boulder gravel	CONGLOMERATE (rounded pieces) / BRECCIA (angular pieces)	No specific name for indurated gravels
6–25 cm	Cobble gravel		
4–64 mm	Pebble gravel		
0.06–2 mm	Sand	SANDSTONE	GREYWACKE
<0.06 mm	Mud (silt & clay)	MUDSTONE	ARGILLITE
All sizes	Calcareous shells	LIMESTONE	MARBLE
<~1 mm	Siliceous shells	CHERT	COAL (hardened plant remains)

X.3 Classification of sedimentary rock.

Sedimentary rocks

Sedimentary rock forms from sediment that accumulates in layers on the floor of the sea, lakes or river valleys. With burial, this loose sediment is compacted and sometimes cemented into rock strata. Special kinds of sedimentary rock are **limestone** and **chert**, which mostly form by the accumulation, compaction and cementation of deposits of the calcareous or siliceous shells of dead marine organisms. Another kind of sedimentary rock is composed of the organic remains of plants, which initially accumulate in swamps as **peat** but with progressively deeper burial become **lignite** and then **coal**.

Clastic sedimentary rocks are those composed of clasts (rock pieces) and are classified based on their grain size (X.3).

Metamorphic rocks

Metamorphic rocks (e.g. **schist**, **marble**, **gneiss**) are sedimentary or igneous types that have been greatly changed by the action of enormous pressures, heat or hot gas and solutions (hydrothermal), usually as a result of deep burial or contact with high-temperature magmas.

How rocks are dated

Relative dating using fossils

Fossils are used to date the sedimentary rocks in which they occur. Geological time has been divided up into a relative time scale of eras, epochs and periods (I.2, page 11) based on the **succession** (evolution) **of fossils** in the strata of progressively younger age. Thus if dinosaur bones are present in the rocks then we know they were deposited during the Age of Dinosaurs (Mesozoic era). This is the fastest and cheapest method of dating rocks. Sometimes fossils can be identified in the field, and the age of the strata are immediately known. The various epochs and smaller units of the geological time scale have been calibrated using absolute dating methods in many places, and thus a relative age can be translated into absolute time.

In New Zealand, different fossil groups are useful for dating strata of different ages (X.4). Extinct groups of larger fossils (**macrofossils**), such as trilobites, graptolites and some brachiopods (lampshells), date Paleozoic rocks. Mollusc fossils are mostly used for dating Permian to Cretaceous strata but can also assist in dating marine Miocene and Pliocene rocks. **Microfossil** groups are more widely used – conodonts are useful in the Paleozoic; spores and pollen date strata of all ages, especially those in freshwater and nearshore sedimentary rocks (because they come from land plants); radiolaria (zooplankton with silica tests) date deep-water marine strata, mostly of Permian–Oligocene age in New Zealand; foraminifera (shell-bearing 'amoeba') date shallow and deep-water marine rocks of late Cretaceous to Pleistocene age.

Absolute dating using radioactive isotopes

An absolute age is the numerical age of the formation or cooling of a mineral or rock given in years. This form of dating is possible because many minerals contain minute radioactive 'clocks' preserved within them. These 'clocks' consist of unstable forms (isotopes) of certain chemical elements which decay over time and transform into more stable isotopes of the same or different elements. The rate of decay is constant for any isotope and is given as the half-life in years. The half-life is the amount of time taken for half the isotope to break down. The half-life of different isotopes varies dramatically between just a few years (^{210}Pb = 22 years), a few thousand years (^{14}C = 5730 years), to millions of years (^{40}K = 1251 Myr). Thus different isotopes are used for dating depending on their age and the type of mineral or rock.

There are two basic ways that radioactive isotopes form and are incorporated into materials used for dating. The first way is during the crystallisation of minerals in igneous or metamorphic rocks. As the parent materials cool and the crystals grow, they include unstable isotopes. From then on these unstable isotopes and their stable daughter isotopes are captured within the crystals, and the amounts of both can

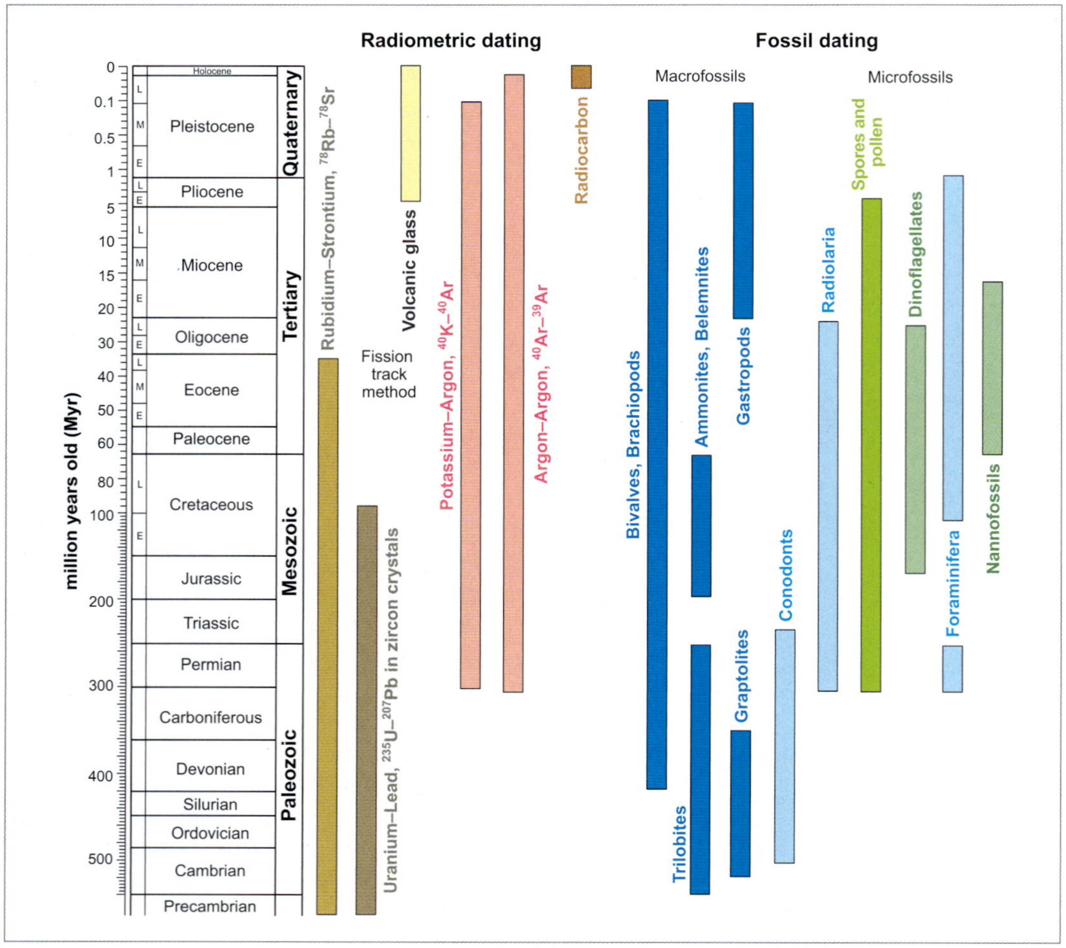

X.4 Methods commonly used for dating New Zealand rocks of different ages.

be measured and the age since crystal formation calculated. The second way that radioactive isotopes are formed is by the bombardment of extra-terrestrial cosmic rays on elements such as carbon (^{14}C) in the atmosphere and beryllium (^{10}Be) and silicon (^{32}Si) in surface rocks.

Radiometric dating is undertaken in sophisticated laboratories in one of several different ways. One method measures the amounts of original unstable and derived (daughter) stable isotopes present with a mass spectrometer. Using the specific half-life and ratio between original and derived isotopes, the time since the mineral was formed can be calculated. In New Zealand, **potassium–argon** (^{40}K–^{40}Ar) isotopes and more recently **argon–argon** (^{40}Ar–^{39}Ar) isotopes have been widely used to date Cretaceous to Pleistocene (100 Myr–12,000 years old) igneous rocks, **uranium–lead** (^{235}U–^{207}Pb) isotopes to date zircon crystals in Paleozoic to early Cretaceous basement rocks (300–100 Myr old), and **radiocarbon** (^{14}C–^{14}N) isotopes to date young Pleistocene and Holocene (0–50,000 years old) organic material such as shell, wood and peat.

An indirect radiometric dating method, called **fission track dating**, determines the density of microscopic damage (called tracks) left in volcanic glass or certain mineral crystals by the fission (radioactive decay) of uranium-238 isotopes. This fission of the nucleus of a single parent uranium atom into two daughter isotopes releases fission fragments which leave trails of damage that are visible and can be counted using a high-powered microscope. Once the uranium content of the material being dated is determined, the decay half-life can be used to calculate the time since the mineral or glass cooled down. In New Zealand, fission track dating has been used on volcanic glass to determine the age of rhyolitic ash deposits of Pliocene–Pleistocene age (5 Myr–10,000 years old).

All dating radiometric methods give numerical ages with different levels of accuracy, which depend on things like the precision of the laboratory measurements and whether the dated crystal, glass or organic matter has suffered from any alteration or weathering since it was formed. Sometimes the mineral clocks have been reset by a period of increased heat and the age determination might date this later event.

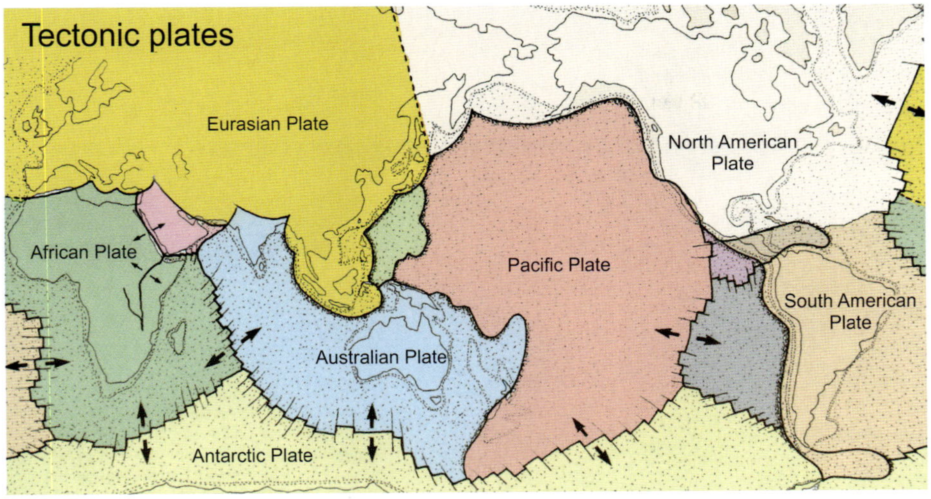

X.5 Map of the major tectonic plates of the world today.

Plate tectonics – an introduction

The Earth's outermost layer, the **lithosphere** (upper mantle plus crust), is broken up into seven or eight major tectonic plates (X.5) and numerous smaller ones. These plates are slowly moving around the Earth's surface in varying directions and colliding with each other. The plates are composed of two kinds of lithosphere. One is dense oceanic lithosphere (50–100 km thick), which consists of thin (5–8-km-thick) **mafic or oceanic crust** (basalt and gabbro) and the more rigid upper part of the **mantle** (peridotite). The second is less dense continental lithosphere (150–250 km thick), which consists of thicker (30–50km) **felsic or continental crust** (mostly granite, metamorphic and sedimentary rocks) plus the more rigid upper mantle.

Many of the Earth's earthquakes and volcanic eruptions, as well as mountain building, occur around the margins of the plates where they interact with the adjacent plates. These plate boundaries can be of three basic types:

1. **Moving apart (divergent boundaries).** Spreading zones often occur within oceanic plates under the ocean, but can also rift apart continental plates (e.g. formation of Zealandia by the opening of the Tasman Sea). As the plates move apart, molten magma from the mantle rises through the gap and erupts on the seafloor as basalt lava flows, forming new oceanic crust (X.6). The presence of rising molten magma at shallow depths raises the temperature of the lithosphere, and it bulges upwards creating a mid-ocean ridge on the seafloor. The erupting magma on the seafloor contributes to building up this mid-ocean **spreading ridge**, which is a lot shallower than the surrounding ocean floor, and sometimes basalt volcanoes form islands along it, such as Iceland. As the plates move apart they raft along the newly formed oceanic crust and lithosphere. The further these move away from the spreading ridge, the cooler, denser and thicker they get, and they gradually subside to the typical depths of the ocean floors (4–5 km).

2. **Moving together (convergent boundaries).** These collisional boundaries commonly occur around the edge of the Pacific Ocean where the oceanic plate collides with a continental plate, e.g. east North Island and south of South Island. In these instances the dense oceanic plate plunges down into the mantle (**subduction**) beneath the edge of the less dense continental plate (X.6). Numerous earthquakes, large and small, occur along these subduction boundaries. A plot of the location (foci) of these earthquakes shows the path of the top of the downward-moving plate. The subducted oceanic plate carries down with it an upper layer of water-saturated marine sediment. With increased temperature and pressure, at about 100 km depth, the water is driven upwards out of the subducted rocks into the overlying upper mantle. The addition of water lowers the melting point of the mantle above the subducting plate, causing partial melting. This hydrated basalt magma rises and collects in large chambers at the base of or within the overlying continental crust.

   In these chambers the composition of the magma may change over time by processes of selective crystallisation of some minerals (**fractional crystallisation**), and incorporation and melting of some of the surrounding more felsic rocks (**assimilation**). Magma from these chambers periodically rises to the surface and erupts as andesite, dacite or rhyolite. These volcanoes form a line above the subducted plate and parallel to the plate boundary, called a **volcanic arc**. Thus the volcanic arcs around the margins of the Pacific Plate together comprise the well-known Pacific Ring of Fire, e.g. Tongariro–Kermadec–Tonga Arc.

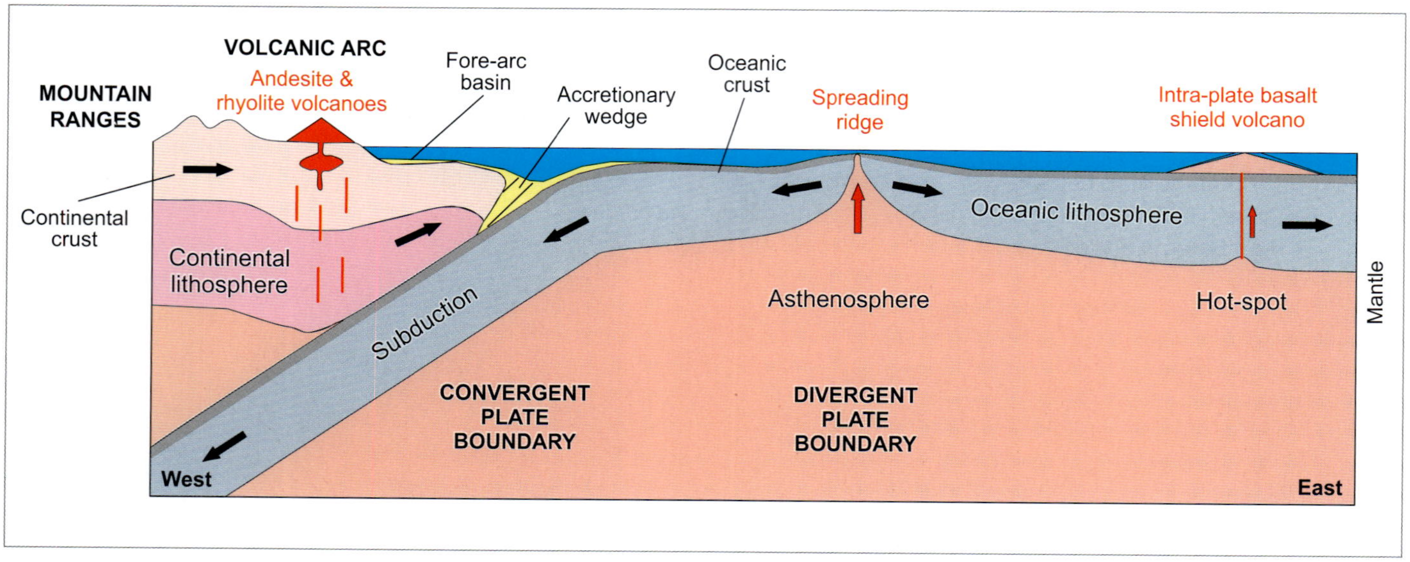

Along these convergent plate boundaries considerable pressure builds up within the rocks in the overlying continental plate, often causing the land to be pushed up as mountain ranges parallel to the plate boundary (e.g. axial ranges of New Zealand). In some places, slices of rock are scraped off the top of the down-going oceanic plate and thrust under the edge of the continental plate. Over time a succession of these slices might be added, forming what is called an accretionary wedge (X.6).

3. **Moving sideways (transform boundaries).** Sometimes a plate boundary might separate two portions of plate of similar density and neither is pushed under or over the other. In most instances the convergence direction between the two plates is at an angle to the boundary itself, and thus it is not difficult for the two plates to be pushed sideways past each other along a major fault. This is happening today through the South Island.

The interior of plates, away from their boundaries, are usually tectonically stable, but sometimes they are pierced by the eruption of basalt sourced direct from partially molten parts of the upper mantle, 50–100 km below. These are called **intra-plate volcanoes** and the deep source beneath is often referred to as a **hot-spot** (e.g. Auckland, Banks Peninsula and Dunedin volcanoes).

⊙ X.6 Schematic cross-section through the South Pacific Ocean showing the mid-ocean spreading ridge and the collisional boundary with oceanic lithosphere subducting down beneath the edge of the continental plate.

A brief history of New Zealand

The rocks and their contents tell us that the Earth was formed about 4500 Myr ago, that New Zealand's history began at least 505 Myr ago (I.2), and that a complicated series of events since then has combined to produce the country we know today. Following its origin, the molten Earth has cooled on the outside to form a solid crust of rock. Much of this crust has been eroded by natural forces; the resulting fragments have been carried by rivers and redeposited, mainly in the sea as sedimentary rocks. Some of the rocks have remelted at depth, then solidified again as they intruded upwards to form plutonic rocks or erupted at the surface as volcanoes. Many of the deeply buried crustal rocks have been greatly changed by the intense pressures and heat at depth, forming metamorphic rocks. There are examples of all these processes in the history of New Zealand's formation (X.7).

Growth on the edge of Gondwana

Between 505 and ~100 Myr ago (most of the Paleozoic and Mesozoic eras) the region of rocks that was to later become the continent of Zealandia, which includes New Zealand, was located along the edge of one of the world's great supercontinents. Initially this was the global supercontinent of Pangaea, but after this split in two, about 200 Myr ago, the Zealandia rocks accumulated along the edge of the Southern Hemisphere supercontinent of Gondwana.

The rocks formed during this 400-Myr period (505–100 Myr ago) are known as New Zealand's basement rocks, and these now underlie all of the country and Zealandia. The

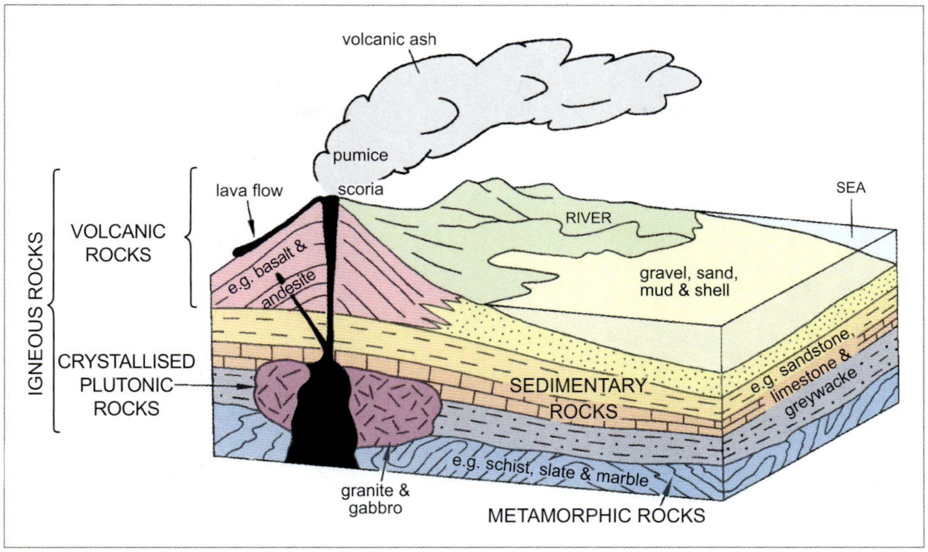

X.7 Schematic cut-away diagram showing how the main groups of rock were formed.

X.8 Terranes cross-section and map.

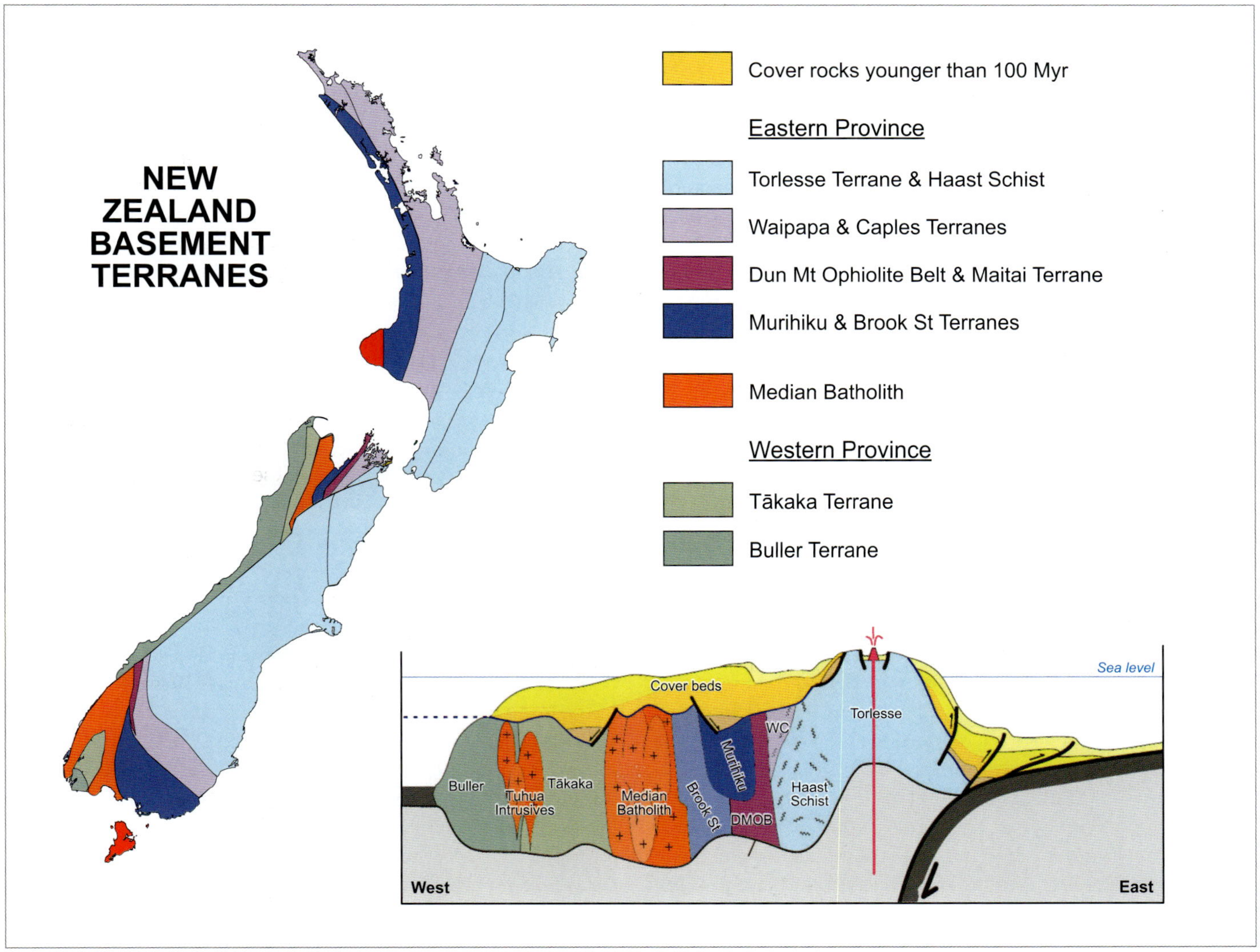

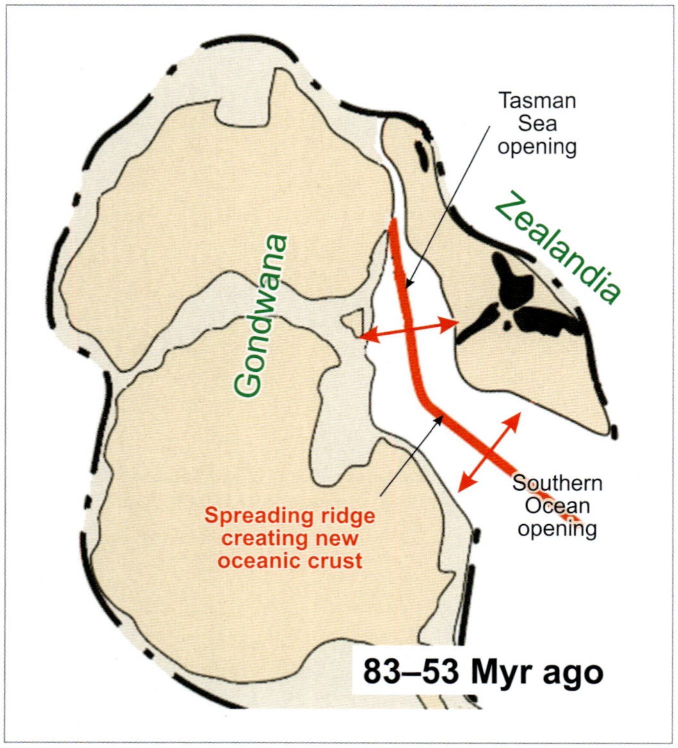

X.9 Zealandia split away from Gondwana by rifting that created the Tasman Sea between 83 and 53 Myr ago.

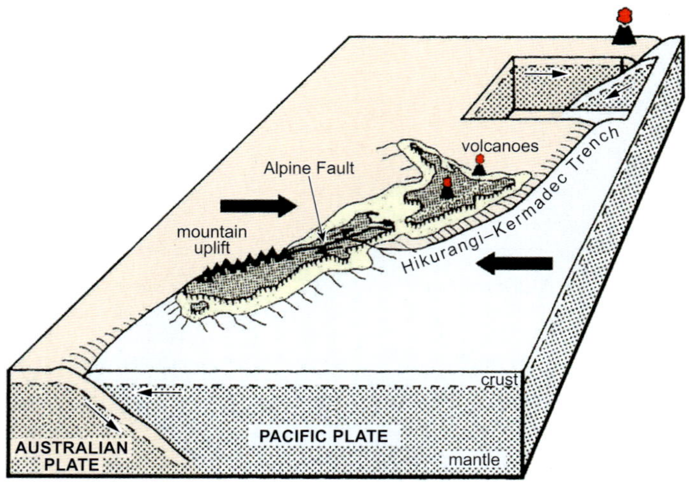

X.10 Cut-away model showing the present plate-tectonic setting of New Zealand located on an obliquely convergent plate boundary.

basement rocks were formed on a collisional plate boundary where oceanic lithosphere on the ancient Pacific Plate was forced down beneath continental lithosphere of the Gondwanan Plate. The convergent forces between the two plates pushed up the margin of the supercontinent to become coastal mountain ranges. Subduction along the boundary pulled down the seafloor offshore, forming a deep trench, and a line (arc) of volcanoes erupted along the coast above the subducting plate. Erosion of the mountain ranges and volcanoes produced large volumes of sand and mud that were transported into the sea and deposited in enormous thicknesses offshore, particularly in the trench and associated submarine fans.

In the collision zone, the edge of Gondwana grew outwards and thicker by two main processes – terrane accretion and magma intrusion. In terrane accretion, or sideways growth, thick sequences of offshore sediment were rafted on top of the west-moving Pacific Plate and collided with the edge of the Gondwanan Plate. As the dense oceanic plate plunged down into the mantle, slices of the less dense pile of overlying sedimentary rock were scraped off the top and thrust under the edge of the continental plate. Over time a succession of these slices was added onto the edge of the continental crust to form an accretionary wedge (The Big Picture 10, page 154).

The direction of collision between these early plates is believed to have been oblique to the margin, rather like it is today. As a result, not only was the Pacific Plate subducted down into the mantle but there was also considerable sideways movement of the plate and its load of rafted sediments, along major transform faults (similar to the present Alpine Fault). As a result, the rafted sedimentary rocks that were added on (accreted) to the edge of Gondwana may have accumulated elsewhere, maybe hundreds – even thousands – of kilometres away, and not directly offshore from this part of the plate boundary. The sporadic movement of the various offshore transform faults resulted in at least nine huge, elongate slabs or belts of unrelated rocks (called terranes) being added to the continent at different times. The oldest terrane belts are in the west and they become progressively younger towards the east where they progressively built out into the ancient Pacific Ocean as they were added (X.8). The oldest terranes are known as the Western Province, and these contain rocks of Cambrian to Devonian age that were added to the edge of Gondwana between 500 and 330 Myr ago (Cambrian–Carboniferous). The more voluminous, younger terranes are known as the Eastern Province, and these were added onto the margin of Gondwana between 160 and 105 Myr ago (middle Jurassic–early Cretaceous).

Subduction along the plate boundary resulted in the periodic generation of magma, some of which rose to the surface and erupted as a volcanic arc, and some of which slowly cooled at depth in large magma chambers forming

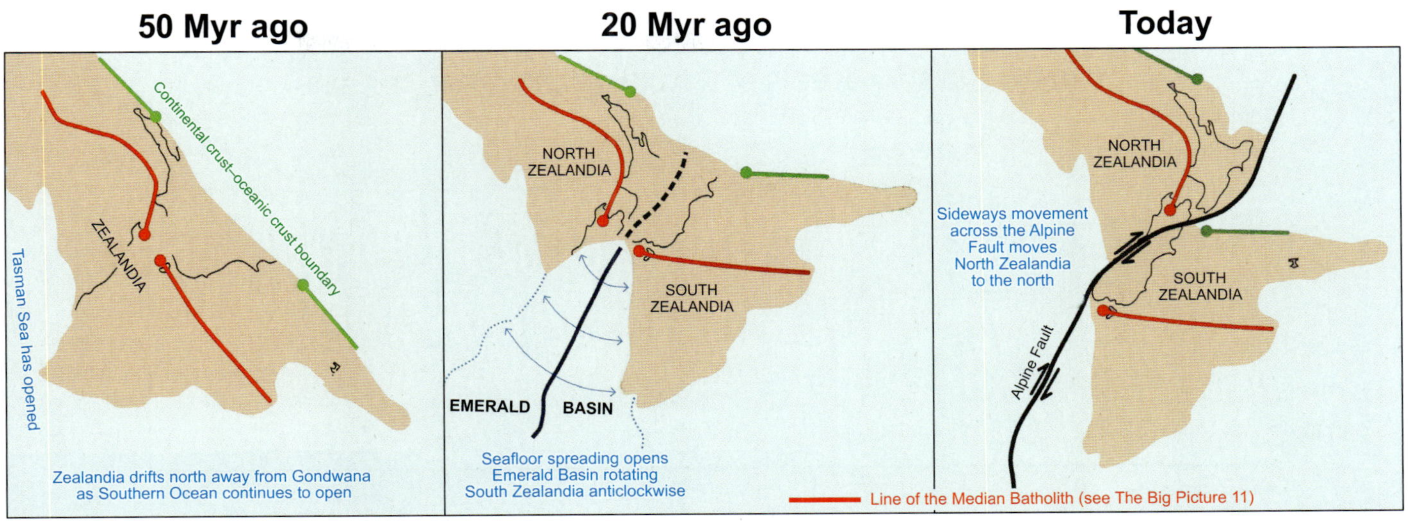

X.11 Tectonic history of Zealandia (brown) over the last 50 Myr. MODIFIED FROM MORTIMER AND CAMPBELL (2014).

plutons of coarsely crystalline granite or diorite. The plutons intruded the western part of the accreted terranes, forming one huge, elongate batholith (Median Batholith) and a number of smaller ones, each composed of numerous plutons. The plutons of the Median Batholith were intruded between the Western and Eastern Provinces (hence its name) over a 400-Myr period (500–105 Myr ago, Cambrian–early Cretaceous; The Big Picture 11, page 207).

Since their formation, most of the basement rocks have been buried sufficiently deeply to be hardened into greywacke and argillite or metamorphosed into schist or gneiss (The Big Picture 17, page 305).

Stretching, break-up and formation of Zealandia

The collisional forces between the plates came to an end about 105 Myr ago (middle Cretaceous). Tectonic movement switched to extension, which started to stretch and thin the continental crust within the southwest side of Gondwana. This reduced the crust's thickness by nearly half and doubled its width. Continued extension created a number of coast-parallel rifts, one of which became dominant and opened up the Tasman Sea between 83 and 53 Myr ago (X.9). As a result, a huge chunk of thinned continental crust split off from Gondwana creating the world's eighth, and smallest, continent, Zealandia.

Erosion, subsidence and immersion

As it split away, Zealandia was mostly rugged land, but over the next 50 Myr the mountains were eroded down to an extensive wave-cut surface, known as the Waipounamu Erosion Surface (The Big Picture 18, page 309). Increased heat flow along the opening Tasman Sea rift had heated the surrounding continental crust, but as Zealandia moved away this crust cooled, became less buoyant and progressively subsided into the underlying mantle. The thinner crust was also less buoyant than continents with thicker crust and contributed to Zealandia's slow subsidence. By about 23 Myr ago (end of the Oligocene) most of Zealandia was under the sea with only a few areas of low-lying land remaining. With almost no eroding land, there was very little sand or mud washed into the sea and the dominant sediment that accumulated on the vast seafloor during the Oligocene (33–23 Myr ago) was the limy shells and skeletons of marine organisms such as molluscs, barnacles, red algae, and plankton. With later burial these vast shell layers have been hardened and cemented into limestone that is widespread around New Zealand today.

Return to plate convergence and subduction and the modern plate boundary

From about 45 Myr ago, a new convergent plate boundary started to slowly develop through the New Zealand sector of Zealandia, and by the start of the Miocene (23 Myr ago) its influence really kicked in. The low-lying Waipounamu Erosion Surface began to buckle and break along faults, with some areas being pushed up to become eroding hills and ranges, supplying sediment to adjacent areas that were pushed down and became valleys and marine depressions. The compressional forces along this new boundary between the Australian and Pacific Plates changed orientation several times in the next 23 Myr, resulting in different areas being pushed up, down or sideways at different times.

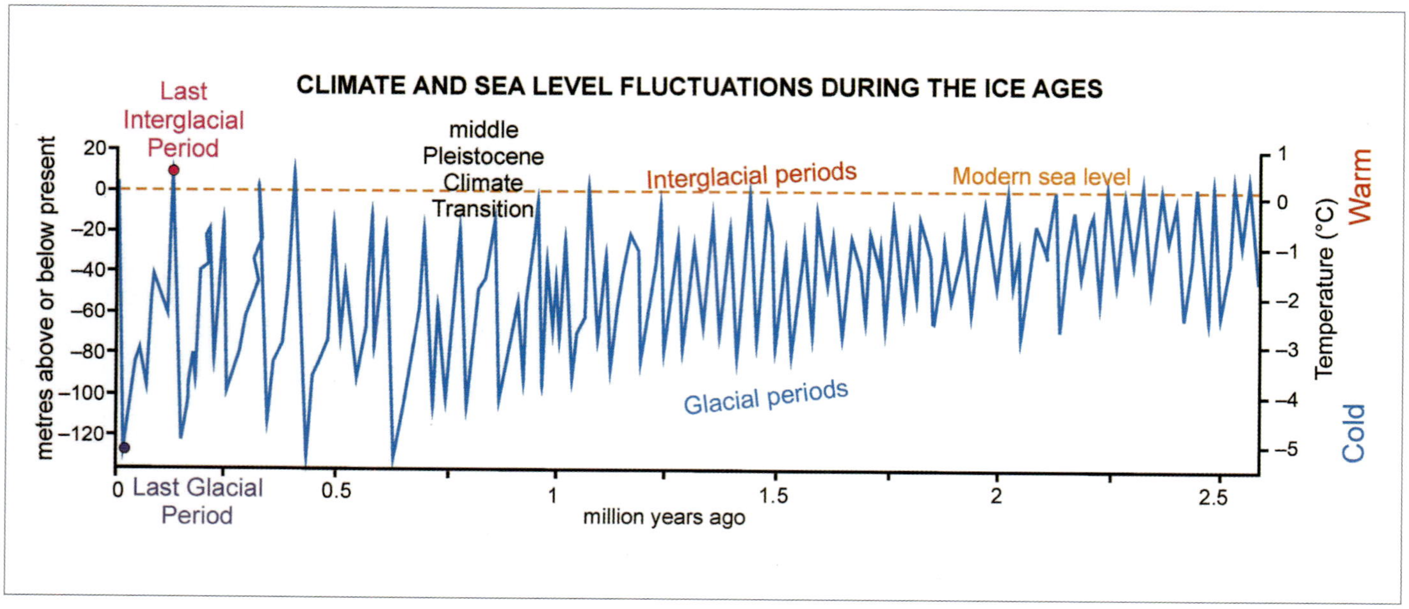

The modern plate boundary (X.10) developed with westward subduction of oceanic lithosphere of the Pacific Plate beneath the edge of the Australian Plate along the Kermadec Trench extending southwards under the North Island. Eastwards subduction of the Australian Plate's oceanic lithosphere beneath the edge of the Pacific Plate developed south of Fiordland along the Macquarie Ridge. These two regions of subduction, with opposite polarity, are linked by a major transform fault, the Alpine Fault (The Big Picture 12, page 248). Dextral sideways movement (far side to the right) along this fault began about 23 Myr ago and speeded up in the last 5–10 Myr (X.11). The oblique compressional forces also resulted in the east side of the Alpine Fault being pushed up, creating the linear Southern Alps mountain chain through much of the length of the South Island.

The west-dipping subduction beneath the northern half of the North Island generated magma that rose to the surface and erupted as migrating chains (volcanic arcs) of mostly andesite and rhyolite volcanoes. These began as the Northland Volcanic Arc in the early Miocene (~23–15 Myr ago; The Big Picture 1, page 23), migrated southwards to the Coromandel and West Taranaki volcanic arcs during the middle Miocene to Pliocene (16–3 Myr ago), and finally shifted (The Big Picture 5, page 90), in the last 2 Myr, to its present location running northwards from Ruapehu through Taupō, Rotorua and offshore into the Bay of Plenty and beyond to the Kermadec Islands (The Big Picture 6, page 97).

Intra-plate (not related to subduction or spreading at a plate boundary) eruptions of mostly basalt volcanoes has occurred in many places east of the South Island and in the northern North Island in the last 13 Myr. Two huge volcano complexes erupted at Dunedin and Christchurch (Banks Peninsula) in the middle and late Miocene (13–6 Myr ago; The Big Pictures 16 [page 298] and 14 [page 266]). Smaller basalt fields have also been active in many places, particularly in the north around Kaikohe–Bay of Islands, Whangārei, Auckland and South Auckland (The Big Picture 3, page 57).

Because the modern New Zealand portion of Zealandia is located astride the plate boundary, it stands high as land, but a much larger portion of the continent is still submerged below sea level. The main axial ranges of both islands are composed of exhumed hard basement rocks brought up from considerable depths by the enormous compressional forces between the plates. Much of the more rolling country is composed of softer and younger sedimentary rocks that accumulated in the more local marine sedimentary depressions and have more recently everted and been partly eroded.

Ice Age cycles of climate and sea level

Astronomically driven cycles of warm and cold climate can be recognised throughout most of Earth's history, but in the last 2.6 Myr (referred to as the Ice Ages) their amplitude increased markedly and resulted in the cold glacial periods developing extensive polar ice caps, especially in the Northern Hemisphere. Up until about 1 Myr ago the peaks and troughs (interglacial and glacial periods) of the climate cycles were 40,000 years apart, but this changed during the middle Pleistocene Climate Transition (1.2–0.7 Myr ago) when the cycles lengthened to 100,000 years. Thus, since 2.6 Myr ago (the Ice Ages) there have been about 42 Ice Age cycles, and the nine most recent of these were 100,000 years long.

X.12 Climate and sea-level curve for the New Zealand region for the Ice Ages (last 2.6 Myr).

X.13 Vegetation cover of New Zealand at the peak of the cold glacial periods of the last 1 Myr. Note the coastline of New Zealand has changed, as sea level was 130 m lower than now.

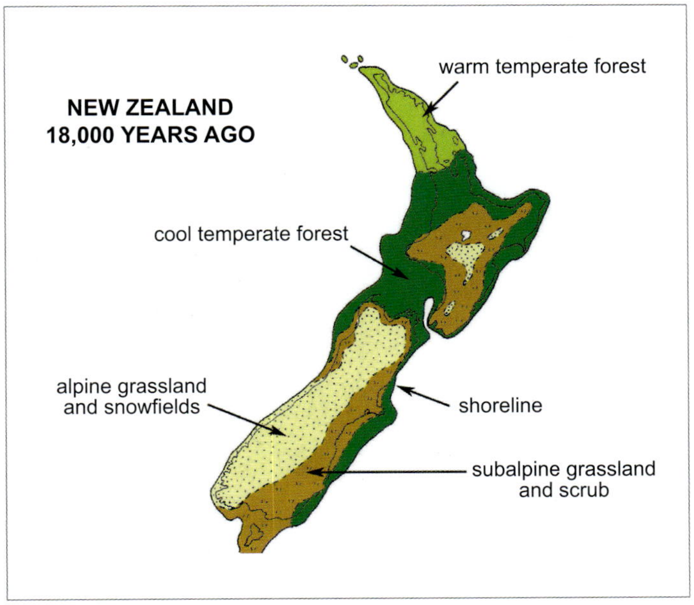

Not only did the climate cycles get longer in the middle Pleistocene but also the cold glacials became increasingly severe, with the development of more extensive polar ice caps. The peaks of cold glacial and warm interglacial periods each lasted only a few thousand years. In the more recent cycles, the climate slowly declined in a saw-tooth-like series of oscillations over a period of 80,000 years after each peak warmth (X.12). After each glacial low, the Earth's climate warmed up much faster, often in less than 10,000 years.

During the intense cold of all the last few glacial intervals, almost all the forest of the South Island and southern North Island was replaced by subalpine grassland and scrub, with snow fields and glaciers developing in the more mountainous areas (X.13). In the northern North Island, forest persisted over all but the highest ranges through the glacial periods, although its composition changed from mixed podocarp–broadleaf in warm times to dominantly southern beech and shrublands during the coldest glacials. The development of extensive ice fields, glaciers and super-glaciers in higher-altitude areas of the South and southern North Islands during the cold intervals resulted in accelerated erosion. Distinctive glacier-carved U-shaped valleys were eroded in the Southern Alps, some of which are currently flooded and form the elongate southern lakes and the fiords on the southwest coast. The vast quantities of sediment eroded during these glacial times built out extensive outwash plains in the South Island (e.g. Canterbury and Southland Plains).

A direct result of the climate cycles, and mimicking them in periodicity and amplitude, were sea-level cycles (X.12). During the more recent, severe glacial peaks, a large amount of the Earth's water was captured on land in polar ice caps, especially in the Northern Hemisphere. This ice was sourced, via evaporation and snow fall, from the oceans, and as a consequence sea level fell. During the most severe glacial periods, such as the Last Glacial peak just 18,000 years ago, sea level fell to as low as 130 m below present, but for most of the Ice Ages sea level fluctuated between 100 and 30 m lower than today. Sea level has only been high and close to the present for about 10% of the time in the last 1 Myr.

These sea-level cycles have had a major impact on the shape of New Zealand and its coastline, especially the rapid rise of sea level that occurred after the end of the Last Ice Age, between 18,000 and 7500 years ago. The rising sea level washed vast quantities of sand up against the land to form beaches, barriers and dunes. Where there was a plentiful supply, whole valleys were filled, or large sand and gravel barriers were created enclosing sheltered estuaries, inlets and shallow harbours. Where there was less sediment available, the incised river valleys were drowned to become our modern embayed coastlines and harbours. The meandering and branching shape of these former river valleys is still recognisable in the extensive headwaters of many northern harbours and the Marlborough Sounds. The extensive inland waterways of Fiordland were created by the drowning of steep-sided U-shaped valleys that had been carved out by glaciers when sea level was much lower during each cold Ice Age period.

Glossary

a'a lava – (pronounced ah-ah) lava flow that moves slowly and has a rubbly or blocky surface

accretion – added on to

accretionary lapilli – small spherical balls of volcanic ash that accrete together in a wet cloud of ash

accretionary wedge – a thick stack of mostly sedimentary rock scraped off the top of the down-going oceanic plate in a subduction zone and added on to the edge of the adjacent continental plate

acidic/felsic igneous rock – a rock that has formed by cooling of liquid magma and contains >63% silica, e.g. granite, rhyolite

alkali feldspar – a group of rock-forming potassium aluminium silicate minerals (often orange-pink in colour) that crystallise out of igneous rocks

allochthon – a large body of rock that has moved from its original site of formation

alluvial fan – a triangle-shaped deposit of alluvium (gravel, sand, mud) often dumped where a stream comes out of the hills and on to a plain

alluvium – sediment deposited by rivers or streams

andesite – a medium- to light-grey-coloured, relatively fine-grained volcanic rock with 52–63% silica composition

antecedent rivers – rivers that retain their course even though the topography around them has altered

anticline – upfold

Ar – symbol for the inert gas element argon

arête – a sharp rocky ridge eroded between two U-shaped glacial valleys

argillite – hardened mudstone, similar to greywacke

ash – mud- and sand-size fragments (<2 mm across) of volcanic and other rock that have been fragmented and erupted into the air from a volcano

asthenosphere – dense, weak layer of semi-molten rock within the mantle at a depth of ~100–400 km beneath the surface

axis, of a fold – a straight line running down the middle of the fold

basalt – dark-grey-coloured, fine-grained volcanic rock with 45–52% silica composition

basic/mafic igneous rock – a rock that has formed by cooling of liquid magma and contains 45–52% silica, e.g. basalt

basin, sedimentary – a large depression of the crust that accumulated a thick deposit of sediment

bathyal – seafloor at continental slope depths of 200–2000 m water depth

bed, bedding – layering in sedimentary rocks

biotite – a black mica mineral that can be separated into thin translucent sheets; commonly found in some igneous and metamorphic rocks

braided river plain – a network of river channels separated by temporary small islands called braid bars

breached crater – a U-shaped crater on a volcanic cone with one of its sides having been taken out by an eruption or lava flow

breccia – rock composed of angular boulders, cobbles or pebbles (fragments >2 mm in diameter)

calcareous – rich in calcium carbonate, $CaCO_3$ (lime)

calcite – mineral composed of calcium carbonate (lime), the main component of limestone

calc-tufa – soft, porous rock consisting of calcium carbonate deposited from springs rich in lime

caldera – large crater formed by collapse during or after an eruption

carbonaceous – containing carbon, usually black plant material

case hardening – a weathering phenomenon that hardens the surface of a rock

cataclasite – a fine-grained rock that has been crushed by grinding movement in a fault zone

cavernous weathering – weathering that results in the formation of hollows on the surface of a rock

chenier – a sand or shell beach ridge that has built up on top of mudflats

chert, cherty – a hard, flinty rock composed of fine-grained silica

chlorite – a green mineral with a soapy feel, formed by metamorphism or hydrothermal alteration

cirque – amphitheatre-like valley formed at the head of a glacier

clast, clastic sedimentary rocks – particles up to boulder size and rocks composed of them

clinopyroxene – dark-coloured group of crystalline silicate minerals

clint – narrow limestone ridge between dissolved gullies

coal measures – sequence of coal layers and associated swamp and alluvial sedimentary rocks

concretion – hard rock formed by precipitation of minerals from solutions passing through the host rocks

conduit – pipe or passageway for conveying fluid, such as magma or thermal water

conglomerate – a sedimentary rock composed of rounded gravel (pebbles, cobbles, boulders)

conodonts – the fossilised teeth of extinct chordates that resemble eels

convergence – moving towards

core stone – the fresh rounded internal rock remaining after spheroidal weathering

country rock – the background rock

cross-section – a vertical slice through the rocks or part of the Earth

crust – the outermost layer of the Earth: average continental crust thickness 30–40 km; average oceanic crust thickness 5–10 km

cuesta – a ridge capped with an inclined, resistant layer with a gentle dip slope on one side and a steep escarpment on the other

cuspate foreland – a triangular (arrow-shaped) coastal lowland formed by longshore drift of sediment from two opposing directions

dacite – a felsic volcanic rock intermediate in composition between rhyolite and andesite (63–68% silica)

debris avalanche – rapid, large-volume flows formed from the collapse of a large sector of a volcanic cone

deflation – process of wind erosion within sand-dune areas

dehydrate – lose a large amount of water

dendritic – branched like a tree

dextral fault – sideways displacement on a fault with the far side moving right

diapir – an intrusive body of mobile material that has squeezed upwards through more brittle rocks

diatoms – microscopic single-celled algae with skeletons made of silica

dike – a narrow sheet-like intrusion of lava or sediment cutting across the rock layers

diorite – speckled black and white, crystalline plutonic rock with 52–63% silica composition

doline – funnel-shaped depression formed by solution of limestone, often referred to as a sinkhole or tomo

dome, volcanic – steep-sided mound formed by extrusion of viscous lava at the surface

ductile metamorphism – involving shearing deformation without breakage

East Coast Allochthon – many oceanic rock types that have been displaced 50 to several hundreds of kilometres on to the East Coast, mostly as large slabs

erratic boulders – rock different from surrounding rocks because it has been transported from its place of origin

exfoliation weathering – process involving the detachment of curved sheets of rock from the surface of an outcrop as a result of the release of pressure within the rock as it reaches the surface

explosion crater – wide, relatively shallow, circular crater remaining after a series of wet explosive eruptions

explosive eruption – a violent wet style of eruption that occurs when molten magma encounters cold water and blasts out a cloud of steam, gas and fragmented rock (also known as phreatomagmatic eruption = steam and magma eruption)

fault – a fracture in rocks along which movement has occurred displacing one side with respect to the other

fault scarp – a step-like feature across the ground surface produced by displacement on a fault plane

fault trace – the visible line across the ground where a fault plane reaches the surface

feldspar – the most common group of silicate minerals

felsic/acidic igneous rock – light-coloured igneous rocks rich in feldspar and silica

fire-fountaining eruption – steady fountaining eruption of fluid magma powered by gas release and the building up of a scoria cone around the vent (also known as lava-fountaining or Hawaiian eruption)

flaggy – rocks that readily split into layers 1–20 cm thick that can be used for flagstone paving

fold – curved layering in rocks

fold axis – the hinge line around which the folding of rock layers occurred

foliated – consisting of thin sheets or laminae

foraminifera – microscopic marine amoeba with chambered shells often found fossilised

fore-arc basin – an elongate depression seaward of a volcanic arc and landward of a subduction trench

fossil – any remains or trace of a plant or animal that has been preserved in rock

frittering – breaking into small pieces

geothermal – heat from hot groundwater or steam

geyser – intermittent jets of boiling water and steam that erupt in geothermal areas

geyserite – an opaline silica deposit (a kind of sinter) that precipitates out around the edges of geysers and boiling springs

glacial flour – fine, silt-size powder produced by mechanical grinding of bedrock by glaciers

glacial period – period during the Ice Ages when global temperatures were much colder, and polar ice caps and extensive mountain glaciers formed

glacier – a slowly moving 'river' or mass of ice

gneiss – (pronounced 'nice') a high-grade, coarsely crystalline, foliated metamorphic rock

Gondwana – Southern Hemisphere supercontinent prior to its break-up about 100–50 Myr ago

granite – a coarsely crystalline plutonic rock consisting mostly of quartz, alkali feldspar and mica

granodiorite – a coarsely crystalline plutonic rock consisting mostly of quartz, plagioclase feldspar, hornblende and mica

graptolites – an extinct group of colonial, pelagic hemichordates that often look like hieroglyphs on the bedding planes of Ordovician slate

greenschist – a metamorphosed rock coloured green by an abundance of chlorite or sometimes epidote

greywacke – hardened sedimentary rock, strictly hardened sandstone; usually basement rocks

grike – a solution fissure

gypsum – calcium sulfate mineral

head scarp – distinct step at the upslope edge of a landslide or slump

honeycomb weathering – a form of cavernous weathering that produces regular honeycomb-like depressions on rock surfaces

horn – a steep-sided peak formed between three or more glacial cirques

hot-spot volcano – a volcano that erupts above an area of hotter, more molten magma in the mantle, away from a plate boundary

hydrate – to take up or combine with water

hydrocarbon – organic compounds of hydrogen and carbon that form oil, natural gas and coal

hydrothermal – related to hot fluid within the Earth's crust

Ice Ages – a time when the Earth's climate cycled back and forth between cold glacial and warmer interglacial periods

igneous rock – rocks formed by the cooling and solidification of molten magma

ignimbrite – a hot, ground-hugging flow of volcanic ash and gas

interglacial period – a period of warmer global temperatures when most polar ice caps and glaciers melted

intermediate igneous rock – a rock that has formed by cooling of liquid magma and contains 52–63% silica

intrusion – magma that has intruded into subsurface rocks

isotope – variants of a chemical element that differ in the number of neutrons in their nuclei

joint – a fracture in rock

K–Ar dating – potassium–argon dating, a method of radiometric dating of the time of crystallisation of igneous rocks and minerals

karst – landscape formed by dissolution of limestone and sometimes basalt rocks; includes sinkholes and caves

klippen (singular **klippe**) – large exotic blocks of rocks that have been transported into their current position surrounded by unrelated rocks

kyr – thousands of years

lahar – a volcanic mud flow consisting of a dense mix of volcanic debris and water

laminated – thinly banded or thinly layered

lapilli – pebble-sized pieces (2–64 mm across) of volcanic and other rock that have been fragmented and erupted from a volcano

lava – molten rock (magma) that has been erupted on to the Earth's surface

limestone – a sedimentary rock composed of >50% calcium carbonate

lithosphere – the outer part of the Earth that is broken up into tectonic plates; comprises the crust and upper mantle

loess – wind-blown fine sediment (usually silt-sized), often yellowish-brown and mostly of glacial flour origin in New Zealand

longshore drift – the overall movement of sediment along the coast by a combination of currents and wind

mafic igneous rock – dark-coloured igneous rock dominated by iron- and magnesium-rich minerals and having 45–52% silica. Also known as basic.

magma – molten rock occurring under the ground; may erupt at the surface (after which it is called lava, but is essentially the same composition as the parent magma)

mantle – part of the Earth's interior between the crust and the core

marble – metamorphosed limestone

metamorphic rock – transformed by heat and/or pressure

meteoric – derived from rain

methane – colourless, odourless, combustible natural gas, CH_4, formed by the breakdown of organic matter

mineral – naturally occurring inorganic substance; the components of rocks

moraine – unconsolidated rocky debris transported then deposited by a glacier as a heap or ridge

mudstone – a rock composed of fine-grained (<0.06 mm) particles (mud)

mylonite – cohesive foliated rock formed by tectonic crushing of the rock in areas of intense deformation, like a fault zone

Myr – millions of years

nannofossils – fossil calcareous shell of a kind of microscopic marine planktonic alga

névé – partially compacted granular snow that feeds a glacier

normal fault – a sloping fault where the hanging wall has moved down with respect to the foot wall; occurs where rocks are being pulled apart (under tension)

Northland Allochthon – many oceanic rock types that have been displaced hundreds of kilometres on to Northland, mostly as large slabs

olivine – a green crystalline mineral composed of magnesium iron silicate that sometimes crystallises in mafic magma underground

ooze – a soft deposit of sediment on the bottom of the sea

orthopyroxene – dark-coloured group of crystalline silicate minerals

outlier – a younger rock formation isolated among older rocks

overthrust – rocks that have been pushed over the top of others along a low-angle thrust fault

oxidise – chemically combine with oxygen, as happens when some minerals are exposed to air

pahoehoe lava – (pronounced pa-ho-ee-ho-ee) lava flow that has a rapidly chilled, rolled-up or ropey, but otherwise smooth, glassy surface

peat – incompletely decomposed black deposit of plant material

peridotite – iron- and magnesium-rich rock that forms much of the mantle and when partly melted produces basalt magma

petrification, petrified – the process whereby organic material is replaced by minerals and turned to stone

phonolite – a fine-grained volcanic rock composed of alkali feldspar and nepheline

phreatic eruption – a steam-powered eruption not involving magma

phreatomagmatic eruption – an explosive volcanic eruption resulting from molten magma mixing with water

pillow lava – finger-like lobes of lava erupted underwater

plagioclase – light-coloured feldspar mineral composed of sodium- and/or calcium-aluminium silicate that is a common constituent of rocks

plankton – small organisms that float or swim in fresh- or saltwater

plate – a rigid, mobile segment of the lithosphere

plutonic – a coarsely crystalline rock that has formed by slow cooling of magma beneath the ground

pothole – a smooth, bowl-shaped hollow carved in the rocky bed of a watercourse

prehnite – a mineral composed of calcium-aluminium silicate, formed in rocks during low levels of metamorphism

proglacial lake – a lake dammed by a terminal moraine barrier and filling the depression left in front of a melting glacier

pumice – solidified, frothy, glassy, volcanic rock formed in violent wet explosive eruptions, usually pale-coloured and rich in silica

pumpellyite – a silicate mineral formed in a rock during low levels of metamorphism

pyroclastic dike – a cross-cutting, sheet-like fissure filled with fragments of volcanic and country rock, broken up in the throat of a volcano by explosive eruptions and injected along the fissure

pyroclastic flow – a hot, fast-moving, gas-rich, ground-hugging flow of pyroclastic debris

quartz – common, inert, variably coloured mineral composed of silica (SiO_2)

radiolaria – microscopic, single-celled plankton having an intricate skeleton of silica, commonly found fossilised in oceanic sedimentary rocks

regression – a geological event or period when sea level falls relative to the land and the shoreline moves seawards

remote sensing – acquisition of subsurface data by remote means such as using seismic waves or measuring gravity or magnetism

reverse fault – a sloping fault in which the hanging wall rocks have been pushed upwards with respect to the foot wall; occurs in zones of compression

rhyolite – light-coloured, fine-grained volcanic rock with >68% silica

rift – a linear zone where the lithosphere is being pulled apart; forming a subsiding rift basin or valley above it

ring plain – a circular plain mostly of lahar deposits that builds up around the foot of a stratovolcanic cone

sandstone – a sedimentary rock composed predominantly of sand grains (0.06–2 mm)

schist – a medium-grade metamorphic, crystalline rock formed by deep burial, heating and recrystallisation so that it breaks along a preferred direction (schistosity)

scoria – lightweight volcanic rock full of holes (vesicular), which was erupted as frothy lava which cooled and solidified to rock; sometimes referred to as cinders

scoria cone – a relatively small, usually steep-sided (c. 30°), volcanic cone made of scoria

scree – an accumulation of loose rocky debris on the slopes or foot of a mountain or cliffs

sedimentary rock – rocks composed of sediment (rock particles and fossil material) and subsequently hardened into a rock

seismic reflection profile – a cross-section of the subsurface rock structure obtained by remote sensing using sound waves bouncing off subsurface layers and features

septarian concretion – a concretion containing angular cracks or cavities, possibly due to shrinkage following dehydration. Also known as turtleback concretion.

serpentinite – green rock made of serpentine, formed by hydrothermal alteration of peridotite

shawl – a curtain-like speleothem that grows downwards from the ceiling of a limestone cave

shield volcano – gently sloping (c. 10°) volcanic cone made of overlapping lava flows erupted from one or more near-central vents

siderite – a pale yellow to brownish, iron-carbonate mineral with high specific gravity (heavy)

silica – silicon dioxide, SiO_2, a common component of many rocks and minerals

siliceous – rich in silica

siltstone – a sedimentary rock composed predominantly of silt grains (0.002–0.063 mm)

sinistral fault – a sideways displacement on a fault with the far side moving left

sinkhole – a depression in the ground with no external drainage, mostly formed by solution of the underlying limestone or marble rock

sinter – a hard siliceous or calcareous deposit precipitated from mineral springs

skarn – a contact metamorphic rock formed by the heat when an igneous intrusion invades a calcareous rock such as limestone

slot canyon – a long, narrow, deep channel with sheer walls; usually eroded into a sedimentary rock

slow slip – long-lived (up to several weeks) shear slip events on subduction thrust faults, rather than sudden ruptures that produce earthquakes

smectite – group of clays that swell up in water

solifluction – a gradual process whereby the surface layers of rock and soil move downslope as a result of freeze and thaw activity, most commonly during glacial periods

speleothem – a mineral deposit in a cave, usually made of lime, such as a stalactite or stalagmite, formed by precipitation out of flowing or dripping carbonate-saturated water

stalactite – icicle-shaped formation hanging from a cave ceiling and produced by precipitation of minerals from dripping water

stalagmite – a mound or column rising from the floor of a cave and produced by mineral precipitation from water, often dripping off the end of a stalactite above

strand plain – a broad belt of sand or fine gravel along a shoreline with well-defined parallel ridges separated by shallow swales

strata – layers of rock

stratovolcano – a steep-sided volcanic cone (usually of andesite) consisting of layers of lava flows and volcanic debris

straw – a narrow, hollow stalactite that water flows down through and grows by precipitation of calcite around the lip

strike – a horizontal line (strike) on a planar surface; used to define the direction of bedding/layering

strike ridge – an elongate hill made of a more erosion-resistant layer and developed along the strike of the bedding

stylobedding – layering within a limestone of alternating resistant bands with softer, thinner mud-rich bands, as a result of stylolite formation; forms pancake rock formations

stylolites – a secondary seam-like structure with a serrated surface that forms by a process of pressure solution often within a limestone

subduction – the process by which the leading edge of one crustal (usually oceanic) plate pushes down beneath the edge of an adjacent (usually continental) plate

submarine canyon – an undersea valley on the seafloor

submarine fan – a delta-like accumulation of sediment deposited at the mouth of a submarine valley

suturing – the process in which one terrane is joined to others

syenite – a coarsely crystalline plutonic rock largely composed of alkali feldspar and minor quartz, hornblende and plagioclase feldspar

syncline – downfold

tafoni (singular **tafone**) – small or large cavities on rock surfaces often producing a honeycomb pattern; formed by weathering, usually involving the growth of salt crystals by evaporation

tarn – a small mountain lake or pond

tectonic – forces within the Earth that cause movements of the crust

tectonic plates – gigantic slabs of lithosphere that form the surface of the Earth and are slowly moving around and jostling each other

terrane – a block of lithosphere (tens to hundreds of kilometres across) broken off one tectonic plate and added to another

terrane suture – boundaries where separate terranes have been stuck together

terrestrial – on land

thrust fault – a low-angle fault with the rocks above having been pushed upwardly across the rocks below

tombolo – a spit of sand or gravel that connects a former island to the mainland

tomo – naturally formed shaft or sinkhole in the ground

tor – a high rock or pile of rocks on top of a hill

trachyte – a light-grey igneous rock mostly composed of small alkali feldspar crystals and no quartz

transcurrent fault – a fault in which the rocks have mostly moved sideways along it

transform fault – a major fracture in the crust with rocks having been displaced sideways (often hundreds of kilometres) on either side of it

transgression – a geological event or period when sea level rises relative to the land and the shoreline moves landwards

travertine – a calcium carbonate (limestone) rock that may form around springs

trilobite – an extinct fossil marine arthropod that lived during the Paleozoic Era

tufa – a porous rock made of calcium carbonate and precipitated out of lime-saturated water

tuff – volcanic ash that has hardened to become rock

tuff ring – a near-circular rampart of bedded volcanic ash (tuff) built up around an explosion crater (sometimes known as a tuff cone when larger)

turbidity current – a dense, turbulent, underwater flow of sediment-laden water that deposits a graded bed of sand (turbidite) on the seafloor

varves – thin annual layers of sediment, often deposited in a glacial lake that ices over in winter

vesicles – holes in a rock

viscous – thick, sticky, resistant to flow, like golden syrup

volcanic arc – a chain of volcanoes erupted above a subduction zone, parallel to a plate boundary

volcanic bomb – a glob of magma ejected into the air while still molten and often acquiring an aerodynamic shape as it cools during flight

weathering – the process of altering rock to clay or wearing it away by long exposure to the air

xenolith – 'foreign rock'; a rock fragment incorporated within rising magma and carried towards the surface with it

younging direction – the direction through a pile of strata in which the layers get younger

Zealandia – the mostly submerged continent centred around New Zealand today

zeolite – group of silicate minerals formed during rock alteration

Further reading and references

The references are divided into three sections: general references, regional references, followed by site-specific references, each listed by year of publication. Entries marked with an asterisk (*) are recommended for general readers.

General New Zealand references

*Cox, G. J., & Hayward, B. W. (1999). *The restless country: Volcanoes and earthquakes of New Zealand*. Auckland: HarperCollins, 64 pp.

*Molloy, L., & Smith, R. (2002). *Landforms: The shaping of New Zealand*. Nelson: Craig Potton Publishing, 160 pp.

*Graham, I. (Ed.). (2008). *A continent on the move: New Zealand geoscience into the 21st century*. Wellington: Geological Society of New Zealand in association with GNS Science Te Pū Ao, 388 pp.

*Crampton, J., & Terezow, M. (2010). *The Kiwi fossil hunter's handbook*. Auckland: Random House, 208 pp.

*Mortimer, N., Campbell, H., & Low, M. (2011). *A photographic guide to rocks and minerals of New Zealand*. Auckland: New Holland, 143 pp.

*Kenny, J. A., & Hayward, B. W. (2013). *On the edge: Celebrating the diversity of New Zealand's coastal landforms* (Guidebook 17). Wellington: Geoscience Society of New Zealand, 48 pp.

*Campbell, H. J., Beu, A., Crampton, J., & Kennedy, L. (2013). *A photographic guide to fossils of New Zealand*. Auckland: New Holland, 139 pp.

*Mortimer, N., & Campbell, H. J. (2014). *Zealandia: Our continent revealed*. Auckland: Penguin, 271 pp.

*Ballance, P. F. (2017). *New Zealand geology: An illustrated guide* (Miscellaneous Publication 148). Wellington: Geoscience Society of New Zealand, 397 pp. Free download from www.gsnz.org.nz/publications-and-webstore/product/73

*Williams, P. (2017). *New Zealand landscape: Behind the scene*. Cambridge, MA, United States: Elsevier, 470 pp.

Schulmeister, J. (Ed.). (2017). *Landscape and Quaternary environmental change in New Zealand*. Paris, France: Atlantis Press, 334 pp.

*Geotrips: www.geotrips.org.nz

*GNS Science Te Pū Ao, New Zealand Geology Web Map: https://data.gns.cri.nz/geology

Regional references

Northland–Auckland

Isaac, M. J. (1996). Geology of the Kaitaia area, geological map 1, 1:250,000. Wellington: Institute of Geological and Nuclear Sciences.

*Cameron, E. K., Hayward, B. W., & Murdoch, G. (1997, 2008). *A field guide to Auckland: Exploring the region's natural and historic heritage*. Auckland: Godwit, 285 pp.

*Homer, L. L., Moore, P. R., & Kermode, L. O. (2000). *Lava and strata: A guide to the volcanoes and rock formations of Auckland*. Wellingon: Landscape Publications in association with the Institute of Geological and Nuclear Sciences, 96 pp.

Edbrooke, S. W. (2001). Geology of the Auckland area, geological map 3, 1:250,000. Wellington: Institute of Geological and Nuclear Sciences.

Edbrooke, S. W., & Brook, F. J. (2009). Geology of the Whangarei area, geological map 2, 1:250,000. Wellington: Institute of Geological and Nuclear Sciences.

*Hayward, B. W. (2017). *Out of the ocean, into the fire: History in the rocks, fossils and landforms of Auckland, Northland and Coromandel* (Miscellaneous Publication 146). Wellington: Geoscience Society of New Zealand, 336 pp.

Coromandel–Waikato

*Chandler, P. (1984). *A trip through time: A guide to landforms, Waitomo Caves–Marokopa Coast*. Waitomo: Waitomo Caves Museum Society, 50 pp.

*Homer, L., & Moore, P. R. (1992). *Vanishing volcanoes: A guide to the landforms and rock formations of Coromandel Peninsula*. Wellington: Landscape Publications, 97 pp.

Edbrooke, S. W. (2005). Geology of the Waikato area, geological map 4, 1:250,000. Wellington: Institute of Geological and Nuclear Sciences.

Bay of Plenty–Taupō

*Houghton, B. F., & Scott, B. (2002). *Geyserland: A guide to the volcanoes and geothermal areas of Rotorua* (Guidebook 13). Wellington: Geological Society of New Zealand, 48 pp.

Nairn, I. A. (2002). Geology of the Okataina Volcanic Centre, sheet parts U15, U16, V15 and V16, geological map 25, 1:50,000. Wellington: Institute of Geological and Nuclear Sciences.

Leonard, G. S., Begg, J. G., & Wilson, C. J. N. (2010). Geology of the Rotorua area, geological map 5, 1:250,000. Wellington: Institute of Geological and Nuclear Sciences.

East Coast–Wairarapa

Mazengarb, C., & Speden, I. G. (2000). Geology of the Raukumara area, geological map 6, 1:250,000. Wellington: Institute of Geological and Nuclear Sciences.

Lee, J. M., & Begg, J. G. (2002). Geology of the Wairarapa area, geological map 11, 1:250,000. Wellington: Institute of Geological and Nuclear Sciences.

Wallace, L., Reyners, M., Cochran, U., Bannister, S., Barnes, P. M., Berryman, K., . . . Power, W. (2009). Characterizing the seismogenic zone of a major plate boundary subduction thrust: Hikurangi Margin, New Zealand. *Geochemistry, Geophysics, Geosystems*, *10*(10). https://doi.org/10.1029/2009GC002610

Lee, J. M., Bland, K. J., Townsend, D. B., & Kamp, P. J. J. (2011). Geology of the Hawke's Bay area, geological map 8, 1:250,000. Wellington: Institute of Geological and Nuclear Sciences.

Clark, K., Howarth, J., Litchfield, N., Cochran, U., Turnbull, J., Howell, A., . . . Wolfe, F. (2019). Geological evidence for past large earthquakes and tsunamis along the Hikurangi subduction margin, New Zealand. *Marine Geology*, *412*: 139–172.

Taranaki–Wellington

Begg, J. G., & Johnston, M. R. (2000). Geology of the Wellington area, geological map 10, 1:250,000. Wellington: Institute of Geological and Nuclear Sciences.

Townsend, D., Vonk, A., & Kamp, P. J. J. (2008). Geology of the Taranaki area, geological map 7, 1:250,000. Wellington: Institute of Geological and Nuclear Sciences.

Marlborough–Nelson

Rattenbury, M. S., Cooper, R. A., & Johnston, M. R. (1998). Geology of the Nelson area, geological map 9, 1:250,000. Wellington: Institute of Geological and Nuclear Sciences.

Rattenbury, M. S., Townsend, D. B., & Johnston, M. R. (2006). Geology of the Kaikoura area, geological map 13, 1:250,000. Wellington: Institute of Geological and Nuclear Sciences.

Westland–Canterbury

*Coates, G. (2002). *The rise and fall of the Southern Alps*. Christchurch: Canterbury University Press, 80 pp.

Nathan, S., Rattenbury, M. S., & Suggate, R. P. (2002). Geology of the Greymouth area, geological map 12, 1:250,000. Wellington: Institute of Geological and Nuclear Sciences.

Cox, S. C., & Barrell, D. J. A. (2007). Geology of the Aoraki area, geological map 15, 1:250,000. Wellington: Institute of Geological and Nuclear Sciences.

Forsyth, P. J., Barrell, D. J. A., & Jongens, R. (2008). Geology of the Christchurch area, geological map 16, 1:250,000. Wellington: Institute of Geological and Nuclear Sciences.

Rattenbury, M. S., Jongens, R., & Cox, S. C. (2010). Geology of the Haast area, geological map 14, 1:250,000. Wellington: Institute of Geological and Nuclear Sciences.

Barrell, D. J. A., Andersen, B. G., & Denton, G. H. (2011). *Glacial geomorphology of the central South Island, New Zealand* (Monograph 27). Wellington: Institute of Geological and Nuclear Sciences, 81 pp.

Otago–Southland

*Turnbull, I. M., & Forsyth, P. J. (1988). *Queenstown: A geological guide* (Guidebook 9). Wellington: Geological Society of New Zealand, 48 pp.

Turnbull, I. M. (2000). Geology of the Wakatipu area, geological map 18, 1:250,000. Wellington: Institute of Geological and Nuclear Sciences.

Forsyth, P. J. (2001). Geology of the Waitaki area, geological map 19, 1:250,000. Wellington: Institute of Geological and Nuclear Sciences.

Turnbull, I. M., & Allibone, A. H. (2003). Geology of the Murihiku area, geological map 20, 1:250,000. Wellington: Institute of Geological and Nuclear Sciences.

*Lee, D., & Forsyth, J. (2008). *Central rocks: A guide to the geology and landscapes of Central Otago* (Guidebook 14). Wellington: Geological Society of New Zealand, 84 pp.

Turnbull, I. M., Allibone, A. H., & Jongens, R. (2010). Geology of the Fiordland area, geological map 17, 1:250,000. Wellington: Institute of Geological and Nuclear Sciences.

Site-specific references

These are a combination of easily accessible material for the general reader (marked with an asterisk *) and more detailed scientific references that have been used as the basis for the site descriptions.

1. Far North

Brook, F. J. (1989). North Cape and Three Kings, geological map of New Zealand, 1:50,000. Wellington: New Zealand Geological Survey, DSIR.

2. Whangaroa Harbour

Hayward, B. W. (1991). Ring plain remnants of a Miocene composite volcano: Wairakau Volcanics at Whangaroa, Northland. *New Zealand Geological Survey Record*, *43*: 65–71.

3. Motu Kōkako / Hole in the Rock, Bay of Islands

Brathwaite, R. L., Isaac, M. J., Challis, G. A., & Brook, F. J. (1990). Tertiary limestone and Zn–Pb mineralised skarn at Motukokako, Cape Brett, northern New Zealand. *Journal of the Royal Society of New Zealand*, *20*: 427–438.

4. Wairere Boulders, Hokianga

Hayward, B. W., & Kenny, J. A. (2011). Karst in basalt. *Geoscience Society of New Zealand Newsletter*, *3*: 12–16.

*Hayward, B. W. (2019). Wairere Boulders. Geotrips: www.geotrips.org.nz/trip.html?id=313

5. Koutu Boulders, Hokianga

Hayward, B. W. (2014). Treasured spheres: Protecting New Zealand's heritage of spherical concretions. *Geoscience Society of New Zealand Newsletter*, *12*: 24–30.

*Hayward, B. W. (2019). Koutu Boulders. Geotrips: www.geotrips.org.nz/trip.html?id=306

6. Tokatoka and Maungaraho, North Kaipara

Black, P. M. (1967). Igneous geology of the Tokatoka district, Northland. *New Zealand Journal of Geology and Geophysics*, *10*: 3–30.

*Hayward, B. W. (2019). Tokatoka volcanic plug. Geotrips: www.geotrips.org.nz/trip.html?id=418

7. Whangārei Heads

*Hayward, B. W. (2019). Taurikura Bay natural jetty. Geotrips: www.geotrips.org.nz/trip.html?id=374

8. Whangaparāoa Head

Sporli, K. B., & Rowland, J. V. (2007). Superposed deformation in turbidites and syn-sedimentary slides of the tectonically active Miocene Waitemata Basin, northern New Zealand. *Basin Research*, *19*: 199–216.

*Isaac, M. J. (2019). Folds and faults at Whangaparaoa. Geotrips: www.geotrips.org.nz/trip.html?id=221

The Big Picture 2: Waitākere Volcano

Hayward, B. W. (1983). Waitakare, sheet Q11, geological map of New Zealand. Wellington: DSIR.

9. Muriwai pillow lava

*Hayward, B. W. (1979). *Ancient undersea volcanoes: A guide to the geological formations at Muriwai, west Auckland* (Guidebook 3). Wellington: Geological Society of New Zealand, 32 pp. Free download from www.gsnz.org.nz/publications-and-webstore/product/124

10. South Piha and The Gap

Hayward, B. W. (1977). Miocene volcanic centres of the Waitakere Ranges, North Auckland, New Zealand. *Journal of the Royal Society of New Zealand*, *7*: 123–141.

*Hayward, B. W. (2019). The Gap, South Piha. Geotrips: www.geotrips.org.nz/trip.html?id=419

11. Whatipu, West Auckland

Williams, P. (1977). Progradation of Whatipu Beach 1844–1976, Auckland, New Zealand. *New Zealand Geographer*, *33*: 84–89.

Blue, B., & Kench, P. (2017). Multi-decadal shoreline change and beach connectivity in a high-energy sand system. *New Zealand Journal of Marine and Freshwater Research*, *51*: 406–426.

12. Rangitoto

*Ballance, P. F., & Smith, I. E. M. (1982). *Walks through Auckland's geological past: A guide to the geological formations of Rangitoto, Motutapu and Motuihe islands* (Guidebook 5). Wellington: Geological Society of New Zealand, 32 pp.

*Wilcox, M. (Ed.). (2008). *Natural history of Rangitoto Island*. Auckland: Auckland Botanical Society, 192 pp.

*Hayward, B. W. (2019). Rangitoto. In B. W. Hayward, *Volcanoes of Auckland: A field guide* (pp 48–61). Auckland: Auckland University Press.

The Big Picture 3: The Auckland Volcanic Field

*Hayward, B. W., Murdoch, G., & Maitland, G. (2011). *Volcanoes of Auckland: The essential guide*. Auckland: Auckland University Press, 234 pp.

*Hayward, B. W. (2019). *Volcanoes of Auckland: A field guide*. Auckland: Auckland University Press, 335 pp.

Hopkins, J. L., Smid, E. R., Eccles, J. D., Hayes, J. L., Hayward, B. W., McGee, L. E., . . . Smith, I. E. M. (2021). Auckland Volcanic Field magmatism, volcanism, and hazard: A review. *New Zealand Journal of Geology and Geophysics, 64*: 213–234.

13. Takapuna Reef, Auckland's North Shore

*Hayward, B. W. (2019). Takapuna fossil forest. In B. W. Hayward, *Volcanoes of Auckland: A field guide* (pp. 77–81). Auckland: Auckland University Press.

14. Wiri Lava Cave, South Auckland

*Hayward, B. W. (2019). Wiri lava cave. In B. W. Hayward, *Volcanoes of Auckland: A field guide* (pp. 302–303). Auckland: Auckland University Press.

15. Pūkorokoro / Miranda Chenier Plain, Firth of Thames

Schofield, J. C. (1960). Sea level fluctuations during the last 4000 years as recorded by a chenier plain, Firth of Thames, New Zealand. *New Zealand Journal of Geology and Geophysics, 3*: 467–485.

Dougherty, A. J., & Dickson, M. E. (2012). Sea level and storm control on the evolution of a chenier plain, Firth of Thames, New Zealand. *Marine Geology, 307–310*: 58–72.

*Hayward, B. W. (2019). Miranda chenier plain. Geotrips: www.geotrips.org.nz/trip.html?id=399

16. Waikawau Valley, Port Waikato

Waterhouse, B. C. (1978). Onewhero, sheet N51, geological map of New Zealand, 1:63,360. Wellington: DSIR.

17. Ngatutura Cliffs, Port Waikato coast

Spratt, P. R., & Rodgers, K. A. (1975). The Ngatutura Volcanics, Southwest Auckland. *Journal of the Royal Society of New Zealand*, 5: 163–178.

Van Niekerk, R. (2016). Reconstructing the complex history of a small volume basaltic volcano (Ngatutura Volcanic Field, New Zealand): The role of subsurface processes and implications for diatreme formation [MSc thesis]. Palmerston North: Massey University.

18. Wairēnga / Bridal Veil Falls, Raglan

*Hayward, B. W. (2019). Wairenga / Bridal Veil Falls. Geotrips: www.geotrips.org.nz/trip.html?id=621

19. Rākaunui Peninsula, Kāwhia

Williams, P. W. (2004). Polygonal karst and paleokarst of the King Country, North Island, New Zealand. *Zeitschrift für Geomorphologie, 136*: 45–67.

20. Waitomo Caves

*Moore, P. R. (1994). *Waitomo Caves* [pamphlet] (Information Series 30). Wellington: Institute of Geological and Nuclear Sciences.

*Hayward, B. W. (2019). Waitomo Glowworm Cave. Geotrips: www.geotrips.org.nz/trip.html?id=609

The Big Picture 4: Limestone and marble karst and caves

*Kenny, J. A., & Hayward, B. W. (2010). *Karst in stone: Karst landscapes in New Zealand – A case for protection* (Guidebook 15). Wellington: Geological Society of New Zealand, 40 pp.

21. Mangapōhue, Waitomo

Williams, P. W. (2004). Polygonal karst and paleokarst of the King Country, North Island, New Zealand. *Zeitschrift für Geomorphologie, 136*: 45–67.

*Crampton, J., & Terezow, M. (2010). Mangapohue: Bridging the waters of time. In J. Crampton & M. Terezow, *The Kiwi fossil hunter's handbook* (pp. 48–53). Auckland: Random House.

22. Te Whanganui-a-Hei / Cathedral Cove, Coromandel Peninsula

Skinner, D. N. B. (1995). Geology of Mercury Bay area, geological map 17, 1:50,000. Wellington: Institute of Geological and Nuclear Sciences.

The Big Picture 5: Coromandel Volcanic Zone

*Homer, L. L., & Moore, P. R. (1992). *Vanishing volcanoes: A guide to the landforms and rock formations of Coromandel Peninsula*. Wellington: Landscape Publications, 97 pp.

*Hayward, B. W. (2017). Coromandel Volcanic Zone. In B. W. Hayward, *Out of the ocean, into the fire* (Miscellaneous Publication 146) (pp. 134–159). Wellington: Geoscience Society of New Zealand.

23. Te Kōhatu-Whakairi-a-Ngātoroirangi / Table Mountain, Thames

Hayward, B. W. (1974). The geology and eruptive history of the Table Mountain area, Coromandel Peninsula. *New Zealand Journal of Geology and Geophysics, 17*: 403–422.

24. Mauao / Mt Maunganui

*Moore, P. R. (1994). *Tauranga Harbour* [pamphlet] (Information Series 25). Wellington: Institute of Geological and Nuclear Sciences.

Briggs, R. M., Hall, G. J., Harmsworth, G. R., Hollis, A. G., Houghton, B. F., Hughes, G. R., . . . Whitbread-Edwards, A. R. (1996). Geology of the Tauranga area, sheet U14, geological map, 1:50,000 (Occasional Report 22). University of Waikato, Department of Earth Sciences, 57 pp.

25. Whakaari / White Island

Hamilton, W. M., & Baumgart, I. L. (1959). *White Island*. DSIR Bulletin 127, 84 pp.

Houghton, B. F., & Nairn, I. A. (1991). The 1976–1982 Strombolian and phreatomagmatic eruptions of White Island, New Zealand: Eruptive and depositional mechanisms at a wet volcano. *Bulletin of Volcanology, 54*: 25–49.

Cole, J. W., Thordarson, T., & Burt, R. M. (2000). Magma origin and evolution of White Island (Whakaari) volcano, Bay of Plenty, New Zealand. *Journal of Petrology, 41*: 867–895.

Scott B. J., & Massiot, C. (2018). White Island. In M. W. Sagar, C. Massiot, & J. L. Hopkins (Eds.), *Field Trip Guides: Geosciences 2018, Napier, New Zealand* (Miscellaneous Publication 151B). Wellington: Geoscience Society of New Zealand, 18 pp.

Kilgour, G., Kennedy, B., Scott, B., Christenson, B., Jolly, A., Asher, C., Rosenberg, M., & Saunders, K. (2021). Whakaari/White Island: A review of New Zealand's most active volcano. *New Zealand Journal of Geology and Geophysics, 64*: 273–295.

26. Mt Tarawera

*Keam, R. F. (1988). *Tarawera: The volcanic eruption of 10 June 1886*. Auckland: Published by the author, 472 pp.

*Leonard, G. (2020). Mount Tarawera. Geotrips: www.geotrips.org.nz/trip.html?id=731

Rowe, M. C., Carey, R. J., White, J. D. L., Kilgour, G., Hughes, E., Ellis, B., Rosseel, J.-B., & Segovia, A. (2021). Tarawera 1886: An integrated review of volcanological and geochemical characteristics of a complex basaltic eruption. *New Zealand Journal of Geology and Geophysics, 64*: 296–319.

*Hayward, B. W. (2019). Mamaku Tors. Geotrips: www.geotrips.org.nz/trip.html?id=648

The Big Picture 7: Caldera volcanoes and ignimbrite eruptions

Barker, S. J., Wilson, C. J. N., Illsley-Kemp, F., Leonard, G. S., Mestel, E. R. H., Mauriohooho, K. & Charlier, B. L. A. (2021). Taupō: An overview of New Zealand's youngest supervolcano. *New Zealand Journal of Geology and Geophysics*, *64*: 320–346.

27. Mamaku Plateau, Rotorua

Kennedy, N. (1994). New Zealand tephrochronology as a tool in geomorphic history of the c. 140 ka Mamaku Ignimbrite Plateau and in related oxygen isotope stages. *Geomorphology*, *9*: 97–115.

29. Waiotapu, Rotorua

Lloyd, E. F. (1959). The hot springs and hydrothermal eruptions of Waiotapu. *New Zealand Journal of Geology and Geophysics*, *2*: 141–176.

*Lynne, B. Y. (2003). *The geothermal guide to Wai-o-tapu Thermal Wonderland*. Auckland: Published by the author, 64 pp.

30. Huka Falls, Taupō

*Wilson, C. J. N., & Houghton, B. F. (2004). *Taupo . . . the volcano* [pamphlet] (Information Series 61). Wellington: Institute of Geological and Nuclear Sciences.

Manville, V., Hodgson, K. A., & Nairn, I. A. (2007). A review of breakout floods from volcanogenic lakes in New Zealand. *New Zealand Journal of Geology and Geophysics*, *50*: 131–150.

32. Red Crater and Ngarotopounamu / Emerald Lakes, Mt Tongariro; 33. Mt Ngāuruhoe; and 34. Crater Lake, Mt Ruapehu

*Williams, K. (1984). *Volcanoes of the south wind: A field guide to the volcanoes and landscape of Tongariro National Park*. Turangi: Tongariro Natural History Society, 128 pp.

*Houghton, B. F., Neall, V. E., & Johnston, D. (1996). *Eruption: Mount Ruapehu awakens*. Auckland: Penguin, 48 pp.

Hobden, B. J., Houghton, B. F., & Nairn, I. A. (2002). Growth of a young, frequently active composite cone; Ngauruhoe volcano, New Zealand. *Bulletin of Volcanology*, *64*: 393–409.

Townsend, D. B., Leonard, G. S., Conway, C. E., Eaves, S. R., & Wilson, C. J. N. (2017). Geology of the Tongariro National Park area, geological map 4, 1:60,000. Wellington: Institute of Geological and Nuclear Sciences, 109 pp.

*Cole, R. (2018). Red Crater, Tongariro. Geotrips: www.geotrips.org.nz/trip.html?id=226

*Thomson, J. (2018). Ngauruhoe Summit Crater. Geotrips: www.geotrips.org.nz/trip.html?id=260

35. Mōhakatino coast, North Taranaki

King, P. R., Scott, G. H., & Robinson, P. H. (1993). *Description, correlation and depositional history of Miocene sediments outcropping along North Taranaki Coast* (Monograph 5). Wellington: Institute of Geological and Nuclear Sciences, 199 pp.

36. Taranaki Maunga / Mt Taranaki

Neall, V. E. (1979). Sheets P19, P20 and P21, geological map of New Zealand, 1:50,000. Wellington: DSIR.

Taranaki Branch of the Geological Society of New Zealand. (1993). *Field guide to Taranaki geology* (Miscellaneous Publication 64). Wellington: Geological Society of New Zealand, 20 pp.

*Neall, V. E. (2003). *The volcanic history of Taranaki* (Occasional Publication 2). Palmerston North: Massey University, Institute of Natural Resources, Soil and Earth Sciences, 18 pp.

Cronin, S. J., Zernack, A. V., Ukstins, I. A., Turner, M. B., Torres-Orozco, R., Stewart, R. B., . . . , Bebbington, M. S. (2021). The geological history and hazards of a long-lived stratovolcano, Mt Taranaki, New Zealand. *New Zealand Journal of Geology and Geophysics*, *64*: 456–478.

37. Pungarehu, Cape Egmont

*Priestley, R., & Cumming, S. (1993). *The volcanoes of Taranaki* [pamphlet] (Information Series 10). Wellington: Institute of Geological and Nuclear Sciences.

Roverato, M., Cronin, S., Procter, J., & Capra, L. (2014). Textural features as indicators of debris avalanche transport and emplacement, Taranaki Volcano. *Geological Society of America Bulletin*, *127*: 3–18.

38. Rangitīkei River, Mangaweka

Milne, J. D. G. (1973). *Map and sections of river terraces in the Rangitikei Basin, North Island, New Zealand*. New Zealand Soil Survey Report 4. Wellington: DSIR.

Litchfield, N. (2003). *Maps, stratigraphic logs, and age control data for river terraces in eastern North Island*. Science Report 2003/31. Wellington: Institute of Geological and Nuclear Sciences, 102 pp.

39. Kai Iwi Beach, Whanganui

Pillans, B. (1990). Late Quaternary marine terraces South Taranaki–Wanganui, geological map 18, 1:100,000. Wellington: New Zealand Geological Survey, DSIR.

Carter, R. M., & Naish, T. R. (1998). A review of Wanganui Basin, New Zealand: Global reference section for shallow marine, Plio–Pleistocene (2.5–0 Ma) cyclostratigraphy. *Sedimentary Geology*, *122*: 37–52.

Dickinson, W., Dunbar, G., Alloway, B., Pillans, B., Beu, A., & Bland, K. (2008). Wanganui: Sea level cycles, tephra and terraces in the Late Cenozoic. *Geoscience Society of New Zealand Conference Field Trip*, *7*: 121–148.

*Crampton, J., & Terezow, M. (2010). Ototoka Beach: A sea-saw sea. In J. Crampton & M. Terezow, *The Kiwi fossil hunter's handbook* (pp. 70–75). Auckland: Random House.

The Big Picture 9: Whanganui Basin

Kamp, P. J. J., & Furlong, K. P. (2006). Neogene plate-tectonic reconstructions and geodynamics of North Island sedimentary basins: Implications for petroleum systems. *2006 New Zealand Petroleum Conference Proceedings*, 14 pp.

40. Te Āpiti / Manawatū Gorge

Milner, A. G. (2017). The Plio–Pleistocene development, uplift and emergence history of the Manawatu Strait, New Zealand [MSc thesis]. Wellington: Victoria University of Wellington.

Rees, C., Palmer, A. G., & Palmer, J. (2018). Quaternary sedimentology and tephrostratigraphy of lower Pohangina Valley, New Zealand. *New Zealand Journal of Geology and Geophysics*, *62*: 171–194.

41. Horoera Point, East Cape

Nelson, C. S., Nyman, S. L., Campbell, K. A., & Rowland, J. R. (2017). Influence of faulting on the distribution and development of cold seep-related dolomitic conduit concretions at East Cape, New Zealand. *New Zealand Journal of Geology and Geophysics*, *60*: 478–496.

42. Mts Hikurangi, Aorangi, Wharekia and Taitai, Ruatōria

Stoneley, R. (1968). A Lower Tertiary décollement on the East Coast, North Island, New Zealand. *New Zealand Journal of Geology and Geophysics*, *11*: 128–156.

Speden, I. G. (1976). Geology of Mt Taitai, Tapuaeroa Valley, Raukumara Peninsula. *New Zealand Journal of Geology and Geophysics*, *19*: 71–119.

43. Waimata, Gisborne

Hayward, B. W., & Francis, D. A. (2013). New Zealand's small mud volcano heritage. *Geoscience Society of New Zealand Newsletter*, *9*: 8–13.

Pettinga, J. R. (2003). Mud volcano eruption within the emergent accretionary Hikurangi margin, southern Hawke's Bay, New Zealand. *New Zealand Journal of Geology and Geophysics*, *46*: 107–121.

44. Kahutara Point / Table Cape, Māhia

Berryman, K. , Clark, K., Cochran, U., Beu, A., & Irwin, S. (2018). A geomorphic and tectonic model for the formation of the flight of Holocene marine terraces at Mahia Peninsula, New Zealand. *Geomorphology*, *307*: 77–92.

45. Lake Waikaremoana

Read, S. A. L., Beetham, R. D., & Riley, P. B. (1991). Lake Waikaremoana barrier: A large landslide dam in New Zealand. In D. Bell (Ed.), *Landslides: Proceedings of the sixth international symposium, 1–14 February 1992, Christchurch* (pp. 1481–1487). Rotterdam: Balkema.

Davies, T. R., McSaveney, M. J., & Beetham, R. D. (2006). Rapid block glides: Slide surface fragmentation in New Zealand's Waikaremoana landslide. *Quarterly Journal of Engineering Geology and Hydrogeology*, *39*: 115–129.

46. Ahuriri Lagoon, Napier

Hull, A. G. (1990). Tectonics of the 1931 Hawke's Bay earthquake. *New Zealand Journal of Geology and Geophysics*, *33*: 309–320.

*Cox, G., & Hayward, B. W. (1999). The Art Deco quake. In G. Cox & B. W. Hayward, *The restless country: Volcanoes and earthquakes of New Zealand* (pp. 54–57). Auckland: HarperCollins.

Hayward, B. W., Grenfell, H. R., Sabaa, A. T., Cochran, U. A., Clark, K. J., Wallace, L., & Palmer, A. S. (2016). Salt-marsh foraminiferal record of 10 large Holocene (last 7500 yr) earthquakes on a subducting plate margin, Hawke's Bay, New Zealand. *Geological Society of America Bulletin*, *128*: 896–915.

47. Te Kauwae-a-Māui / Cape Kidnappers

Beu, A. G., & Grant-Taylor, T. L. (1975). *Geology and fossils of the Cape Kidnappers area*. Wellington: Department of Lands and Survey, 19 pp.

Black, T. M. (1992). Chronology of the Middle Pleistocene Kidnappers Group, New Zealand, and correlation to global oxygen isotope stratigraphy. *Earth and Planetary Science Letters*, *109*: 573–584.

*Hull, A. G. (1993). *Cape Kidnappers* [pamphlet] (Information Series 22). Wellington: Institute of Geological and Nuclear Sciences.

*Thomson, J. (2018). Clifton Cliffs to Cape Kidnappers. Geotrips: www.geotrips.org.nz/trip.html?id=182

48. Te Mata Peak, Havelock North

*Crampton, J., & Terezow, M. (2010). Te Mata Peak: A giant resting on a bed of shells. In J. Crampton & M. Terezow, *The Kiwi fossil hunter's handbook* (pp. 88–95). Auckland: Random House.

49. Rangiwhakaoma / Castlepoint

*Homer, L. L., & Moore, P. R. (1989). Castlepoint. In L. L. Homer & P. R. Moore, *Reading the rocks. A guide to geological features of the Wairarapa coast* (pp. 18–21). Wellington: Landscape Publications.

*Crampton, J., & Terezow, M. (2010). Castle Point: A faulted fortress. In J. Crampton & M. Terezow, *The Kiwi fossil hunter's handbook* (pp. 102–109). Auckland: Random House.

Buckeridge, J. S., Beu, A. G., & Gordon, D. P. (2018). Depositional environment of the early Pleistocene Castlepoint Formation, New Zealand: A canyon fill in-situ. *New Zealand Journal of Geology and Geophysics*, *61*: 524–542.

50. Pūtangirua Pinnacles, Cape Palliser

*Homer, L. L., & Moore, P. R. (1989). Putangirua Pinnacles. In L. L. Homer & P. R. Moore, *Reading the rocks. A guide to geological features of the Wairarapa coast* (pp. 50–53). Wellington: Landscape Publications.

*Thomson, J. (2017). Putangirua Pinnacles. Geotrips: www.geotrips.org.nz/trip.html?id=52

51. Cape Turakirae, Wellington south coast

*Homer, L., & Moore, P. R. (1989). Turakirae Head. In L. L. Homer & P. R. Moore, *Reading the rocks. A guide to geological features of the Wairarapa coast* (pp. 58–61). Wellington: Landscape Publications.

McSavenay, M. J., Graham, I. J., Begg, J. G., Beu, A. G., Hull, A. G., Kim, K., & Zondervan, A. (2006). Late Holocene uplift of beach ridges at Turakirae Head, south Wellington coast, New Zealand. *New Zealand Journal of Geology and Geophysics*, *49*: 337–358.

*Thomson, J. (2018). Turakirae Head raised beaches. Geotrips: www.geotrips.org.nz/trip.html?id=249

52. Wellington Fault

Begg, J., Langridge, R., Van Dissen, R., & Little, T. (2008). *Field Trip 1 – Wellington Fault: Neotectonics and earthquake geology of the Wellington–Hutt Valley segment* (Miscellaneous Publication 12B). Geological Society of New Zealand, 67 pp.

Langridge, R., Van Dissen, R., Rhoades, D., Villamor, P., Little, T., Litchfield, N., . . . Clark, D. (2011). Five thousand years of surface ruptures on the Wellington Fault, New Zealand: Implications for recurrence and fault segmentation. *Bulletin of the Seismological Association of America*, *101*: 2088–2107.

*Thomson, J. (2018). Wellington Fault Trace. Geotrips: www.geotrips.org.nz/trip.html?id=49

53. Marlborough Sounds

Hayward, B. W., Grenfell, H. R., Sabaa, A. T., Kay, J., Daymond-King, R., & Cochran, U. (2010). Holocene subsidence at the transition between strike-slip and subduction on the Pacific–Australian plate boundary, Marlborough Sounds, New Zealand. *Quaternary Science Reviews*, *29*: 648–661.

Nicol, A. (2011). Landscape history of the Marlborough Sounds, New Zealand. *New Zealand Journal of Geology and Geophysics*, *54*: 195–208.

*Calkin, T. (2019). Cullen Point Lookout. Geotrips: www.geotrips.org.nz/trip.html?id=673

54. Wairau Bar and Waikārapi Lagoons

Pickrill, R. A. (1976). The evolution of coastal landforms of the Wairau Valley, New Zealand. *New Zealand Geographer*, *32*: 17–29.

Brown, L. J. (1981). Late Quaternary geology of the Wairau Plain, Marlborough, New Zealand. *New Zealand Journal of Geology and Geophysics*, *24*: 477–490.

Davidson, P., & Wilson, S. (2011). Regional Hydrogeology. In P. Hamill (Ed.), *Groundwaters of Marlborough* (pp. 58–71). Marlborough: Marlborough District Council. Retrieved from www.marlborough.govt.nz

56. Mead Stream, Chalk Range, Clarence Valley

Strong, C. P., & Hollis, C. J. (2009). *Climatic and biotic events of the Paleogene, 12–15 January 2009, Wellington, New Zealand: Conference field trip guides* (Miscellaneous Series 17). Wellington: Institute of Geological and Nuclear Sciences, 141 pp.

58. Te Taero a Kereopa / Nelson Boulder Bank

Johnston, M. R. (2001). Nelson Boulder Bank, New Zealand. *New Zealand Journal of Geology and Geophysics*, *44*: 79–88.

*Warren, K. (2009). *Rolling Stones: Nelson's Boulder Bank – its place in our history and hearts*. Nelson: Nikau Press, 320 pp.

59. Tokangawhā / Split Apple Rock, Kaiteriteri

*Calkin, T. (2020). Toka Ngawhā / Split Apple Rock. Geotrips: www.geotrips.org.nz/trip.html?id=700

The Big Picture 11: The Median Batholith

Mortimer, N., Tulloch, A. J., Spark, R. N., Walker, N. W., Ladley, E., Allibone, A., Kimbrough, D. L. (1999). Overview of the Median Batholith, New Zealand: A new interpretation of the geology of the Median Tectonic Zone and adjacent rocks. *Journal of African Earth Sciences*, *29*: 257–268.

60. Abel Tasman coast

*Calkin, T. (2020). Kaiteriteri Beach. Geotrips: www.geotrips.org.nz/trip.html?id=676

*Pedley, K. (2018). Anapai and Little Anapai. Geotrips: www.geotrips.org.nz/trip.html?id=523

61. Tākaka Hill

Grindley, G. W. (1971). Takaka, sheet S8, geological map of New Zealand, 1:63,360, including notes. Wellington: DSIR.

Grindley, G. W. (1980). Cobb, sheet S13, geological map of New Zealand, 1:63,360, including notes. Wellington: DSIR.

62. Rāwhiti Cave, Tākaka

*Calkin, T. (2019). Rawhiti Cave. Geotrips: www.geotrips.org.nz/trip.html?id=698

63. Cobb Valley, northwest Nelson

*Cooper, R. A. (1984). *Cobb Valley: A geological guide* (Guidebook 6). Wellington: Geological Society of New Zealand, 48 pp.

*Cooper, R. (2019). Cobb Valley roche moutonnée. Geotrips: www.geotrips.org.nz/trip.html?id=97

64. Te Waikoropūpū, Tākaka

Williams, P. W. (2018). Evidence presented before a Special Tribunal considering an application for a Water Conservation Order in respect of Te Waikoropūpū Springs, 26 pp. Retrieved from www.epa.govt.nz

*Calkin, T. (2019). Te Waikoropūpū Springs/Pūpū Springs. Geotrips: www.geotrips.org.nz/trip.html?id=687

*Department of Conservation. (n.d.). Department of Conservation interpretation panels for Te Waikoropūpū Springs, 15 pp. Retrieved from www.doc.govt.nz/globalassets/documents/parks-and-recreation/places-to-visit/nelson-marlborough/te-waikoropupu-interpretation-panels.pdf

65. Onetahua / Farewell Spit

Tribe, H., & Kennedy, D. (2010). The geomorphology and evolution of a large barrier spit: Farewell Spit, New Zealand. *Earth Surface Processes and Landforms*, *35*: 1751–1762.

*Calkin, T. (2019). Farewell Spit. Geotrips: www.geotrips.org.nz/trip.html?id=674

66. Wharariki Beach, northwest Nelson

*Pedley, K. (2020). Wharariki Beach/Archway Islands. Geotrips: www.geotrips.org.nz/trip.html?id=440

67. Mt Owen, northwest Nelson

*Thomson, J. (2019). Ngarua marble caves. Geotrips: www.geotrips.org.nz/trip.html?id=593

*Pedley, K. (2019). Mt Owen massif. Geotrips: www.geotrips.org.nz/trip.html?id=391

68. Longford Syncline, Murchison

Holden, A. M. (1983). Fossil *Nothofagus* from the Longford Formation, Murchison, New Zealand. *Journal of the Royal Society of New Zealand*, *12*: 65–77.

Suggate, R. P. (1984). Mangles Valley, sheet M29AC, geological map of New Zealand, 1:50,000. Wellington: DSIR.

Lihou, J. C. (1993). The structure and deformation of the Murchison Basin, South Island, New Zealand. *New Zealand Journal of Geology and Geophysics*, *36*: 95–105.

*Thomson, J. (2019). Nuggety Creek. Geotrips: www.geotrips.org.nz/trip.html?id=37

69. Devil's Dining Table, Matiri Range

Suggate, R. P. (1984). Mangles Valley, sheet M29AC, geological map of New Zealand, 1:50,000. Wellington: DSIR.

70. Ōpārara Arch, Karamea

*Pedley, K. 2019. Oparara Basin arches. Geotrips: www.geotrips.org.nz/trip.html?id=555

71. Punakaiki / Pancake Rocks

*Laird, M. G., & Coates, G. (1988). *Guide to the pancake rocks, Punakaiki* [pamphlet] (Information Series 70). Wellington: Institute of Geological and Nuclear Sciences.

*Pedley, K. 2019. Pancake Rocks. Geotrips: www.geotrips.org.nz/trip.html?id=436

The Big Picture 12: Alpine Fault and Southern Alps

*Coates, G. (2002). *The rise and fall of the Southern Alps*. Christchurch: Canterbury University Press, 80 pp.

Cochran, U. A., Clark, K. J., Howarth, J. D., Biasi, G. P., Langridge, R. M., Villamor, P., . . . Vandergoes, M. J. (2017). A plate boundary earthquake record from a wetland adjacent to the Alpine Fault in New Zealand refines seismic hazard estimates. *Earth and Planetary Science Letters*, *464*: 175–188.

Lang, K. A., Glotzbach, C., Ring, U., Kamp, P. J. J., & Ehlers, T. A. (2020). Linking orogeny and orography in the Southern Alps of New Zealand: New observations from detrital fission-track thermochronology of the Waiho-1 borehole. *Earth and Planetary Science Letters*, *552*. https://doi.org/10.1016/j.epsl.2020.116586

72. Gaunt Creek, Franz Josef

Cooper, A. F., & Norris, R. J. (1994). Anatomy, structural evolution and slip rate of a plate boundary thrust: The Alpine Fault at Gaunt Creek, Westland, New Zealand. *Geological Society of America Bulletin*, *106*: 627–633.

*Thomson, J. (2018). Alpine Fault at Gaunt Creek. Geotrips: www.geotrips.org.nz/trip.html?id=57

73. Kā Roimata o Hine Hukatere / Franz Josef Glacier

*Chin, T., & Coates, G. (1994). *The Franz Josef and Fox Glaciers* [pamphlet] (Information Series 24). Wellington: Institute of Geological and Nuclear Sciences.

Alexander, D., Davies, T., & Schulmeister, J. (2014). Formation of the Waiho Loop terminal moraine, New Zealand. *Journal of Quaternary Science*, *29*: 361–369.

*Thomson, J. (2018). Franz Josef Glacier. Geotrips: www.geotrips.org.nz/trip.html?id=245

74. Kaikōura Peninsula

*Coates, G., Browne, G. H., & Coates, F. (1993). *The Kaikoura Peninsula* [pamphlet] (Information Series 8). Wellington: Institute of Geological and Nuclear Sciences.

*Thomson, J. (2019). Kaikoura Peninsula clifftop walk. Geotrips: www.geotrips.org.nz/trip.html?id=188

The Big Picture 13: The 2016 Kaikōura Earthquake

Berryman, K., Hamling, I., Kaiser, A., & Stahl, T. (2018). Introduction to the Special Issue on the 2016 Kaikōura Earthquake. *Bulletin of the Seismological Society of America*, *108*: 1491–1495.

Litchfield, N. J., Villamor, P., Van Dissen, R. J., Nicol, A., Barnes, P. M., Barrell, J. A., . . . Zinke, R. (2018). Surface rupture of multiple crustal faults in the 2016 Mw 7.8 Kaikōura, New Zealand, Earthquake. *Bulletin of the Seismological Society of America*, *108*: 1496–1520.

Duffy, B. (2020). A geometric model to estimate slip rates from terrane rotation above an offshore, listric thrust fault, Kaikōura, New Zealand. *Tectonophysics, 786*, 228460. https://doi.org/10.1016/j.tecto.2020.228460

75. The Wall of Waiau and Leader River

Massey, C., Townsend, D., Rathje, E., Allstadt, K. E., Lukovic, B., Kaneko, Y., . . . Villeneuve, M. (2018). Landslides triggered by the 14 November 2016 Mw 7.8 Kaikōura Earthquake, New Zealand. *Bulletin of the Seismological Society of America, 108*: 1630–1648.

*Pedley, K. (2020). Leader River landslide dam. Geotrips: www.geotrips.org.nz/trip.html?id=560

*Pedley, K. (2020). The Wall of Waiau. Geotrips: www.geotrips.org.nz/trip.html?id=559

76. Kura Tāwhiti / Castle Hill

Gage, M. (1970). Late Cretaceous and Tertiary rocks of Broken River, Canterbury. *New Zealand Journal of Geology and Geophysics, 13*: 507–559.

Bradshaw, J. D. (1975). The folds at Castle Hill (Canterbury) and their bearing on Kaikouran deformation style in the Canterbury Basin. *Journal of the Royal Society of New Zealand, 5*: 209–217.

*Pedley, K. (2019). Kura Tāwhiti/Castle Hill. Geotrips: www.geotrips.org.nz/trip.html?id=390

The Big Picture 14: Banks Peninsula Volcanoes

Weaver, S., Sewell, R., & Dorsey, C. (1985). *Extinct volcanoes: A guide to the geology of Banks Peninsula* (Guidebook 7). Wellington: Geological Society of New Zealand, 48 pp.

*New Zealand Geological Survey. (1987). Scenery and geology from the Summit Road, Christchurch [pamphlet]. Wellington: New Zealand Geological Survey.

Sewell, R. J., Weaver, S. D., & Reay, M. B. (1992). Geology of Banks Peninsula, geological map 3, 1:100,000. Wellington: Institute of Geological and Nuclear Sciences.

Hampton, S. J., & Cole, J. W. (2009). Lyttelton Volcano, Banks Peninsula, New Zealand: Primary volcanic landforms and eruptive centre identification. *Geomorphology, 104*: 284–298.

Timm, C., Hoernle, K., Van Den Bogaard, P., Bindeman, I., & Weaver, S. (2009). Geochemical evolution of intraplate volcanism at Banks Peninsula, New Zealand: Interaction between asthenospheric and lithospheric melts. *Journal of Petrology, 50*: 989–1023.

77. Ōnawe Peninsula, Akaroa Harbour

*Weaver, S., Sewell, R., & Dorsey, C. (1985). *Extinct volcanoes: A guide to the geology of Banks Peninsula* (Guidebook 7). Wellington: Geological Society of New Zealand, 48 pp.

Sewell, R. J., & Weaver, S. D. (1990). Akaroa west, sheet N36AC, geological map of New Zealand, 1:50,000. Wellington: DSIR.

*Pedley, K. (2018). Hilltop lookout. Geotrips: www.geotrips.org.nz/trip.html?id=433

*Pedley, K. (2018). Ōnawe Peninsula. Geotrips: www.geotrips.org.nz/trip.html?id=389

78. Rakaia River

*BRaid. Our braided rivers [website]. https://braidedrivers.org/rivers

79. Aoraki / Mt Cook

*Chin, T. J., McSaveney, M. J., & McSaveney, E. R. (1992). *The Mount Cook rock avalanche of 14 December 1991* [pamphlet]. Wellington: DSIR.

*Thomson, J. (2018). Plateau Hut and Glacier Dome. Geotrips: www.geotrips.org.nz/trip.html?id=450

*Thomson, J. (2019). Hooker Lake walk. Geotrips: www.geotrips.org.nz/trip.html?id=685

80. Haupapa / Tasman Glacier, Mt Cook

*Chin, T. J., & Coates, G. C. (1993). *The glaciated landscape of Mount Cook* [pamphlet]. Wellington: Institute of Geological and Nuclear Sciences.

*Thomson, J. (2017). Tasman Glacier terminal ice face. Geotrips: www.geotrips.org.nz/trip.html?id=139

*Thomson, J. (2017). Tasman Glacier terminal moraine. Geotrips: https://www.geotrips.org.nz/trip.html?id=147

*Thomson, J. (2018). Tasman Glacier white ice. Geotrips: www.geotrips.org.nz/trip.html?id=432

81. Pareora dolines

Gair, H. S. (1959). The Tertiary geology of the Pareora district, South Canterbury. *New Zealand Journal of Geology and Geophysics, 2*: 265–296.

Jennings, J. N. J. (1975). Doline morphology as a morphogenetic tool: New Zealand examples. *New Zealand Geographer, 31*: 6–28.

82. Clay Cliffs, Ōmārama

*Pedley, K. (2020). Omarama clay cliffs. Geotrips: www.geotrips.org.nz/trip.html?id=448

83. Cape Wanbrow, Ōamaru

*Forsyth, J. (1994). *The Oamaru volcano* [pamphlet] (Information Series 32). Wellington: Institute of Geological and Nuclear Sciences.

Kaulfuss, U., Nemeth, K., & White, J. D. L. (2012). *Field guide: Miocene subaerial to subaqueous monogenetic volcanism in Otago, New Zealand* (Miscellaneous Publication 131C). Wellington: Geoscience Society of New Zealand, 57 pp.

*Thomson, J. (2018). Boatmans Harbour pillow lavas. Geotrips: www.geotrips.org.nz/trip.html?id=73

84. Moeraki Boulders

*Forsyth, J., & Coates, G. (2007). *The Moeraki boulders* [pamphlet] (Miscellaneous Series 9). Wellington: Institute of Geological and Nuclear Sciences.

*Mortimer, N. (2017). Moeraki Boulders. Geotrips: www.geotrips.org.nz/trip.html?id=71

85. Matanaka sea caves, Waikouaiti

Barth, N. (2013). Caves of Otago. *New Zealand Speleological Bulletin, 11*(206): 122–159.

The Big Picture 16: Dunedin Volcano

*Forsyth, J., & Coates, G. (1989). *The Dunedin volcano* [pamphlet]. Wellington: New Zealand Geological Survey.

*Mortimer, N. (2019). Organ Pipes, Mt Cargill track. Geotrips: www.geotrips.org.nz/trip.html?id=118

Scott, J. M., Pontesilli, A., Brenna, M., White, J. D. L., Giacalone, E., Palin, J. M., & Le Roux, P. J. (2020). The Dunedin Volcanic Group and a revised model for Zealandia's alkaline intraplate volcanism. *New Zealand Journal of Geology and Geophysics, 63*: 510–529.

86. Pikiwhara / Sandymount, Otago Peninsula

*Forsyth, J., & Coates, G. (1989). *The Dunedin volcano* [pamphlet]. Wellington: New Zealand Geological Survey.

87. Tunnel Beach, Dunedin

*Bland, K. (2019). Tunnel Beach. Geotrips: www.geotrips.org.nz/trip.html?id=587

88. Upper Taieri River valley

Craw, D., Craw, L., Burridge, C. P., Wallis, G. P., & Waters, J. M. (2016). Evolution of the Taieri River catchment, East Otago, New Zealand. *New Zealand Journal of Geology and Geophysics, 59*: 257–273.

*Craw, D. (2020). The circuitous path of the Taieri River in east Otago [website]. University of Otago, Department of Geology:

www.otago.ac.nz/geology/research/environmental-geology/geomorphology/taieri-river-path.html

The Big Picture 18: The Waipounamu Erosion Surface and Otago's basin and range topography

Landis, C. A., Campbell, H. J., Begg, J. G., Mildenhall, D. C., Paterson, A. M., & Trewick, S. A. (2008). The Waipounamu Erosion Surface: Questioning the antiquity of New Zealand land surface and terrestrial fauna and flora. *Geological Magazine*, *145*: 173–197.

89. Kopuwai / Old Man Range, Alexandra

*Lee, D., & Forsyth, J. (2008). *Central rocks: A guide to the geology and landscapes of Central Otago* (Guidebook 14). Wellington: Geological Society of New Zealand, 84 pp.

*Thomson, J. (2019). Tors on the Old Man Range. Geotrips: www.geotrips.org.nz/trip.html?id=278

The Big Picture 19: The former Wakatipu Super Glacier

Brodie, J. W., & Irwin, J. (1970). Morphology and sedimentation in Lake Wakatipu, New Zealand. *New Zealand Journal of Marine and Freshwater Research*, *4*: 479–496.

*Lee, D., & Forsyth, J. (2008). *Central rocks: A guide to the geology and landscapes of Central Otago* (Guidebook 14). Wellington: Geological Society of New Zealand, 84 pp.

*Mortimer, N. (2018). Glacial erratic, Frankton marina. Geotrips: www.geotrips.org.nz/trip.html?id=133

Sutherland, J. L., Carrivick, J. L., Shulmeister, J., Quincey, D. J., & James, W. H. M. (2019). Ice-contact proglacial lakes associated with the Last Glacial Maximum across the Southern Alps, New Zealand. *Quaternary Science Reviews*, *213*: 67–92.

92. Tokata / Nugget Point

*Crampton, J., & Terezow, M. (2010). Roaring Bay: Echoes from Gondwana. In J. Crampton & M. Terezow, *The Kiwi fossil hunter's handbook* (pp. 174–178). Auckland: Random House.

93. Tumu Toka / Curio Bay, The Catlins

Pole, M. (2001). Repeated flood events and fossil forests at Curio Bay (Middle Jurassic), New Zealand. *Sedimentary Geology*, *144*: 223–242.

*Crampton, J., & Terezow, M. (2010). Curio Bay: Curioser and curioser. In J. Crampton & M. Terezow, *The Kiwi fossil hunter's handbook* (pp. 180–185). Auckland: Random House.

*Thomson, J. (2019). Curio Bay. Geotrips: www.geotrips.org.nz/trip.html?id=25

94. Key Summit, Milford Road

*Blattner, P. (1988). *A guide to the Routeburn Track, NZ* [pamphlet]. Wellington: DSIR.

95. Piopiotahi / Milford Sound

Blattner, P., & Coates, G. (1989). *A guide to Milford Sound, NZ* [pamphlet]. Wellington: New Zealand Geological Survey.

Dykstra, J. L. (2012). The Post-LGM Evolution of Milford Sound, Fiordland, New Zealand [PhD thesis]. Christchurch: University of Canterbury.

97. Green Lake, Fiordland

Hancox, G. T., & Perrin, N. D. (2009). Green Lake landslide and other giant and very large postglacial landslides in Fiordland, New Zealand. *Quaternary Science Reviews*, *28*: 1020–1036.

98. Awarua / Bluff coastal ventifacts

Bishop, D. G., & Mildenhall, D. C. (1994). The geological setting of ventifacts and wind-sculpted rocks at Mason Bay, Stewart Island and their implications for late Quaternary paleoclimates. *New Zealand Journal of Geology and Geophysics*, *37*: 169–180.

Jennings, C. G. (2009). The use of Southland argillite in New Zealand prehistory: Distribution, chronology and form [MA thesis]. Dunedin: University of Otago.

99. Kaka-kura / Gog and Tu-pouri / Magog, Stewart Island

Allibone, A. H., & Tulloch, A. J. (2004). Geology of the plutonic basement rocks of Stewart Island, New Zealand. *New Zealand Journal of Geology and Geophysics*, *47*: 233–256.

Sawyer, L., Smith, K., & Hayward, B. W. (2020). Granite fluting, basins and tafone, southern Stewart Island. *Geocene*, *23*: 7–10.

100. Ohira Bay, Chatham Island

Campbell, H. J., Andrews, P. B., Beu, A. G., Maxwell, P. A., Edwards, A. R., Laird, M. G., . . . Hayward, B. W. (1993). *Cretaceous–Cenozoic geology and biostratigraphy of the Chatham Islands, New Zealand* (Monograph 2). Wellington: Institute of Geological and Nuclear Sciences, 269 pp.

Thanks

This book presents our modern understanding of the geology and landforms of many parts of New Zealand that has been built up over nearly two centuries from the research of hundreds of individuals, all of whom I acknowledge as having contributed to this book. I thank many colleagues who have made more specific and significant contributions through sharing field work experiences and ideas: especially the late Peter Ballance, Mike Isaac, Les Kermode and Margaret Morley; Fred Brook, Kate Clark, Ursula Cochran, Hugh Grenfell, Glenys Hayward, Jill Kenny, Phil Moore, Peter Crossley and many members of the Auckland Geology Club. For advice and discussion away from the field, I additionally thank: Nicolas Barth, Marco Brenna, Roger Briggs, Hamish Campbell, Murry Cave, Chris Hollis, Mike Johnston, Jeff Mauk, Nick Mortimer, Vince Neall, James Scott, Ian Smith, Bernhard Sporli and Paul Williams. I also acknowledge the voluntary contributions of hundreds of recent geoscientists to the New Zealand Geopreservation Inventory of the Geoscience Society of New Zealand, which identifies and documents the outstanding geoheritage sites in the country. The Inventory listings have influenced my selection of the sites presented in this book. For providing helpful initial reviews of the project I thank Hamish Campbell and Simon Nathan.

The majority of diagrams and maps have been drafted by me, but I am indebted to GNS Science for making freely available their digital QMap, which was utilised to derive the geological bases for many of the maps in this book. For allowing me to include original or modified versions of their maps or figures I thank Kyle Bland, Greg Browne, Kate Clark, Alan Cooper, Louise Cotterall, Geoff Cox, Dave Craw, Hugh Grenfell, James Scott and the late Margaret Morley. I am deeply indebted to my colleagues Alastair Jamieson and Lloyd Homer for the fabulous aerial photographs that adorn the book. For permission to use their land-level photographs I am grateful to Nicolas Barth, Hamish Campbell, Jonathan Carr, Murry Cave, the late Roger Cooper, Peter Crossley, Egon Eberle, Lloyd Esler, Graham Hancox, Chris Hollis, Jill Kenny, Mike Johnston, Nicola Litchfield, Kate Pedley, Lee Sawyer, Ken Smith, Julian Thomson, Michael Turek, Rose Turnbull, Paul Williams and Murchison Helitours (all acknowledged in the relevant captions). Unattributed photographs were taken by me. I particularly acknowledge GNS Science, for allowing me to reproduce many wonderful aerial photographs in their collections, mostly taken by Lloyd Homer, but also by Kyle Bland, Karen Britten, Bruce Mountain, Brad Scott, Dougal Townsend and Paul van der Werff. I also acknowledge the exceptional photographers who have placed some of their masterpieces on Wikimedia Commons which I have been able to include in this book, specifically Michal Klajban, Geoff Leeming, Marian Predescu and C. Rodliffe. Specific helicopter aerial photography taken by Alastair Jamieson and myself for this project was sponsored by generous grants from the Royal Society of New Zealand's Charles Fleming Publication Fund, and we thank the pilots, Pete Darling and Charlie Rowe of Helicopters Nelson, and Ryan Trounce and Jamie Barrow of Heliflite Ardmore, for their great skill and tolerance during these missions. We also acknowledge Michael Horton's commitment to New Zealand's natural world, and thank him for the generous grant that helped enable the new photography for this book.

Nick Mortimer and Paul Williams are thanked for their encouraging and helpful reviews of the draft manuscript. The AUP team of Sam Elworthy, Katharina Bauer, Sophia Broom and Lauren Donald, editor Teresa McIntyre, designer Carolyn Lewis and proofreaders Mike Wagg and Kate Stone are thanked for their friendly encouragement, patience and support throughout the latter stages of this book's preparation. My thanks also go to Tim Nolan for designing the site maps and his illustration input as well as to Susan Brookes for creating the index.

Finally, I thank my wife, Glenys, and daughters Kathryn, Jessica and Clare for their support, encouragement and especially tolerance during the many decades that this book has been gestating.

Bruce W. Hayward

Index

Page numbers are bolded where images also occur

A

B

C